CRC Handbook of Animal Models of Pulmonary Disease

Volume I

Editor
Jerome O. Cantor, M.D.
College of Physicians and Surgeons
Columbia University
New York, New York

CRC Press
Taylor & Francis Group
Boca Raton London New York

CRC Press is an imprint of the
Taylor & Francis Group, an **informa** business

First published 1989 by CRC Press
Taylor & Francis Group
6000 Broken Sound Parkway NW, Suite 300
Boca Raton, FL 33487-2742

Reissued 2018 by CRC Press

Library of Congress Cataloging-in-Publication Data

Handbook of animal models of pulmonary disease / [edited by] Jerome O. Cantor.
p. cm.
Includes bibliographies and index.
ISBN 0-8493-2978-7 (v. 1). ISBN 0-8493-2979-5 (v. 2)
1. Lungs—Diseases—Animal models. I. Cantor, Jerome O.
[DNLM: 1. Disease Models. Animal. 2. Lung diseases—etiology.
WF 600 H236]
RC756.H287 1989
616.2' 4027—dc19
DNLM/DLC
for Library of Congress 88-24203

A Library of Congress record exists under LC control number: 88024203

ISBN 13: 978-1-315-89186-6 (hbk)
ISBN 13: 978-1-351-07096-6 (ebk)

Visit the Taylor & Francis Web site at http://www.taylorandfrancis.com and the
CRC Press Web site at http://www.crcpress.com

PREFACE

Many of us can recall the frustration which accompanied our first attempts to reproduce an animal model of disease. The most basic procedures seemed to require an impossible amount of skill. Aside from the practical difficulties, questions may have arisen about the model itself. How closely did it resemble the human disease? Was it well-suited for the studies being performed?

Hopefully, this book will provide some assistance in dealing with the problems of choosing and setting up an animal model of lung disease. Each chapter is designed to familiarize the investigator with the characteristics of a particular model, the steps involved in implementing it, and the procedures for evaluating morphological, physiological, or biochemical changes. The book emphasizes methodology, but it should not be viewed solely as a "how-to" manual. Much of the material presented concerns recent experimental findings and ideas for future investigations.

Since these volumes are the result of a collaborative effort, I wish to thank each of the contributors. The care and enthusiasm expressed in their work reinforce the belief that research is a meaningful and humane endeavor.

Jerome O. Cantor, M.D.
New York City

FOREWORD

These are landmark volumes. For the first time a compendium of papers devoted to models of pulmonary disease has been drawn together into one publication. It is of even more significance that these volumes reflect the research which has resluted in the need for such animal models. The lung is a major link in the cardiovascular system of mammalian organisms and fulfills the critical function of gas exchange. These functions require absolute architectural integrity to fulfill its physiological purpose. While the functional capacity of the lung is great, any pathological mechanism which can produce a generalized injury and anatomical remodeling of the lung is a critical threat to survival of the organism. Thus, proliferative lesions which destroy alveolar spaces and alter the alveolar capillary interface or increase lung retractile force destroy lung functional capacity for circulation, gas exchange, and the energy of breathing. Similarly, mechanisms which cause alveolar destruction and specific loss of tissue components of lung structure such as lung elastin cause major functional aberrations of lung tissue inducing loss of lung retractile force with consequent loss of airway patency. Similarly, injury which induces changes in alveolar capillary permeability alters water transport in the lung and interferes with gas exchange and lung retractile force. It does not matter critically that such models may not be perfect replications of interstitial fibrosis or pulmonary emphysema or the acute respiratory distress syndrome as seen in man. What does matter is that the pathogenic mechanisms involved may bear a significant resemblance to the mechanisms which are at play in human disease. The hope of all of these models is that the precise biochemical, cellular, or immunological mechanisms which may cause tissue proliferation, tissue destruction, increased pulmonary vascular resistance, or changes in alveolar capillary fluid transport can be understood in terms of the molecular and cellular factors which initiate lung injury and perpetuate lung remodeling.

Simulations of human lung disease in terms of the initiation and progression of lung injury and anatomical remodeling can be of enormous help in investigating and even defining the chemical and cellular agents which produce these lesions in man. In this regard, the more we understand about the lesions and processes producing interstitial fibrosis, emphysema, or altered capillary permeability in animals, the easier it will be to conceive critical experiments in man to understand the development of human disease. Ultimately, understanding that is gained in animals should lead the investigator to critical studies on lavaged fluid, inflow and outflow blood perfusates, as well as tissue. Animal models of lung disease are therefore a critical link in understanding the basis of the human disease but in addition, in the future, can be the experimental means for developing new therapeutic modalities which can ultimately prevent or control such pathological states in man.

This much needed publication is worthy of the interest and attention of individuals working in lung research but may also serve as a useful guide to investigators studying pathological states in other organ systems.

Gerard M. Turino, M.D.
John H. Keating, Sr., Professor of Medicine
College of Physicians and Surgeons
Columbia University
and
Director, Department of Medicine
St. Luke's-Roosevelt Hospital
New York, New York

THE EDITOR

Dr. Jerome Cantor is Associate Clinical Professor of Pathology at the College of Physicians and Surgeons of Columbia University and a Research Associate at St. Luke's-Roosevelt Medical Center. He received his undergraduate degree at Columbia College and attended medical school at the University of Pennsylvania. After completing a research fellowship at the Roche Institute of Molecular Biology, he performed an internship and residency in pathology at Columbia-Presbyterian Medical Center. Dr. Cantor obtained additional training in lung pathology at the Armed Forces Institute of Pathology and is a diplomate of the American Board of Pathology. He has authored numerous scientific papers on the role of connective tissue in pulmonary fibrosis and emphysema and is currently interested in the factors which modulate elastin synthesis following lung injury.

To Linda

CONTRIBUTORS

Ian Y. R. Adamson, Ph.D.
Professor
Department of Pathology
University of Manitoba
Winnipeg, Manitoba, Canada

C. Redington Barrett, M.D.
Associate Professor of Clinical Medicine
College of Physicians and Surgeons
Columbia University
Director of ICU and Associate Director of Pulmonary Division
Department of Medicine
St. Luke's-Roosevelt Hospital
New York, New York

David J. P. Bassett, Ph.D.
Associate Professor
Department of Environmental Health Sciences
The Johns Hopkins University
Baltimore, Maryland

Arnold R. Brody, Ph.D.
Head
Pulmonary Pathology Laboratory
National Institute of Environmental Health Sciences
National Institutes of Health
Research Triangle Park, North Carolina

Jerome O. Cantor, M.D.
Associate Clinical Professor of Pathology
College of Physicians and Surgeons
Columbia University and
Research Associate
St. Luke's-Roosevelt Hospital
New York, New York

Laurie J. Carpenter-Deyo, Ph.D.
Research Associate
Department of Biochemistry and Biophysics
Oregon State University
Corvallis, Oregon

Joseph M. Cerreta, Ph.D.
Associate Professor
Department of Pharmaceutical Sciences
College of Pharmacy and Allied Health Professions
St. John's University
Jamaica, New York

Robert A. Durr
Department of Medicine
Duke University Medical Center
Durham, North Carolina

Kent J. Johnson, M.D.
Associate Professor
Department of Pathology
University of Michigan Medical School
Ann Arbor, Michigan

Yoon T. Kim, M.D.
Active Staff
Department of Radiation Therapy
Resurrection Hospital
Chicago, Illinois

Jerold A. Last, Ph.D.
Professor of Medicine
Department of Internal Medicine
University of California
Davis, California

Deng F. Liau, Ph.D.
Assistant Professor
Department of Pathology
College of Physicians and Surgeons
Columbia University
New York, New York

Daniel H. Matulionis, Ph.D.
Associate Professor
Department of Anatomy and
Neurobiology
University of Kentucky
Lexington, Kentucky

Lila H. Overby, B.S.
Chemist
Laboratory of Cellular and Molecular
Pharmacology
National Institute of Environmental
Health Sciences
National Institutes of Health
Research Triangle Park, North Carolina

Richard E. Parker, Ph.D.
Research Assistant Professor
Department of Pulmonary Medicine
Vanderbilt School of Medicine
Nashville, Tennessee

Robert A. Roth, Ph.D.
Professor
Department of Pharmacology and
Toxicology
Michigan State University
East Lansing, Michigan

Stephen F. Ryan, M.D.
Professor of Clinical Pathology
Columbia University
College of Physicians
and Surgeons
Chief Anatomic Pathologist
Department of Pathology
St. Lukes-Roosevelt Hospital
New York, New York

Lun A. Thet, M.D.
Assistant Professor of Medicine
Department of Medicine
Duke University Medical Center
Durham, North Carolina

Peter A. Ward, M.D.
Professor and Chairman
Department of Pathology
University of Michigan Medical School
Ann Arbor, Michigan

William F. Ward, Ph.D.
Professor
Department of Radiology
Northwestern University Medical School
Chicago, Illinois

Jeffrey S. Warren, M.D.
Assistant Professor
Department of Pathology
University of Michigan Medical School
Ann Arbor, Michigan

TABLE OF CONTENTS

Volume I

TABLE OF CONTENTS

Volume II

Acute Parenchymal Injury

ANIMAL MODELS OF HYPEROXIC LUNG INJURY

Robert A. Durr and Lyn A. Thet

INTRODUCTION

The potential for adverse effects from exposure to hyperoxia was suspected by Priestly more than 2 centuries ago: " . . . for, as a candle burns outs much faster in dephlogisticated than in common air, so we might, as may be said, live out too fast, and the animal powers be too soon exhausted in this pure kind of air."[1]

In 1899, Smith showed that a high concentration of oxygen was, in fact, toxic.[2] Since that time, interest in the adverse effects from exposure to hyperoxic conditions has increased dramatically. This has been stimulated in recent years by the development of high-altitude air and space travel and the advent of more advanced techniques of critical care medicine with the ability to deliver high concentrations of oxygen for prolonged periods of time. Many of the early modern studies (from the 1930s to about 1960) attempted to determine the extent of the toxicity; determine the hormonal, chemical, and environmental factors influencing it; and characterize the qualitative pathologic changes resulting from hyperoxia.[3] More recent studies have focused on the etiology, quantitative pathology, cellular biochemistry, prevention, and repair of oxygen toxicity.

Although many organ systems are affected by hyperoxia, most research efforts have centered on the respiratory system. This is likely due to the fact that the lung has, at least at normobaric conditions, the greatest susceptibility to damage from hyperoxia. Although this susceptibility may be due to some inherent characteristic of lung tissue itself, it may also be a result of its direct interface with the atmospheric environment and the fact that, especially in critical care settings, the lung is exposed to higher oxygen tensions than any other internal organ. From a clinical perspective, this is of major importance because hypoxia has been implicated as a contributing factor in adult respiratory distress syndrome (and the pulmonary fibrosis which follows it) and in bronchopulmonary dysplasia in newborns. It is this clinical relevance, combined with the fact that the degree of oxygen toxicity is reproducible and can be controlled by varying either the oxygen tension or the duration of exposure, that makes hyperoxic lung injury an important and appropriate model for studying acute as well as chronic respiratory distress.

The mechanisms through which hyperoxia induces cellular damage are not completely known. However, oxygen-derived free radicals (superoxide anion and hydroxyl radical), hydrogen peroxide, and epoxides probably play a key role. Epoxides, which are peroxidation products of lipids such as linoleate, are produced by alveolar leukocytes and have been postulated to be directly toxic to alveolar cells.[4] The evidence in support of oxygen free radicals as mediators of hyperoxic injury comes from several sources. In rats rendered tolerant to 100% oxygen by prior exposure to 85% oxygen, an increase in superoxide dismutase activity, an enzyme which converts O_2- to H_2O_2, has been demonstrated.[5] This increase in enzyme activity does not occur in animals in which such tolerance in not induced. Mitochondrial and lung-slice oxygen radical production in rats exposed to 85% oxygen for 7 d is increased[6] and intraperitoneal administration of superoxide dismutase[7] or intravenous injection of liposome-entrapped superoxide dismutase and catalase[8] have been shown to be partially protective against oxygen toxicity. Additional evidence for the role of free radicals in hyperoxia-induced lung damage is reviewed by Jamieson et al.[9] The relative role of the neutrophil in hyperoxic lung damage is somewhat more controversial. Although perhaps not necessary for the development of toxicity, neutrophils probably serve to magnify the direct toxic effects of hyperoxia on the lung.[10-13]

The purpose of this chapter is to discuss the use of hyperoxia as a method of experimentally inducing acute and chronic lung injury. Techniques for inducing lung injury in both small and large animal models will be discussed, the former represented by the rat and the latter by the baboon. The exposure and evaluation techniques described are easily adaptable to other similarly sized animals. The specific exposure techniques described in detail are those currently in use in research laboratories at the Duke University Medical Center. Additional techniques are also discussed and referenced.

IMPLEMENTING THE MODEL

Effective utilization of animal models of pulmonary oxygen toxicity requires an awareness of interspecies, intraspecies, and age-dependent variations in the response of the lung to hyperoxia.[14] Historically, this has been characterized primarily in terms of mean survival times for various species exposed to high oxygen concentrations, although morphologic and morphometric differences between species also occur.[15-17]

As for all animal models of respiratory disease, a reliable source of healthy animals and appropriate holding areas are mandatory. Discussion of specific animal care needs with vivarium and veterinary personnel is often helpful. Because many animals are susceptible to viral and bacterial respiratory infections, appropriate screening and care in preventing disease before, during, and after the exposure period are important. Maintenance of positive pressure or laminar flow ventilatory systems and the use of high-efficiency air filters are ideal. However, this may not be practical for large animals. Noxious waste products, particularly ammonia which is a respiratory irritant, are a potentially important problem but can be easily resolved by ensuring frequent removal of waste material and by providing adequate ventilation.

Rat Model

A chamber used for exposing rats to normobaric hyperoxic conditions is shown in Figures 1A and B. The chamber, measuring (37 × 47 × 41 cm) with an internal volume of 71 l, is constructed of polystyrene and is large enough to allow exposure of six rats weighing 250 to 350 g each. Separate, sealable openings are provided for animal access, cleaning, delivery, monitoring and exhaust of gases, temperature monitoring, food, and water (Figure 1). In the chamber, animals are supported on a stainless-steel wire mesh grid which allows waste material to pass through to an absorbent, deodorized, and neomycin-impregnated animal cage board (Shepard specialty Products, Kalamazoo, MI) below.

Prior to each use, chambers are cleaned with disinfectant, dried, and ventilated for several hours using gases at exposure concentrations. This allows titration and equilibration of the gases and removal of any disinfectant vapors remaining after cleaning. The relative concentrations of various gases are controlled using flow meters (Aalborg Instruments, Monsey, NY), with a total flow rate sufficient to provide five to seven volume changes per cage per hour. Using these parameters, the CO_2 concentration can be maintained at less than 0.6%.[16] Oxygen concentration in the chamber is continuously monitored during the exposure period using an oxygen monitor (Beckman Instruments, Palo Alto, CA) and recorded on a strip chart recorder. Daily changes of the absorbant, deodorized pad are made through the separate sealable door below the animal compartment (Figure 1). Opening this door for brief periods results in only a moderate drop in oxygen concentration inside the chamber. Experimental concentrations can be reattained quickly by flushing the chamber with 100% oxygen until the desired level of hyperoxia is reached. A concentration of greater than 98% can be reached within approximately 15 to 20 min.

In the laboratory of Duke University Medical Center, studies on the effects of hyperoxia in adult rats are generally done using specific-pathogen-free male Sprague-Dawley rats

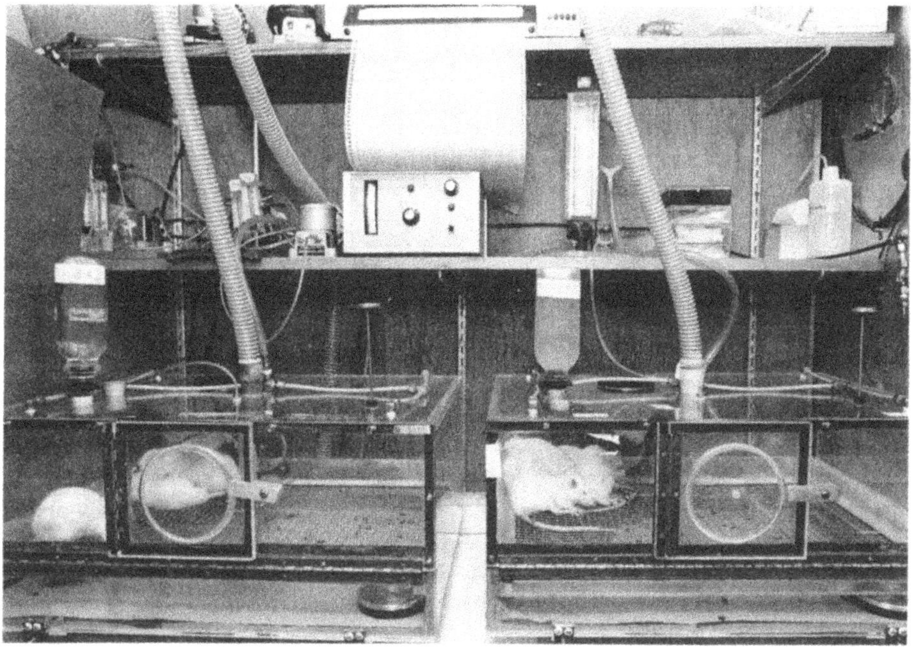

FIGURE 1A. Set-up for exposure of small animals to hyperoxia. Two exposure chambers are shown, with six adult rats in each. See Figure 1B and text for description and schematic representation of the system.

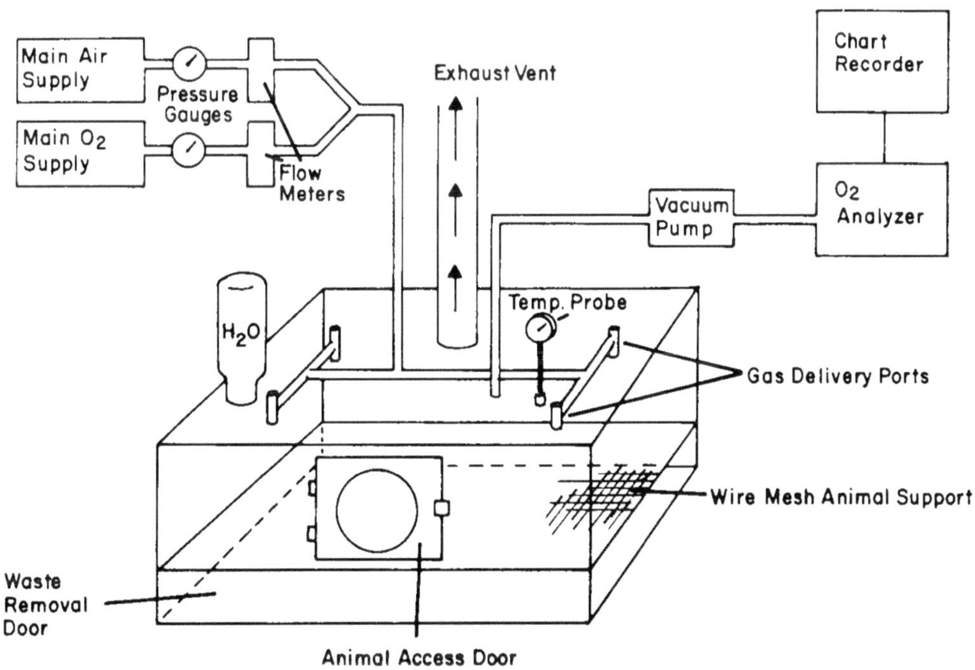

FIGURE 1B. A schematic drawing of the exposure chamber and monitoring system shown in Figure 1A. The arrangement of the gas supply, flow meters, oxygen analyzer, and continuous chart recorder is clearly illustrated. Where large variations in pressure of the main air or oxygen supply sources are present, accurate control of oxygen concentration is achieved by using a series of two to three step-down valves installed between the pressure gauges and the flow meters.

(Charles River Laboratories, Wilmington, MA) weighing approximately 250 to 350 g each. These animals are screened for, and free from, multiple viral, bacterial, mycoplasmal, and parasitic agents using serologic, culture, and pathologic techniques. The adult rat is quite susceptible to oxygen toxicity. The combined mortality of rats exposed to 100% O_2 for 60 h and returned to ambient air is 20 to 25%. The exact mortality rate depends on the age and weight of the animals.

For studies on the effects of hyperoxia in newborns, pregnant female rats are obtained from Zivic Miller Laboratories (Allison Park, PA). More than one lactating mother with pups should be available during the course of the experiment. Mothers should be alternated daily between control litters exposed to ambient air and hyperoxia-exposed litters to avoid adverse effects on the level of maternal care caused by the development of oxygen toxicity in the mother rats. After various durations of exposure of either adults or newborns, the animals are removed from the chamber and may be used either immediately or following a period of recovery.

For morphologic and morphometric studies, animals are anesthetized using an intraperitoneal injection of 85 mg/kg of sodium methohexital (Brevital, Eli Lilly and Company, Indianapolis). The trachea is cannulated, a pneumothorax is induced by puncturing the diaphragm, and the lungs are inflation-fixed with an appropriate fixative. For most routine microscopic techniques, 2% glutaraldehyde in 0.01 *M* cacodylate buffer, pH 7.4, instilled at a filling pressure of 20 to 25 cm of water, will give good results. When using glutaraldehyde fixative, the degree of inflation of the fixed lung is dependent on several variables including the inflation pressure, the diameter and length of the tubing and tracheal cannula, and the characteristics of the fixative itself.[18] The cannula and tubing used should be the widest and shortest practical for the size of the animal, although in 20-gauge or smaller cannulas, the system resistance is independent of tubing length.[18] Ligatures are placed around the trachea and catheter to insure that a good seal is made and to prevent loss of pressure and leakage of fixative. Adequate time for fixation of the tissues should be allowed (approximately 20 to 30 min for 2% glutaraldehyde) before simultaneously removing the cannula and ligating the trachea. After fixation, the tissue is prepared for examination by either light or electron microscopy using standard techniques.[16,19,20]

For biochemical studies, animals are anesthetized as above and sacrificed by transection of the abdominal vessels. Prior to removal of the lungs from the thorax, perfusion of the pulmonary vasculature to remove blood or alveolar lavage to obtain intra-alveolar cells, fluid, and surfactant may be performed. Perfusion of the lungs is performed by first incising the left atrium, then perfusing the lung with iced phosphate-buffered saline at physiologic pH through a cannula inserted into the right ventricle (Figure 2). In adult animals, this can easily be done using a 16- to 18-gauge intravenous catheter cut to a length of approximately 2 cm. In newborn animals as young as 1 d old, a 22- to 25-gauge needle inserted directly into the right ventricle just proximal to the pulmonary artery works well. Approximately 10 ml of perfusate for adult animals, or 5 ml in newborns, is sufficient to completely blanch the lung tissue. Using a syringe, the perfusion should be performed avoiding excessive pressure to prevent retrograde flow or rupture of the heart or pulmonary artery.

Alveolar lavage is performed by inserting a cannula through a small incision in the exposed trachea (Figure 3). An intravenous catheter of appropriate gauge cut to a length of 2 cm works well. Ligatures are used around the trachea to prevent loss of the lavage fluid (Figure 3). When using a syringe, excessive force during infusion and aspiration of the lavage fluid should be avoided to prevent either rupturing the lung during infusion or collapsing the airways during aspiration. Experience indicates that single lavage volumes of approximately 2 ml/100 g body weight are generally adequate to fully inflate the lungs and lavage the alveolar space. Five to ten lavages are usually sufficient for optimal recovery of cells and secretions. Following perfusion and/or lavage, the lungs are excised and either used im-

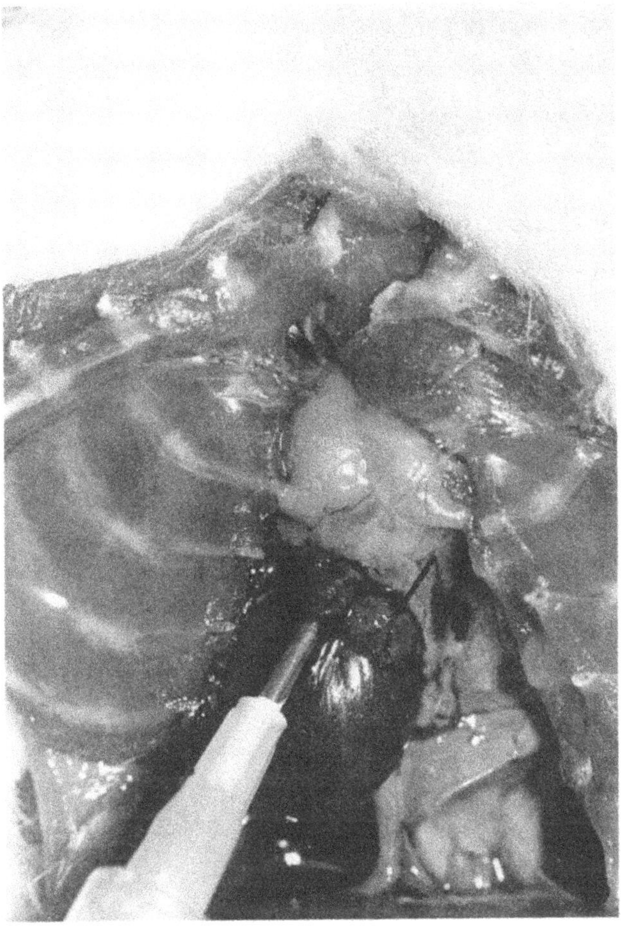

FIGURE 2. Close-up view of technique for performing lung perfusion in rats. The head of the rat is at the top of the photo. A 16-gauge catheter attached to a syringe is shown inserted into a small incision made in the right ventricular wall just proximal to the pulmonary artery. Grasping the myocardial tissue with forceps at the point of entry or placing ligatures around the pulmonary artery and catheter prevent leakage of perfusate. An incision (arrow) has been made in the left atrium to allow efflux of blood and perfusate.

mediately for biochemical assays or cell isolation or stored using techniques appropriate for the planned future assays.

Baboon Model

Adult, colony-raised baboons are obtained from the Southwest Foundation for Biomedical Research (San Antonio). On arrival, the animals are quarantined on a research farm for a minimum of 2 weeks. Screening for tuberculosis and other active respiratory and nonrespiratory infections is performed during this period. Utilization of colony-raised baboons lessens the likelihood of lung mites which may be common in feral animals. For all morphometric and biochemical studies, hematoxylin-eosin-stained control and experimental lung sections are also examined under a light microscope for evidence of active pulmonary infection. Data obtained from animals suspected of active infection are rejected.

A chamber used for exposing large animals such as the baboon is shown in Figure 4. This chamber, which measures 127 × 184 × 151 cm and has an approximate volume of

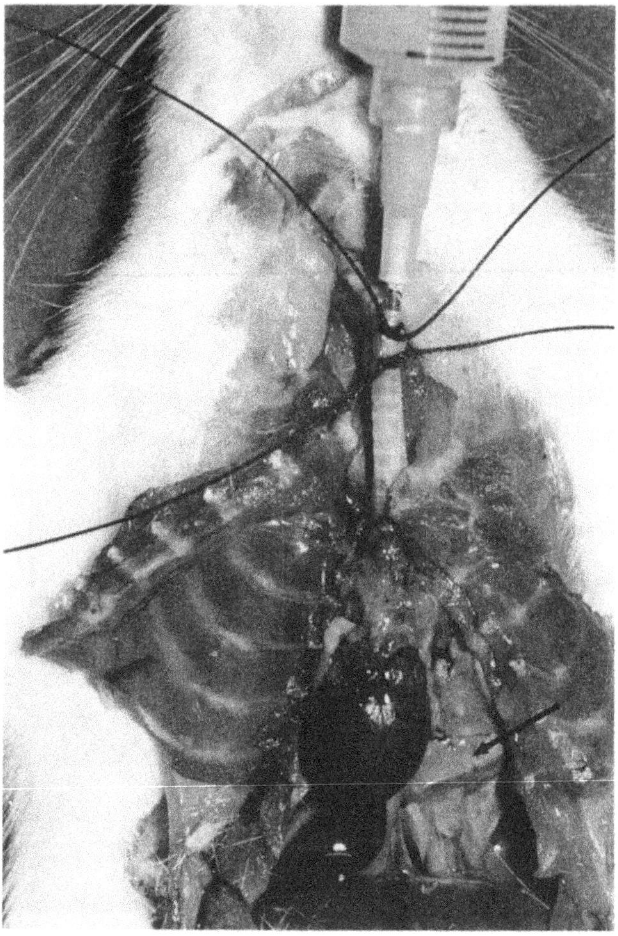

FIGURE 3. Photograph showing technique for performing alveolar lavage in rats. The thorax and neck are opened with a median incision, exposing the thoracic contents and trachea. A 16-gauge catheter is shown inserted into a small incision in the trachea. Ligatures are used to prevent leakage of lavage fluid. The lungs (arrow) in this photograph are deflated.

3528 l, is made of plexiglass sheets attached to a wooden frame using screws and a caulk sealant. The chamber is provided with sealable ports for the delivery, sampling, and exhaust of gases and is large enough to accept a holding cage in which the animal is actually kept. This obviates the need for anesthesia, sedation, intubation, or restraint of the animal during at least the first 80 h or exposure and the infectious, pharmacologic, and traumatic complications from such measures. After placing the caged animal in the chamber, the front wall and gasket are replaced and sealed with wing nuts and occlusive duct tape.

During the intrachamber exposure period, food and water are provided *ad libitum*. Adequacy of food and water is assessed every 12 h and provided as needed. Waste material is collected in a disinfected, wood-shaving-lined metal pan which is removed after 36 h and replaced with another. After initially placing the animal in the chamber, or following food, water, or waste pan changes, the chamber is flushed with 100% oxygen at high flow rates until the desired oxygen concentration is reached. Oxygen concentrations of greater than 98% can be achieved in approximately 30 to 40 min. The rate of gas flow is then decreased to about 30 l/min for the remainder of the exposure period.

In baboons, exposure to 100% oxygen for 72 to 80 h results in obvious signs of clinical

FIGURE 4. Oxygen exposure chamber for large animals such as baboons. A holding cage containing an adult male baboon (arrow) is inside the chamber. An oxygen probe and monitor are shown (arrow heads). The details for connection of flow meters, gas analyzers, and chart recorders are similar to those shown in Figure 1B.

illness including cough, tachypnea, anorexia, and emesis. If longer exposure periods are planned, sedation and more direct supportive measures are indicated to both prevent premature death as well as dehydration and discomfort for the animal. At this point, or when surgical procedures or physiologic measurements are made, the baboons are given a tranquilizing dose of 10 to 15 mg/kg of ketamine intramuscularly and removed from the chamber. An intravenous line is started and the animals are intubated with a #7 French® endotracheal tube. Additional sedation and neuromuscular blockade is accomplished with approximately 0.5 mg/kg of diazepam and 2 mg of pancuronium bromide (Pavulon, Organon Pharmaceuticals, West Orange, NJ) intravenously. These doses are repeated approximately every 2 h. While intubated and sedated, the baboons are turned, given chest percussion, and endotracheally suctioned every 2 h. Hydration is maintained by intravenous administration of a 5% dextrose, 0.9% sodium chloride solution with 20 meq/l of potassium chloride at 50 to 100 ml/h. Ventilation is supported with a volume-cycled ventilator using a tidal volume of 10 to 15 mg/kg body weight, a respiratory rate of approximately 10 to 12 breaths per minute, and positive end-expiratory pressure of 3 cm of water. Adequacy of ventilation is monitored with blood gases and the above parameters are adjusted to maintain a pCO_2 of

32 to 38 mmHg. To prevent infection during the period of mechanical ventilation, animals are given penicillin (1 million units intravenously every 6 h), polymyxin (20,000 units orally once a day and 20,000 units intratracheally every 4 h), and gentamicin (5 to 7 mg/kg/d intratracheally in six divided doses).

When morphometric studies are planned, animals may be used as their own controls by performing a surgical right middle lobectomy prior to oxygen exposure. Following lobectomy, animals are allowed to recover for a minimum of 4 weeks before beginning the oxygen exposure. After excision, the control right middle lobe bronchus is cannulated, inflation-fixed with an appropriate fixative at a filling pressure of 20 to 25 cm of water, and processed as described previously for the rat model. When hemodynamic or pulmonary physiologic studies are planned, the sedated animals are intubated, a Swan-Ganz catheter is placed inn the pulmonary artery via the femoral or axillary vein, and an arterial line is inserted into the femoral or axillary artery. The various physiologic studies performed are discussed below. For biochemical determinations, control samples obtained from the right middle lobe or post-oxygen-injury lung tissue samples are used immediately for a given biochemical assay or frozen at −70°C for future use.

EVALUATING THE MODEL

Microscopic Techniques

Both light and electron microscopic techniques are used to qualitatively and quantitatively evaluate the pathologic characteristics of hyperoxic lung injury and its subsequent repair, as well as to assess the efficacy of any therapeutic intervention. Morphometric determination of parameters such as the alveolar epithelial and capillary surface areas; epithelial, endothelial, and interstitial volumes; thickness of the interstitium; and calculation of individual cell numbers are particularly useful and performed using techniques described by Weibel and others.[16,19-21] Data obtained may be presented in several manners. For volume and cell number determinations, data may be reported as absolute total volume or number per lung of a given compartment or cell type or expressed in terms of volume or number per unit of lung volume. The disadvantage of the latter method is that the total lung volume is variable and dependent on many factors including the compliance of the lung; the pressure, flow rate, and nature of the fixative; and whether fixation is accomplished *in vivo* or *ex vivo*.[18] This variability makes comparison of different sets of data difficult; it does not occur when data are reported in absolute terms. Calculation of absolute volumes or cell numbers does, however, require determination of the proportion of the total displacement lung volume that is alveolar lung tissue.

Qualitative and nonmorphometric quantitative microscopy studies (i.e., immunocytochemistry or autoradiography) can also be applied to the hyperoxic injury model. The individual techniques available and specific technical difficulties associated with these techniques are, however, beyond the scope of this chapter.

Physiologic Techniques

A variety of physiologic parameters have been used to assess the lung injury resulting from hyperoxia. Spirometry, lung volumes, pulmonary gas exchange, and control of ventilation have all been studied and many of the findings have been reviewed.[3] In rabbits and larger animals, pulmonary artery and pulmonary capillary wedge pressures, cardiac output, lung lymph flow, lymph protein concentration, capillary filtration coefficient, and arterial blood gas tensions have all been studied.[22-25] For physiologic studies on baboons in laboratories at Duke University, animals are first sedated and intubated as described above. Swan-Ganz and arterial catheters are inserted to allow measurement of pulmonary artery, right atrial and systemic pressures, thermodilution cardiac output, and for obtaining blood samples. In addition, the stroke volume, ejection fraction, and cardiac output are measured using gated radionuclide angiography.[26] Total lung volume is determined using a helium dilution method. A 10% helium:air mixture is introduced into the lungs of the intubated animal at functional residual capacity and the lungs are then inflated to total lung capacity, defined as the volume at 40 cm H_20 pressure. Ten ventilatory cycles are completed, the final concentration of helium is measured using a helium analyzer (Warren-Collins, Boston), and that measurement is used to calculate the total lung volume using a standard formula. To minimize variability in the level of sedation and the side effects thereof, all physiologic measurements are performed at the midpoint between sedative doses.

Biochemical Techniques

The number of biochemical tests that can be performed to evaluate the effects of hyperoxia on the lung is almost limitless. However, a few tests are more standard and are therefore useful for assessing the degree of acute lung damage, its chronic effects, and any potential alterations in the usual course resulting from therapeutic intervention.

Because the pulmonary alveolar capillary bed is a major site of hyperoxia-induced lung injury in all species, measurement of pulmonary capillary permeability can be an important means of assessing toxicity. This has been done *in vivo* by measuring the accumulation of intravenously administered ^{125}I-albumin,[27] ^{57}Co-cyanocobalamin,[27] and ^{125}I-polyvinyl pyrrolidone[28] and by determining the levels of intra-alveolar albumin in bronchoalveolar lavage fluid using and ELISA technique.[20,29] Measurement of wet-to-dry lung weight ratios, lung lymph flow, and lymph protein content have also been performed.[22,23] A more sophisticated technique of assessing capillary permeability is to determine the permeability-surface-area product for the lung.[25]

The late sequalae of hyperoxic lung injury include the development of pulmonary fibrosis.[19,30-33] An integral part of this fibrotic process is an increase in total lung collagen. As such, assessment of changes in lung collagen metabolism offers a convenient means of evaluating the degree of lung injury and the potential efficacy of any therapeutic interventions aimed at limiting the injury or its complications. Close approximation of lung collagen is most simply and reproducibly done by measurement of tissue hydroxyproline content using the method of Kivrikko et al.[34] as described by Berg.[35] Although this amino acid is also a component of elastin and the C1q component of complement, these have been estimated to contribute less than 10% of the total lung hydroxyproline.[36]

Measurement of the absolute rate of collagen synthesis provides more detailed information regarding collagen metabolism but is more complex due to the effects of precursor availability and collagenolysis. Collagen synthesis may be determined by measuring the incorporation of radiolabeled proline into the hydroxyproline of procollagen or collagen[37] and can be preformed either *in vivo* or *in vitro*. Appropriate interpretation of results requires measurement of the specific activity of the radiolabeled precursor as well as determination of the rate of degradation of newly synthesized collagen containing radiolabeled hydroxyproline. For collagen, the optimal precursor to study is somewhat controversial. Determination of the specific activity of tRNA proline pools is perhaps ideal[38] but requires the use of relatively large amounts of isotope and is technically more demanding. Determination of the specific activity of free proline pools may be equally appropriate and is more easily done using a combination of photometric and radiolabeling techniques.[35,38] Although release of newly incorporated radiolabed hydroxyproline resulting from collagenolysis cannot readily be measured, it has been estimated to result in a less than 2.5% underestimation of synthesis in the normal rabbit lung.[36]

Levels of activity of lung superoxide dismutase and catalase have also been used to assess the response to hyperoxia.[5-7,9,16,40] Such measurements are particularly useful in evaluating therapeutic modalities directly aimed at increasing the levels of these enzymes.[7,40]

Many other biochemical alterations have been observed in the hyperoxic model of lung injury, including changes in levels of lung ornithine decarboxylase and polyamines,[41] hyaluronidase,[42] fibronectin,[19] glycosaminoglycans,[43] histamine,[44] serotonin,[45] angiotensin,[23] and prostaglandins.[46] While these changes are very important to the understanding of hyperoxic lung injury and repair, they are of less general utility in evaluating the model itself.

EXPERIMENTAL FINDINGS

In any animal model as widely studied as hyperoxic lung injury, there is inevitably an extensive body of published scientific literature. Because the subject of pulmonary oxygen toxicity was thoroughly reviewed by Clark and Lamberston in 1971,[3] only a few of the significant contributions since that time will be discussed. This is not intended to be an exhaustive review of the subject, but rather an introductory, updated summary of the current understanding of hyperoxic lung injury and repair.

Morphology and Cell Kinetics

In a recent review, Crapo grouped the cellular and structural responses of the lung to both lethal (acute) and sublethal (chronic) oxygen concentrations into overlapping stages or phases.[47] In lethal oxygen toxicity, initiation, inflammatory, and destructive phases occur. Some of the morphologic changes in these phases are summarized in Figure 5 and below.

The initiation phase of lethal oxygen toxicity, during which the production of oxygen free radicals begins to exceed the ability of the cell to metabolize them, is characterized by a relative lack of significant morphologically demonstrable changes. Increases in pericapillary fluid and the accumulation of increased numbers of intracapillary neutrophils and platelets mark the onset of the inflammatory phase. Microscopic evidence of parenchymal cellular damage begins to increase rapidly, possibly as a result of amplification of direct alveolar damage by inflammatory cell products and metabolites,[31] although studies on the role of neutrophils in hyperoxic lung injury are somewhat contradictory.[11,13] The destructive phase of lethal hyperoxic lung injury overlaps the inflammatory phase and is characterized by destruction of the capillary endothelial cells and proportional loss of capillary surface area. In rats, this phase leads directly to death if exposure continues and, although total interstitial and inflammatory cell number does increase, there is no evidence of a fibrotic response.

In sublethal oxygen toxicity, initiation, inflammatory, and destructive phases also occur (Figure 6). The onset and duration of these phases and the specific morphologic changes depend in part on the concentration of oxygen used. In contrast to lethal hyperoxic exposure, a fibrotic phase is described. This phase, which overlaps the destructive phase, is associated with an increase in interstitial fibroblasts and monocytes and an increase in the noncellular interstitium.[16] Similar morphologic changes have also been reported in newborn mice exposed to 80% oxygen for up to 6 weeks.[17]

The cellular and morphologic events occurring during the repair of oxygen-induced lung injury have been reviewed by Clark and Lambertson[3] and more recently by Thet.[31] Although both acute and chronic oxygen exposure are associated with increases in the number of interstitial fibroblasts and myofibroblasts, the duration of increases in myofibroblast number, the volume of the extracellular matrix,[20,31] and the amount of lung collagen[19,40] are greater following chronic hyperoxia.[20] These changes probably reflect the cumulative effect of concurrent long-term injury and repair.

Physiologic Changes

Pulmonary function testing has revealed a decrease in the total lung capacity, expired minute volume diffusing capacity to carbon monoxide, lung compliance, and inspiratory flow volumes and rates in humans exposed to greater than 75% oxygen at 1.0 atm and above

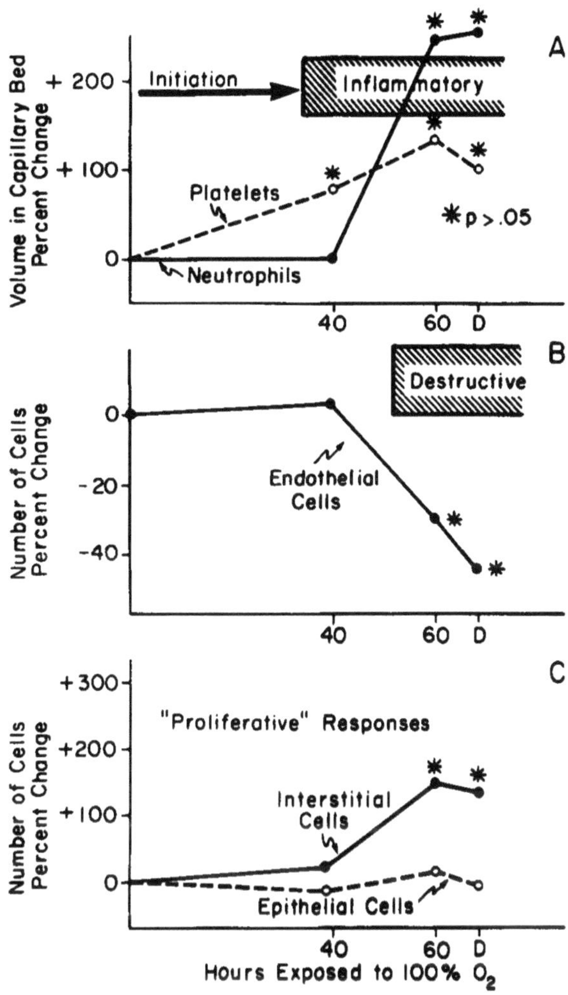

FIGURE 5. Major morphologic changes in the lungs of rats associated with each major phase of injury during exposure to a lethal level of hyperoxia. Panel A shows changes in the total volume of platelets and neutrophils found in the alveolar capillary bed. Panels B and C correlate the timing of changes in the numbers of endothelial, interstitial, and epithelial cells. (Reproduced with permission from the *Annual Review of Physiology*, Vol. 48, © 1986 by Annual Reviews Inc.)

of ambient pressure.[3] Although identical measurements are difficult to duplicate in animals due to their inability to perform many of the ventilatory maneuvers necessary, comparable changes in function have been demonstrated. Rats exposed to 60% oxygen for 7 d demonstrate an increase in the pressure/volume relationship of their excised lungs, suggesting a decrease in lung compliance.[18] In rabbits, however, exposure to 100% oxygen for 64 h did not alter lung compliance or total lung capacity, although both parameters decreased by 30% during the first 24 h following exposure.[48]

The susceptibility of the pulmonary vascular system to hyperoxia can also be detected physiologically. Increases in lung water content, lymph flow, and lymph protein clearance have been demonstrated in sheep during[23] and following[22] exposure to 100% oxygen. These changes are associated with an increase in the alveolar-arterial oxygen differences and loss of the normal pulmonary vasoconstrictive response to hypoxia.

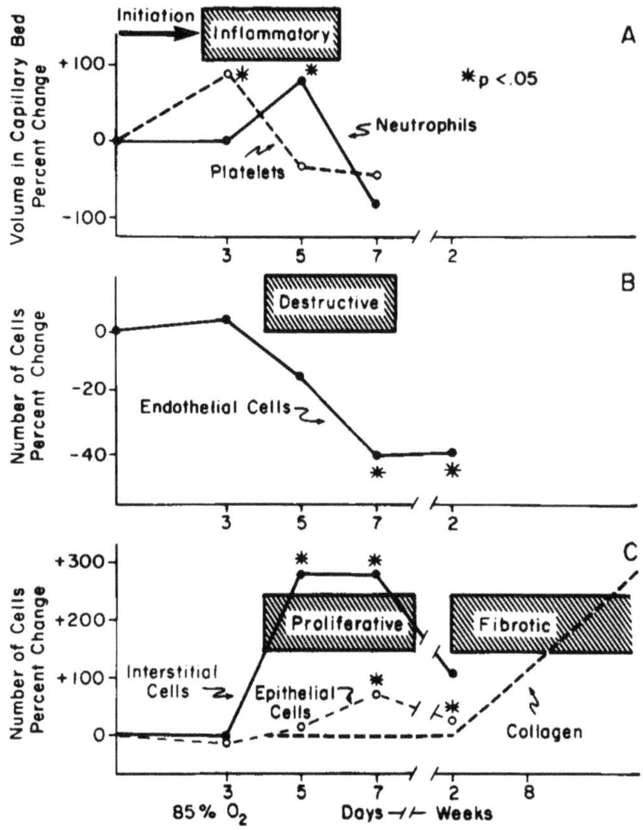

FIGURE 6. Major morphologic changes in the lung associated with each major phase of injury during exposure to a sublethal level of hyperoxia. (Reproduced with permission from the *Annual Review of Physiology*, Vol. 48, © 1986 by Annual Reviews Inc.)

Surprisingly few cardiovascular hemodynamic alterations in response to hyperoxia have been reported. No significant difference in mean pulmonary arterial pressure, left atrial pressure, cardiac output, or mean systemic arterial pressure was observed in dogs[44] or sheep,[23] except in the immediate preterminal period. In baboons, however, a decrease in the cardiac output, as measured by a thermodilution method, was observed following exposure to 98% oxygen. This has been corroborated by a decrease in ejection fraction measured by radionuclide angiography.[26]

Etiology and Biochemical Changes

It is generally accepted that oxygen-derived free radicals are mechanistically important in the production of lung tissue damage due to oxygen toxicity.[9,49] Many of these free radicals are undoubtedly produced by the lung cells themselves. Although controversial, some evidence exists to suggest that neutrophils and platelets may act to augment or magnify this injury,[10-12] probably through the release of oxygen-derived free radicals, proteases, vasoactive substances such as serotonin and histamine, or production of lipid and arachidonic acid metabolites such as the epoxides and leukotrienes.[4,12,50]

The biochemical changes occurring during the repair period following hyperoxic injury are the subject of much recent research. Many of the changes occurring in the metabolism of collagen, fibronectin, glycosaminoglycans, and polyamines have recently been reviewed by Thet.[31]

Prevention and Therapeutic Intervention

The prevention of oxygen toxicity and the therapeutic modulation of its long-term sequelae are also major areas of interest. One approach has focused on the efficacy of exogenously supplied superoxide dismutase or catalase in preventing or ameliorating oxygen toxicity. Liposome-encapsulated superoxide dismutase and/or catalase have been found to protect against oxygen toxicity *in vivo* when administered intravenously,[51] intratracheally,[40] and intraperitoneally[7] and *in vitro* when administered to cultured endothelial cells.[52]

Several other agents have been studied in the hope that they might favorably alter oxygen toxicity or its sequelae. Unfortunately, these agents have to date been nonefficacious, deleterious, or of uncertain clinical utility. α-Tocopherol (vitamin E), an antioxidant capable of inhibiting lipid peroxidation, had no effect on lung lymph flow, lymph protein concentration, or cardiopulmonary hemodynamics when administered to lambs exposed to pure oxygen.[53] *In vitro* administration of *N*-acetylcysteine, another antioxidant, protects cultured lung epithelial cells from polymorphonuclear cell-mediated oxidant injury,[54] although *in vivo* studies are lacking. However, in rats exposed to 95% oxygen intratracheal insufflation of erythrocytes, a source of readily oxidizable glutathione, resulted in a significant increase in survival[55] and suggests a potential clinical role for such agents.

The effects of glucocorticoids on pulmonary oxygen toxicity appear to be dependent on the time of administration. Administration of glucocorticoids to rats early in the course of oxygen exposure is associated with diminished survival,[56,57] although there is some evidence suggesting that later administration may be protective.[56] Antifibrotic agents such as the proline analogues have been shown to inhibit the accumulation of collagen following oxygen-induced injury in rats, although the effect was nonspecific because increases in noncollagen protein were also noted.[32] The functional implications of such inhibition are not known.

FUTURE DIRECTIONS AND SUMMARY

Although much is known about the pathologic and physiologic changes in the lung resulting from exposure to hyperoxia, a large number of unanswered questions remain regarding the mechanisms, biochemical changes, prevention, and treatment of hyperoxic lung injury. Further definition of the changes in composition of the extracellular matrix, basement membrane, and cell surface during hyperoxic injury and its repair are important areas for future research. Definition of these changes, their functional role, and the molecular mechanisms controlling them will not only add to our understanding of oxidant-induced lung injury and repair but might potentially identify points at which the process could be beneficially interrupted.

Another potentially fruitful area of research is the improvement of methods of directly limiting the damage caused by oxygen-derived free radicals. At present, increasing levels of activity of antioxidant enzymes[51,52] or administration of exogenous scavengers of free radicals[54,55] offer the greatest potential clinical utility. Additional study of the effects of acute and chronic systemic administration of conjugated or liposome-encapsulated antioxidant enzymes on the cellular and immune function of the host[58] is needed before extensive human trials can be considered.

Although the pulmonary fibrosis following hyperoxic lung injury is not severe, its prevention may potentially lead to functional improvement. Unfortunately, the factors which differentiate the process of collagen synthesis and reorganization during normal wound healing and that of excessive or uncontrolled fibrosis are not known. Nonselective inhibition of collagen metabolism during the repair period has the potential risk of disrupting not only the abnormal collagen deposition associated with fibrosis, but also the normal or reparative process, thus resulting in emphysematous changes in alveolar structure. This phenomenon has been demonstrated in rats given β-aminoproprionitrile during and after hyperoxic injury.[59]

In summary, animal models of pulmonary oxygen toxicity provide a practical, reproducible, and controllable means of inducing both acute and chronic lung injury. As such, they are useful not only for the investigation of oxidant-mediated lung injury and repair but also the more specific and clinically relevant problems of adult respiratory distress syndrome and bronchopulmonary dysplasia

ACKNOWLEDGMENTS

The authors thank Dr. Phil Fracica and Dr. James Crapo for providing details of the techniques used in their laboratory for exposure of primates to hyperoxia. The authors also thank Mr. Roberto Soto for typing and proofreading this manuscript. R. A. Durr is a Thomas H. Davis Fellow of the American Lung Association of North Carolina and L. A. Thet is an Established Investigator of the American Heart Association. Much of the authors' research cited in this chapter has been supported by P01 HL31992 from the National Heart, Lung and Blood Institute.

REFERENCES

1. **Priestly, J.,** *Experiments and Observations on Different Kinds of Air,* Vol. 2, London, 1775.
2. **Smith, J. L.,** The pathological effects due to increase of oxygen tension in the air breathed, *J. Physiol. (London),* 24, 19, 1899.
3. **Clark, J. M. and Lamberston, C. J.,** Pulmonary oxygen toxicity: a review, *Pharmacol. Rev.,* 23, 37, 1971.
4. **Ozawa, T., Hayakawa, M., Takamura, T., Sugiyama, S. Suzuki, K., Iwata, M., Taki, F., and Tomita, T.,** Biosynthesis of leukotoxin 9,10-epoxy-12 octadecanoate, by leukocytes in lung lavages of rat after exposure to hyperoxia, *Biochem. Biophys. Res. Commun.,* 134, 1071, 1986.
5. **Crapo, J. D. and Tierney, D. F.,** Superoxide dismutase and pulmonary oxygen toxicity, *Am. J. Physiol.,* 226, 1401, 1974.
6. **Freeman, B. A. and Crapo, J. D.,** Hyperoxia increases oxygen radical production in rat lungs and lung mitochondria, *J. Biol. Chem.,* 256, 10986, 1981.
7. **McLennan, G. and Autor, A. P.,** Effect of intraperitoneally administered superoxide dismutase on pulmonary damage resulting from hyperoxia, in *Pathology of Oxygen,* Autor, A. P., Ed., Academic Press, New York, 1982, 85.
8. **Turrens, J. F., Crapo, J. D., and Freeman, B. A.,** Protection against oxygen toxicity by intravenous injection of liposome-entrapped catalase and superoxide dismutase, *J. Clin. Invest.,* 73, 87, 1984.
9. **Jamieson, D., Chance, B., Cadenas, E., and Boveris, A.,** The relation of free radical production to hyperoxia, *Annu. Rev. Physiol.,* 48, 703, 1986.
10. **Fox, R. B., Hoidal, J. R., Brown, B. M., and Repine, J. E.,** Pulmonary inflammation due to oxygen toxicity: involvement of chemostatic factors and polymorphonuclear leukocytes, *Am. Rev. Respir. Dis.,* 123, 521, 1981.
11. **Shasby, D. M., Fox, R. B., Hanada, R. N., and Repine, S. E.,** Reduction of the edema of acute hyperoxic lung injury by granulocyte depletion, *J. Appl. Physiol.,* 52, 1237, 1982.
12. **Repine, J. E.,** Pulmonary oxygen toxicity: current assessment of the contributions of oxygen metabolites and neutrophils, in *Physiology of Oxygen Radicals,* Taylor, A. E., Matalon, S., and Ward, P. A., Eds., American Physiological Society, Bethesda, MD, 1986, 119.
13. **Raj, J. U., Hazinski, T. A., and Bland, R. D.,** Oxygen-induced lung microvascular injury in neutropenic rabbits and lambs, *J. Appl. Physiol.,* 58, 921, 1985.
14. **Frank, L., Bucher, J. R., and Roberts, R. J.,** Oxygen toxicity in neonatal and adult animals of various species, *J. Appl. Physiol. Respir. Environ. Exercise Physiol.,* 45, 699, 1978.
15. **Adamson, I. Y. R., Bowden, D. H., and Wyatt, J. P.,** Oxygen poisoning in mice. Ultrastructural and surfactant studies during exposure and recovery, *Arch. Pathol.,* 90, 463, 1970.
16. **Crapo, J. D., Barry, B. E., Foscue, H. A., and Shelburne, J. D.,** Structural and biochemical changes in rat lungs occurring during exposure to lethal and adaptative doses of oxygen, *Am. Rev. Respir. Dis.,* 122, 123, 1980.

17. **Pappas, C. T. E., Obara, H., Bensch, K. G., and Northway, W. H.,** Effect of prolonged exposure to 80% oxygen on the lung of the newborn mouse, *Lab. Invest.*, 48, 735, 1983.
18. **Hayatdavoudi, G., Crapo, J. D., Miller, F. J., and O'Neil, J. J.,** Factors determining degree of inflation in intratracheally fixed rat lungs, *J. Appl. Physiol. Respir. Environ. Exercise Physiol.*, 48, 389, 1980.
19. **Thet, L. A., Parra, S. C., and Shelburne, J. D.,** Sequential changes in lung morphology during the repair of acute oxygen induced lung injury in adult rats, *Exp. Lung Res.*, 11, 209, 1986.
20. **Durr, R. A., Dubaybo, B. A., and Thet, L. A.,** Repair of chronic hyperoxic lung injury: changes in lung ultrastructure and matrix, *Exp. Mol. Pathol.*, in press.
21. **Weibel, E. R. and Elias, H.,** *Quantitative Methods in Morphology*, Springer-Verlag, New York, 1967.
22. **Erdmann, A. J., Huttemeier, P. C., Landolt, C., and Zapol, W. M.,** Pure oxygen breathing increases sheep lung microvascular permeability, *Anesthesiology*, 58, 153, 1983.
23. **Newman, J. H., Loyd, J. E., English, D. K., Ogletree, M. C., Fulkerson, W. J., and Brigham, K. L.,** Effects of 100% oxygen on lung vascular function in awake sheep, *J. Appl. Physiol Respir. Environ. Exercise Physiol.*, 54, 1379, 1983.
24. **Harabin, A. L., Homer, L. D., and Bradley, M. E.,** Pulmonary oxygen toxicity in awake dogs: metabolic and physiologic effects, *J. Appl. Physiol. Respir. Environ. Exercise Physiol.*, 57, 1480, 1984.
25. **Taylor, A. E. and Townsley, M. I.,** Assessment of oxygen radical tissue damage, in *Physiology of Oxygen Radicals*, Taylor A. E., Matalon, S., and Ward, P. A., Eds., American Physiological Society, Bethesda, MD, 1986, 19.
26. **Fracica, P. J., Knapp, M. J., Jafri, A., Seifer, F. D., Coleman, R. E., Wolfe, W. G., and Crapo, J. D.,** Myocardial oxygen toxicity in primates, *Am. Rev. Respir. Dis.*, 135 (Suppl. 4), A11, 1987.
27. **Matalon, S. and Egan, E. A.,** Interstitial fluid volumes and albumin spaces in pulmonary oxygen toxicity, *J. Appl. Physiol. Respir. Environ. Exercise Physiol.*, 57(6), 1767, 1984.
28. **Valimaki, M., Kivisaari, J., and Niinikoski, J.,** Permeability of alveolar capillary membrane in oxygen poisoning, *Aerosp. Med.*, 45, 370, 1974.
29. **Dubaybo, B. A., Durr, R. A., and Thet, L. A.,** Paraquat induced lung fibrosis: changes in lung fibronectin and collagen after different degrees of paraquat lung injury, *J. Toxicol. Environ. Health*, in press.
30. **Valimaki, M., Juva, K., Rantanen, J., Ekfors, T., and Niinikoski, J.,** Collagen metabolism in rat lung during chronic intermittent exposure to oxygen, *Aviat. Space Environ. Med.*, 46, 684, 1975.
31. **Thet, L. A.,** Repair of oxygen induced lung injury, in *Physiology of Oxygen Radicals*, Taylor A. E., Matalon, S., and Wood, P. A., Eds., American Physiological Society, Bethesda, MD, 1986, 87.
32. **Riley, D. J., Kerr, J. S., and Shin, Y. Y.,** Effect of proline analogs on oxygen toxicity — induced pulmonary fibrosis in the rat, *Toxicol. Appl. Pharmacol.*, 75, 554, 1984.
33. **Chvapil, M. and Peng, Y. M.,** Oxygen and lung fibrosis, *Arch. Environ. Health.*, 30, 528, 1975.
34. **Kivrikko, K. I., Laitinerr, O., and Prockop, D. J.,** Modification of a specific assay for hydroxyproline in urine, *Anal. Biochem.*, 19, 249, 1967.
35. **Berg, R. A.,** Determination of 3- and 4-hydroxyproline, *Methods Enzymol.*, 82, 372, 1982.
36. **Laurent, G. J.,** Rates of collagen synthesis in lung, skin and muscles obtained *in vivo* by amplified method using [^{3}H] proline, *Biochem. J.*, 206, 535, 1982.
37. **Peterofsky, B., Chojkier, M., and Bateman, J.,** Determination of collagen synthesis with tissue and cell culture systems, in *Immunochemistry of the Extracellular Matrix*, Vol. 2, Furthmayr, H., Ed. CRC Press, Boca Raton, FL, 1982, 19.
38. **Rannels, D. E., Low, R. B., Yondale, T., Volkin, E., and Longmore, W. J.,** Use of radioisotopes in quantitative studies of lung metabolism, *Fed. Proc.*, 41, 2833, 1982.
39. **Troll, W. and Lindsley, J.,** A photometric method in the determination of proline, *J. Biol. Chem.*, 215, 655, 1955.
40. **Padmanabhan, R. V., Gudapaty, R., Liener, I. E., Schwartz, B. A., and Hoidal, J. R.,** Protection against pulmonary oxygen toxicity in rats by intratracheal administration of liposome encapsulated superoxide dismutase or catalse, *Am. Rev. Respir. Dis.*, 132, 164, 1985.
41. **Thet, L. A., Parra, S. C., and Shelburne, J. D.,** Repair of oxygen induced lung injury in adult rats. The role of ornithine decarboxylase and polyamines, *Am. Rev. Respir. Dis.*, 129, 174, 1984.
42. **Thet, L. A., Howell, A. C., and Han, G.,** Changes in hyaluronidase activity associated with lung growth, injury and repair, *Biochem. Biophys. Res. Commun.*, 117, 71, 1983.
43. **Durr, R. A. and Thet, L. A.,** Changes in glycosaminoglycan content following acute hyperoxic lung injury in adult rats, *Am. Rev. Respir. Dis.*, 133 (Suppl. 4), A110, 1986.
44. **Januszkiewicz, A. J., Huntrakoon, M., Wilson, P., and Faiman, M. D.,** Isolated perfused lung histamine release, lipid peroxidation and tissue superoxide dismutase from rats exposed to normobaric hyperoxia, *Toxicology*, 39, 37, 1986.
45. **Mais, D. E., Lahr, P. D., and Rosin, T. R.,** Oxygen induced lung toxicity. Effect of serotinin disposition and metabolism, *Toxicol. Appl. Pharmacol.*, 64, 221, 1982.
46. **Toivonen, H., Hartiala, J., and Bakhle, Y. S.,** Effects of high oxygen tension on the metabolism of vasoactive hormones in isolated perfused lungs, *Acta Physiol. Scand.*, 111, 185, 1981.

47. **Crapo, J. D.,** Morphologic changes in pulmonary oxygen toxicity, *Annu. Rev. Physiol.*, 48, 721, 1986.
48. **Holm, B. A., Notter, R. H., Siegle, J., and Matalon, S.,** Pulmonary physiologic and surfactant changes during injury and recovery from hyperoxia, *J. Appl. Physiol.*, 59, 1402, 1985.
49. **Freeman, B. A. and Crapo, J. D.,** Biology of disease. Free radicals and tissue injury, *Lab. Invest.*, 47, 412, 1982.
50. **Iwata, M., Takagi, K., Satake, T., Sugiyama, S., and Ozawa, T.,** Mechanism of oxygen toxicity in rat lungs, *Lung*, 164, 93, 1986.
51. **Turrens, J. F., Crapo, J. D., and Freeman, B. A.,** Protection against oxygen toxicity by intravenous injection of liposome-entrapped catalase and superoxide dismutase, *J. Clin. Invest.*, 73, 87, 1984.
52. **Freeman, B. A., Young, S. L., and Crapo, J. D.,** Liposome mediated augmentation of superoxide dismutase in endothelial cells prevents oxygen injury, *J. Biol. Chem.*, 258, 12534, 1983.
53. **Hansen, T. N., Hazinzki, T. A., and Bland, R. D.,** Vitamin E does not prevent oxygen induced lung injury in newborn lambs, *Pediatr. Res.*, 16, 583, 1982.
54. **Simon, L. M. and Suttorp, J.,** Lung cell oxidant injury: decrease in oxidant mediated cytotoxicity by N-acetyl cysteine, *Eur. J. Respir. Dis.*, 66 (Suppl. 139), 132, 1985.
55. **VanAsbeck, B. S., Hoidal, J., Vercellotti, G. M., Schwartz, B. A., Moldow, C. F., and Jacob, H. S.,** Protection against lethal hyperoxia by tracheal insufflation of erythrocytes: role of red cell glutathione, *Science*, 227, 756, 1985.
56. **Koizumi, M., Frank, L., and Massaro, D.,** Oxygen toxicity in rats. Varied effects of dexamethasone treatment depending on duration of hyperoxia, *Am. Rev. Respir. Dis.*, 131, 907, 1985.
57. **Halpern, P., Teitelman, U., and Lanir, A.,** Effect of methylprednisolone on normobaric pulmonary oxygen toxicity in rats, *Respiration*, 48, 153, 1985.
58. **McDonald, R. J., Berger, E. M., White, C. W., White, C. G., Freeman, B. A., and Repine, J. E.,** Effect of superoxide dismutase encapsulated in liposomes or conjugated with polyethylene glycol on neutrophil bactericidal activity in vitro and bacterial clearance in vivo, *Am. Rev. Respir. Dis.*, 131, 633, 1985.
59. **Riley, D. J., Kramer, M. J., Kerr, J. S., Chae, C. U., Yu, S. Y., and Berg, R. A.,** Damage and repair to lung connective tissue in rats exposed to toxic levels of oxygen, *Am. Rev. Respir. Dis.*, 135, 441, 1987.

OZONE-INDUCED LUNG INJURY

David J. P. Bassett

INTRODUCTION

The effects of ozone exposure on lung morphology, physiology, and biochemistry have been extensively investigated during the past 20 years. It is the vast literature associated with these studies that provides the basis of future and ongoing investigations into the mechanisms of acute lung injury, the control of epithelial repair, and the processes that lead to the development of irreversible lung diseases such as interstitial fibrosis and cancer.

In vitro studies have suggested that ozone exposure causes cytotoxicity by oxidative damage of membrane lipids and cellular sulfhydryls and by inactivation of key enzymes involved in cell energy metabolism. Short-term exposures of a few hours to relatively high ozone concentrations of greater than 2 ppm result in epithelial lung damage, edema formation, and inflammatory cell infiltration.[1] This type of exposure protocol provides a model of acute lung epithelial injury that is useful for investigating the pathogenesis of adult respiratory distress syndrome and for testing possible preventative procedures. Similar exposures to ozone are also being used to investigate in experimental animals the possible role that inflammatory cell infiltration might have in airway smooth muscle responsiveness to pharmacological agents.[2,3]

Longer exposures of 2 to 3 d to ozone concentrations of less than 1 ppm have been shown to result in alveolar type I cell damage in the centriacinar regions of the lung, which is followed by proliferation of alveolar type II cells associated with the initial stages of epithelial repair.[4,5] Alterations of bronchiolar epithelium have also been characterized by a loss of ciliated cells and a flattening of the nonciliated Clara cells.[6,7] Because the alveolar and bronchiolar epithelia return to a normal structure by 7 to 9 d postexposure,[8,9] these ozone exposures provide a reversible model to study the processes associated with epithelial damage and repair. Prolonged exposures for longer than 1 week to ozone concentrations greater than 0.5 ppm have been shown to result in increased lung collagen content, persistence of inflammatory cells, and an altered morphology associated with fibrogenesis.[10,11] Therefore, by altering the ozone concentration and the duration of exposure, it is possible to obtain model systems for investigating several different stages of lung epithelial injury and repair.

A detailed description of all the methods that have been or could be used to study lung injury is obviously beyond the scope of this review. A description of methods for establishing models of lung injury based on ozone exposures is therefore presented with reference to and discussion of procedures and assays useful to investigators interested in using ozone for the study of lung injury. Relatively straightforward methods for evaluating the models are presented with a discussion of how they can best be used to interpret changes in lung biochemistry and pathology. It is hoped that the following information will provide a framework onto which new approaches using modern developments in histochemistry, molecular biology, and immunology can be applied to the study of pulmonary diseases resulting from the inhalation of toxic substances.

IMPLEMENTATION OF THE MODEL

Animals

The choice of animal species and strain obviously depends on the intended use of the model. For example, mice have been used primarily for studying interactions of ozone with infectious agents[12] and guinea pigs for studying the effects of ozone on airway reactivity.[2,3]

The majority of data on which the development of ozone models of lung injury has been based has been obtained using rats. On the other hand, Mustafa and associates have suggested that mice might offer a more sensitive model for studying the pulmonary effects of ozone exposure.[13,14] These authors demonstrated greater increases in tissue weight, DNA and sulfhydryl contents, and glucose-6-phosphate dehydrogenase activity in the lungs of Swiss-Webster mice following 5 d of continuous exposure to 0.45 ppm ozone, compared with similar measurements made in Long-Evans, Wistar, and Sprague-Dawley rats.[13] It should be noted that these different strains of rat demonstrated similar responses to the ozone exposure. Similar comparisons between rats of different ages, using a 3-d continuous exposure to 0.8 ppm ozone, suggested that older rats (60 to 90 d) are more responsive to ozone exposure, demonstrated by larger increases in lung activity of NADPH-linked dehydrogenases of glucose-6-phosphate, 6-phosphogluconate, and isocitrate.[14,15] Although mice have the added advantage of being available with specific genetic deficiencies, the size of the young adult rat lung (60 to 90 d old) permits several measurements to be made on the same lung.

The choice of rat strain often depends on the ability of investigators to obtain animals that are specific-pathogen-free. A procurement and management program should be adopted that minimizes the development of diseases often found in animal colonies that include Sendai virus, rodent coronaviruses, and mycoplasmal infection.[16] Although several suppliers now have certified virus-titer-free colonies, continual monitoring of animals entering the facility and during experiments is recommended. Animals should be shipped from suppliers with virus-free colonies in filtered boxes, with minimal time being spent in transit. Because most suppliers test their colonies only on a quarterly basis, more frequent serological testing of batches of animals on entering the facility is often required.[16] Animals should be kept for up to 2 weeks in high-efficiency particulate-filtered air, isolated from other batches of animals to prevent possible cross-infection before exposure. This quarantine period ensures that animals have adjusted to their new surroundings and any serological test results have been received. Because most animal beddings either create dusts that affect baseline phagocytic cell number or result in induction of microsomal enzyme systems, suspended wire caging is recommended. Routine storage of blood serum samples taken from rats at all stages of exposure is recommended so that tests can be made if infection is suspected, bearing in mind that at least 2 weeks are required before the antibodies to a particular infecting virus are detectable in the serum. Measurements of weight gain and routine pathology and bronchoalveolar lavage of test animals at the beginning and end of the quarantine period also act as additional monitors of the health status of the animals.

Ozone Exposures

In order to ensure that all animals receive the same concentration of ozone, exposures are best carried out in stainless-steel inhalation chambers where the animals are individually caged with free access to food and water as described in detail elsewhere.[17] Alternatively, small numbers of animals can be exposed to ozone in glass chambers provided that the volume of the rats does not exceed 5 to 10% of the chamber volume and good air circulation is maintained. In all cases, the air flow through the chambers should exceed 15 to 20 changes per hour to ensure that there is uniform concentration of ozone throughout the chamber and that ammonia concentrations are kept to a minimum. Air supplied to both control and ozone chambers should be maintained at 20 to 25°C with relative humidity of 50 to 65% and be charcoal and high-efficiency particulate filtered. The chambers should be cleaned and the cages rotated to different positions on a daily basis.

Ozone can be generated by passing oxygen at a slow rate of 0.5 to 2 l through ozone generators that use either electrical discharge or ultraviolet (UV) light (Orec Corporation, Phoenix). Smaller generating systems can be made by mounting UV lights (Hamamatsu

Corporation, Middlesex, NJ) in a pickle jar covered with aluminum foil. Such a system will provide a stable output but is not easily adjusted for different ozone concentrations. Ozone is then mixed with air before entering the inhalation chamber at sufficient dilution to ensure that chamber oxygen concentration is not significantly elevated. Equilibration time normally depends on the flow rate and the volume of the chamber given by the formula $t = (V \times F)/4.5$, where t is the time in minutes required to reach 95% concentration, V is the volume of the chamber in liters, and F is the flow in liters per minute. Additional time might be required for the ozone to equilibrate with exposure equipment and animal fur. Best results are therefore obtained with large flow rates and small animal to chamber volume ratios.

Several methods are available for the measurement of chamber ozone concentration, including chemical methods based on the reaction of ozone with solutions containing potassium iodide and on UV photometry. The original chemical reference method based on ozone absorption in a neutral-buffered potassium iodide solution has been replaced by the use of a boric-acid-buffered potassium iodide (1% KI,0.1 *M* H_3BO_4) method that has been shown to give consistent agreement with UV photometry.[18] Chamber air is passed directly through impingers (0.5 l/min) containing the boric-acid-buffered potassium iodide solution (10 ml) for a fixed period of time (10 min). The iodine produced by ozone interaction with the potassium iodide is measured spectrophotometrically as I_3^- at 352 nm and compared with standard iodine solutions.[18] Direct measurements of chamber concentration are possible using a Mast® oxidant meter (Mast Instrument Company, Davenport, IO) that measures iodine production electrochemically. Alternatively, ozone concentrations can be determined by UV photometry (e.g., Dasibi Environmental Corporation, Glendale, CA and PCI Corporation, West Caldwell, NJ). The absorbance of ozone at 250 nm forms the basis of present standard reference methods and can be used in combination with a known ozone source (Model 1008PC, Dasibi Corporation) for calibration of other instruments and procedures.

EVALUATION OF THE MODEL

The literature describing methods for evaluating ozone injury to the lung is very extensive and includes physiological, morphological, and biochemical techniques. In this section, a generalized system for routine evaluation of lung injury is described with reference to more complicated procedures that have been employed by investigators to answer specific questions. As previously noted, ozone exposures of the lung result in cytotoxicity, edema formation, inflammatory cell infiltration, epithelial cell proliferation, changes in tissue enzyme activities, and alterations in connective tissue distribution and content. It is therefore the purpose of this section to outline some of the standard procedures used to quantitate these events.

Bronchoalveolar Lavage

Relatively simple procedures have been developed for early detection of pulmonary injury based on the measurements of enzymes and protein in bronchoalveolar lavage fluid.[19-21] By bronchoalveolar lavage of Syrian hamster lungs with increasing concentrations of the detergent Triton® X-100, Henderson and co-workers showed elevated lavage fluid lactate dehydrogenase activity (LDH) that was identified by isoenzyme analysis as being derived from lung cells.[19] By comparing different pneumotoxicants, Roth further demonstrated that lavage fluid LDH measurements are a useful early indicator of lung cell damage.[21] Evaluation of the inflammatory response resulting from lung injury can also be assessed by differential counting of the cells recovered by repeated lavages of the lungs.[22]

Animals are anesthetized by intraperitoneal injection with 50 mg/kg sodium pentobarbital and the trachea is then cannulated. If lavage fluid analysis is to include determination of recoverable inflammatory cells, serial lavages using a total volume of about 50 ml of 0.1

M phosphate-buffered isotonic saline, pH 7.4 containing 3 m*M* EDTA (PBS) at room temperature, is suitable for 250-g rats.[22] Mauderly suggests that the volume for each lavage be calculated on the basis of 23 ml/kg mean body weight.[23] Cell-free samples for analysis of enzymes and protein are prepared and stored on ice by centrifugation (500 × *g*) of the combined fluids obtained from the first two lavages.[21] The serial lavage procedure is then continued until the total volume of PBS has been used to isolate the inflammatory cells from the lung. The cells recovered from centrifugation of all lavages are then combined and washed by resuspension and centrifugation in PBS. The resulting cell pellets are resuspended in PBS, counted, and differentially stained for analysis of macrophage, lymphocytes, and neutrophil numbers.[22]

Lactate dehydrogenase activity is measured by standard procedures based on changes in NADH in the presence of excess pyruvate.[24] Although lavage fluid protein can readily be determined by the Lowry procedure,[25] albumin measurements using bromocresol green represent a more specific indicator of changes in vascular permeability.[26] Although results are normally represented as activity or content per milliliter of lavage fluid, the total volume of recovered lavage fluid should be recorded as an indicator of fluid retention associated with extensive damage to the lung.

Morphology

Various lung fixation methods can be employed that include both intratracheal and intravascular infusion with formaldehyde and glutaraldehyde solutions. For morphometric studies, the degree of inflation of the lung during fixation is usually kept constant between tissues by maintaining the intratracheal infusion pressure at 20 to 25 cm of H_2O. The disadvantage of tracheal fixation is that the position of inflammatory cells and edema fluid within the airways might be disrupted. Perfusion fixation might offer a suitable alternative provided that the inflation volume is standardized between tissues. Using standard methods for perfusing isolated lungs,[27] the inflation volume is set to a constant airway pressure of 10 cm of H_2O and the Krebs-Ringer bicarbonate perfusion medium is switched to fixative containing 1% formaldehyde-1% glutaraldehyde, 1% sucrose, and 0.1 *M* sodium cacodylate, pH 7.6. Perfusion with the fixative is continued for at least 15 min, at which time the lung is transferred to a beaker of fixative prior to embedding, sectioning, and staining. It should be noted that although glutaraldehyde is suitable for subsequent electron microscopy, formaldehyde appears to be more suitable for morphometric studies where uneven fixation and shrinkage artifacts must be minimized.

Because of the complexity of the lung and the relatively nonuniform distribution of lesions associated with ozone exposures, sampling techniques for morphometric studies must be carefully designed.[28] The upper airways should be consistently sampled at specifed distances distal to the carina. In order to obtain an overall distribution of ozone-induced lesions between large and small airways, transverse sections of the whole lung are prepared for light microscopy. Samples for electron microscopy can then be prepared from proximal and distal regions of these tissue slices by use of a 20-gauge punch, as described in more detail by Crapo et al.[29]

Although paraffin embedment with hemotoxylin and eosin staining is satisfactory for routine light microscopic examinations, glycol methacrylate embedment (JB-4, Polysciences, Warrington, PA) with Giemsa staining provides superior sections that can be used to distinguish cell types of both airway and alveolar epithelium.[30] Histochemical staining can also be used to identify specific cell populations by enzyme activity localization.[31] The disadvantage of methacrylate embedment is that the area of tissue examined is limited, although larger embedding blocks are now available. Specific staining for connective tissue in paraffin-embedded sections is achieved by trichrome staining and in methacrylate sections by staining with 0.5% Sirius Red in a saturated solution of picric acid.[32]

Morphometric methods provide a quantitative measure of changes in lung structure.[7,28,29,33] Because the methods can be very time consuming, detailed morphometry has been used primarily to answer specific questions. A strategy for studying ozone-induced lesions might therefore involve the use of paraffin-embedded sections to determine the area of lung adversely affected by the exposure and methacrylate sections for characterizing alveolar lesions by determining, for example, the number of alveolar macrophages present in affected alveoli. Electron microscopic examination would then be used to quantitate changes in interstitial spaces, cellular dimensions, and the numbers of intracellular organelles.[8,10,29,34]

Biochemical Analyses

Whole tissue measurements of protein, DNA, and collagen have been used extensively as a means of quantitating the observed changes in pathology resulting from ozone exposures. Measurements of key metabolic enzymes involved in energy generation and in cellular antioxidant defense mechanisms have been used as indicators of altered lung metabolism. The following describes a scheme that is suitable for analyzing rat lung tissue for determination of these biochemical parameters.

The lungs are first perfused with either isotonic saline or Krebs-Ringer bicarbonate buffer in order to remove as much blood from the vasculature as possible.[27] The lungs are kept in ice-cold saline prior to being trimmed of extra parenchymal tissue, blotted, and then weighed. The tissue is divided into two representative portions of the whole lung. One third of the lung is used for the determination of dry to wet ratio and the remainder is used for the preparation of an homogenate. The first portion, following freeze-drying to a constant weight, can subsequently be used for determinations of hydroxyproline as an indicator of lung collagen content.[35,36] The second weighed portion is homogenized in approximately four volumes of ice-cold 0.15 *M* sucrose, 0.15 *M* mannitol, 1 m*M* EDTA, and 2 m*M* trizma buffer pH 7.4 as previously described.[37] Based on the use of about 750 mg of tissue, homogenizing medium is then added to adjust the volume to exactly 6 ml before splitting into fractions for separate analyses for DNA and enzyme activities.

To precipitate proteins and nucleic acids, which are then separated by centrifugation (10,000 × *g*), 1 ml of homogenate is added to 5 ml of ice-cold 0.5 *M* perchloric acid. The resulting pelleted samples, following an additional wash with 10 ml of ice-cold 0.5 *M* perchloric acid, can be stored frozen at −80°C until analysis. The DNA assay involves the extraction of the nuclear protein by resuspending the pellet in a known volume (10 ml) of 0.5 *M* perchloric acid and heating in a water bath at 90°C for 15 min.[38] The solution is then centrifuged (10,000 × *g*) and the supernatant assayed for DNA content using methods based on its color reaction with diphenylamine read at 600 nm.[39] The remaining pellet can be used for determination of protein following alkaline digestion according to standard assay procedures.[25]

Although isolation of both mitochondrial and microsomal subcellular fractions requires large volumes of homogenate, estimation of mitochondrial content is possible by direct measurement of oxygen utilization of the homogenate in the presence of succinate and excess ADP.[15,40] Using an oxygen electrode (Yellow Springs Instrument Company, Yellow Springs, OH), 0.1- to 0.2-ml portions of the lung homogenate are added to 1 ml of air-saturated buffer (100 m*M* sucrose, 100 m*M* mannitol, 10 m*M* KCl, 5 m*M* $MgCl_2$, 1 m*M* Na_2HPO_4, 25 m*M* trizma, pH 7.5) in the presence of 1 m*M* ATP, 5 m*M* glucose, and hexokinase (units per milliter).[37,40] Succinate oxidase activity is then determined by measurement of the increase in oxygen utilization resulting from the addition of 20 μmol sodium succinate to the electrode.

Cytosolic enzymes are measured on postmitochondrial fractions prepared by centrifugation of homogenate samples at 15,000 × *g* for 20 min at 4°C. The resulting supernatant can then be subdivided into aliquots and frozen at −80°C prior to analysis. Enzyme activities that have been measured in ozone-exposed lungs include the NADPH-linked dehydrogenases of glucose-6-phosphate, 6-phosphogluconate and isocitrate, and the antioxidant glutathione-linked reductase and peroxidase.[13,15,41,42] It is recommended that although the dehydrogenases appear to lose less than 5% activity on freezing, glutathione reductase and glutathione peroxidase activity should be determined the same day that the homogenates are prepared. It has been found that 0.1- to 0.2-ml aliquots of supernatant in assay volumes of 2 to 2.5 ml are adequate for assay procedures that use standard techniques based on the appearance of disappearance of NADPH measured spectrophotometrically at 340 nm.[13,15,24,41,42] Lung superoxide dismutase and catalase activities have also been determined in ozone-exposed lungs as previously described.[42,43]

Use of Whole Tissue Preparations

Although measurement of subcellular fraction enzyme activities provides an indication of the potential capacity of the lung to carry out certain metabolic functions, whole tissue preparations are often used for assessing the actual rates at which these functions are proceeding within the intact organ. These functions include energy generation,[13,44] synthesis of NADPH required for fatty acid synthesis and maintenance of antioxidant defenses, and processes that can mainly be attributed to specific cell types that include serotonin uptake and the synthesis of dipalmitoylphosphatidylcholine and collagen. The use of isolated perfused lungs, slices, or minces permits the measurement of metabolic fluxes under conditions where multienzyme pathways remain intact under the influence of normal or altered cellular control mechanisms. The relationships between cells and physiological structures are also maintained. Although purified lung epithelial and endothelial cell populations have been extensively used for investigating specific metabolic functions of the lung, their usefulness is limited in studies involving altered lung pathology because the isolation procedure does not necessarily provide a representative population of the cells as they exist *in vivo*.

Lung intermediary metabolism has been studied primarily in isolated lung slices and perfused organ preparations using tritium and carbon-14-labeled substrates. Total glucose utilization can be readily determined by the appearance of ^3H_2O on perfusion or incubation with [5-^{3}H] D-glucose, being derived from the enolase and triose-phosphate isomerase reactions of the Emden-Meyerhof glycolytic pathway, as previously described.[44,45] Measurements of glucose catabolism to carbon dioxide by mitochondria and the pentose-phosphate cycle have been demonstrated by measurements of $^{14}CO_2$ production on separate incubations with [U-^{14}C], [1-^{14}C], and [6-^{14}C] D-glucose.[45] Failure of the mitrochondria to oxidize reducing equivalents generated in the cytosol has been indicated by increases in the ratio of lactate to pyruvate production by the lung.[27] Mitochondrial metabolism can be separately evaluated by measurements of $^{14}CO_2$ production on incubation with either [U-^{14}C] palmitate[46] or [1-^{14}C] pyruvate.[44]

In these studies of lung intermediary metabolism, endogenous pools of substrates and metabolic intermediates are rapidly equilibrated with exogenously added radiolabeled precursors, giving linear rates of product formation. On the other hand, similar studies to investigate lung synthetic activity, which have included measurements of DNA, lipid, and protein synthesis, have required consideration of the possibility that endogenous precursor pool sizes might be sufficiently large to lower the specific activity of the exogenous radiolabeled substrate.[47] Although increasing the amount of exogenously added radioactive substrate will shorten the time required for radioisotope equilibration with endogenous substrate pools, measurements of synthetic rate from both endogenous and exogenous substrate sources often require estimates of the intracellular precusor specific activity.[47] Lung DNA synthesis has been determined by incubation with tritiated thymidine, protein metabolism with radiolabeled amino acids, and phospholipid synthesis by use of radiolabeled fatty acids and choline.

Studies that have used ozone exposures as a model of lung fibrogenesis have examined changes in both collagen content and collagen biosynthesis. Last and co-workers have used lung tissue minces consisting of 2-mm^3 pieces incubated *in vitro* with radiolabeled proline for 0, 2, and 3 h prior to protein precipitation in trichloroacetic acid.[11,48-50] Collagen synthesis was indicated by measurements of radiolabel incorporation into collagen hydroxyproline, following acid hydrolysis of trichloroacetic acid-insoluble material and oxidation of the hydroxyproline to pyrrole according to the methods of Juva and Prockop.[51] A recent review by Laurent demonstrates the relatively high turnover rates of certain tissue collagens and emphasizes the need to flood the intracellular pools with exogenous proline when measuring tissue rates of collagen synthesis.[52]

EXPERIMENTAL FINDINGS

Acute Lung Injury

Exposure of rats to 1.5 to 2.0 ppm ozone for 2 and 4 h provides a model of acute lung oxidant injury that demonstrates different time courses for the appearance of each inflammatory cell type and a certain degree of reversibility.[53] Bronchoalveolar lavage of lungs isolated from rats 24 h after ozone exposure demonstrated that neutrophils rapidly appeared in numbers proportional to the duration of exposure, giving values of 0.10 ± 0.01 (control), 0.27 ± 0.10 (2 h), and 0.78 ± 0.11 (4 h) million cells per lung (±SEM, n = 5). The duration of exposure could also be correlated with changes in vascular permeability, indicated by lavage fluid albumin concentrations of 106 ± 32, 299 ± 41, and 484 ± 55 μg/ml (±SEM, n = 4) for control, 2- and 4-h ozone-exposed lungs, respectively. Although neutrophils were no longer present in significant numbers by 3 d postexposure, lavage fluid albumin concentrations were still elevated but at diminished levels of 192 ± 20 (2 h) and 356 ± 78 μg/ml (4 h). Lymphocytes were observed to be significantly increased by ozone exposure at 24 h, with further enhancements at 3 d postexposure, giving values of 0.03 ± 0.01 (control), 0.18 ± 0.06 (2 h), and 0.35 ± 0.10 (4 h) million cells per lung. In contrast, infiltration of monocytes into the lung exhibited a slower response, giving no changes at 24 h but significant increases at 3 d postexposure from a control value of 0.67 ± 0.07 to 2.25 ± 0.46 (2 h) and 2.55 ± 0.82 (4 h) million cells per lung (±SEM, n = 5). These changes in inflammatory cell number and albumin concentration could be correlated with pathological observations of edema fluid accumulation and hypercellularity, which in part could also be accounted for by proliferation of alveolar type II epithelium. By 8 d postexposure, the structure of the lung appeared to be returning to normal, as indicated by control numbers of lavaged inflammatory cells and normal levels of lavage fluid albumin.[53]

Guth and associates investigated the sensitivity of different methods of evaluating early lung damage by broncholaveolar lavage, following short-term exposures of rats to ozone concentrations between 0.12 and 0.96 ppm.[20] These authors used the recovery of [^{3}H]-labeled albumin in lung lavage fluid 1 h following injection of 10 μCi via the tail vein as an index of permeability. Although the recovery of radiolabeled albumin was observed to be proportional to ozone concentration when measured in rats after 6- and 24-h exposures to 0.4 ppm ozone, measurement of lavage fluid protein was considered to be a more sensitive index in those experiments that employed longer exposures with concentrations as low as 0.12 ppm.[20] These authors also examined several lavage fluid enzyme activities that included lactate dehydrogenase, acid phosphatase, and *N*-acetyl-β-D-glucosamidase and concluded that they represented the least sensitive indices of lung ozone damage.[20] Lavage fluid protein measurements have also be used in studies by Hu et al.,[54] who demonstrated significant increases following a 72-h exposure of guinea pigs to ozone concentrations of 0.26 to 1 ppm. This latter group of authors noted that if ozone exposures of 0.5 ppm were reduced to 3 h, no protein accumulation could be detected unless the time of lavage was delayed by 10 to 15 h postexposure.[54] These two studies therefore suggest that recovery of radiolabeled albumin is a suitable index of vascular permeability in short-term ozone exposures of a few hours, while measurement of lavage fluid protein content requires a certain amount of time before significant accumulations can be detected.

Lung Epithelial Damage and Repair

The damage and repair of terminal bronchiolar and alveolar epithelia following ozone exposures of less than 1 ppm have been extensively investigated. Early studies have demonstrated that continuous exposure results in lesions in the centriacinar region of the lung that appear to be maximal within 3 to 5 d from the onset of exposure.[5,10] Morphological changes have been correlated with increases in certain lung enzyme activities that include succinate oxidase, glucose-6-phosphate dehydrogenase, and glutathione peroxidase.[13,15,41,55] Changes in microsomal enzymes have also been reported.[55,56] Autoradiographic techniques based on injection with tritiated thymidine have demonstrated that within 2 to 3 d from the onset of exposure, damaged terminal bronchiolar epithelial cells are being replaced by nonciliated secretory cells[6,7] and damaged alveolar type I cells are being replaced by proliferating type II cells.[5] The repair of the rat lung bronchiolar and alveolar epithelia is completed by transformation of nonciliated cells to new ciliated and nonciliated cells[6] and by transformation of type II cells to new type I cells[5] 6 to 9 d after initial 2- to 3-d exposures to ozone concentrations of less than 1 ppm.[8,9] Plopper and co-workers have also demonstrated in rats[9] that the newly repaired alveolar epithelium is just as susceptible to the damaging effects of ozone as unexposed control animals when reexposed 6 d after an initial 3-d continuous exposure to 0.8 ppm ozone. It should be noted that lung enzyme activities had returned to normal at this time point.[41]

On the other hand, Evans and co-workers have suggested that tolerance to ozone can be demonstrated at earlier times following the initial exposure period.[8] These authors proposed that tolerance to ozone reexposure, observed 3 d following an initial exposure to 0.5 ppm for 2 d, was a result of increased thickness of newly formed type I cells and not a result of increased activity of enzymes involved in cellular antioxidant defenses, as proposed by other authors.[42,55] They argue that by representing the enzyme data as per milligram DNA, the previously observed increases in superoxide dismutase, catalase, and glutathione peroxidase[42] observed following 7-d exposure to 0.8 ppm ozone and associated with tolerance to oxygen exposure could be accounted for by increased cellularity. However, it should be noted that 4 d after an initial 3-d exposure to 0.75 ppm ozone, glucose-6-phosphate dehydrogenase and glutathione peroxidase activities per milligram DNA are significantly enhanced compared with air-exposed control lungs.[57] These latter data suggest that newly formed epithelium might have an increased ability to detoxify hydroperoxides, which based on previous reports might be expected to be diminished by 6 d postexposure.[8,9,41] However, other investigators have been unable to correlate changes in lung antioxidant enzymes with the development of ozone tolerance.[43,58]

Fibrogenesis

Although centriacinar lesions, induced by ozone exposures of rats to 0.8 ppm, appear to be diminished by continued exposure beyond 3 to 5 d, morphological alterations have been shown to persist through 90 d of exposure.[10] Ozone-exposed lungs exhibited enhanced numbers of inflammatory cells within proximal alveoli and thickening of the air-blood barrier, associated with a corresponding increase in the volume density of the alveolar septal interstitium from 44 to 56%. In a separate investigation, Last and Greenberg demonstrated that continuous exposure of rats to 0.5 ppm for up to 6 months was associated with increased lung collagen measured as hydroxyproline content.[11] These data could be correlated with increased rates of collagen synthesis at all time points between 3 and 180 d of exposure when measured by incubation of lung minces with radiolabeled proline. Although lung collagen content remained elevated 2 months postexposure, the rates of collagen synthesis had returned to control values.[11] Increases in collagen synthesis have also been used as an indicator of early fibrogenesis in studies that used ozone exposures of 0.8 to 1.5 ppm for as short as 7 d.[48,49] Hesterberg and Last demonstrated that daily treatment of rats with 50 mg/kg body weight of the steroid methylprednisolone during exposure[48] prevented the two-fold increase in collagen synthesis observed 7 d following continuous exposure to 1.2 ppm ozone. Further studies by Reiser and Last demonstrated by CNBr peptide mapping an increase in the ratio of newly synthesized type I to type III collagen in rat lungs following similar exposures to ozone.[50]

DISCUSSION

This brief review has demonstrated how ozone exposure of experimental animals can provide useful models of lung alveolar and terminal bronchiolar injury that are well characterized by both biochemical and histological methods. Although a certain amount of specialized equipment is required for implementation of the model, ozone exposure has certain advantages when compared with those models of lung injury that use single administrations of damaging agents by either intraperitoneal or intratracheal routes. Ozone exposure results in reproducible lesions with predictable distribution within the lung which can be readily altered by changing the concentration and duration of exposure. Raising the concentration results in increased penetration of ozone to distal regions of the lung and increased severity of the resulting lesions, while prolonging exposures results in transition from reversible damage to the development of irreversible fibrosis. Different stages of lung oxidant damage and repair can therefore be studied using the same agent.

Although the description of bronchoalveolar lavage in this review was restricted to methods for evaluating lung injury resulting from ozone exposures, measurements of proteinases, peptides, and chemoattractants are also possible. Such measurements, when used in combination with determinations of inflammatory cell infiltration and histological evaluations, provide the basis for studying the early events leading to pulmonary fibrosis.[59] It should be noted that quantitative measurements of inflammatory cell infiltration by bronchoalveolar lavage would be limited following the development of consolidated lesions. Significant morphometric differences between macrophages obtained by lavage and those observed in the centriacinar region of ozone-exposed lungs have also been reported.[34] These data further demonstrate that cells recovered by lavage might not accurately represent the population of cells present within the lung.

Rat exposures to 1.5 ppm ozone increasing from a few hours to 7 d provide a suitable model to investigate the individual role each cell type has in the initial inflammatory response and the subsequent increase in collagen synthesis associated with the development of fibrosis. Identification of points of reversibility should also be possible. The possible role of the neutrophil in initial derangement of the lung extracellular matrix, the role of macrophage arachidonic acid metabolites, the involvement of lymphocyte cell populations, and the relationship between epithelial repair and fibroblast proliferation represent some important areas for future investigation.[59,60]

The development of agents or procedures to protect the lung from the development of progressive lung damage brought about by exposures to toxic substances requires an understanding of the factors that control the repair of both alveolar and conducting airway epithelia. Haschek and Witschi have suggested that type II cells, during the proliferative stages of alveolar epithelial repair, are particularly susceptible to oxidant damage and when exposed to oxidants permit interstitial cells to proliferate relatively unchecked.[61] In contrast, newly formed alveolar epithelium appears to be relatively resistant to further oxidant injury.[8,61] Ozone exposures to less than 1 ppm for 2 to 3 d therefore appear to provide a suitable model for future investigations of both alveolar and bronchiolar epithelial proliferative and repair processes. Studies to determine the possible biochemical determinants of susceptibility and tolerance to oxidant exposure during epithelial repair have mainly involved measurements of enzymes involved in antioxidant defense mechanisms. Although representing enzyme content on a per milligram DNA basis provides additional information, the use of such measurements is limited for these types of investigation without more detailed knowledge of the relative distribution of antioxidant enzymes between different lung cell populations. On the other hand, evaluation of the ability of the intact lung to carry out certain homeostatic functions, which include the maintenance of reduced glutathione and of cellular energy supply, offers an alternative method for further evaluating biochemical mechanisms of tolerance and susceptibility to oxidant stress. Further investigations into how changes in cell morphology might be involved in tolerance mechanisms, together with studies to determine how surfactant and mucus might be involved in oxidant protection, are also indicated.

Finally, recent studies demonstrating that ozone exposure alters the extent and pattern of lung metabolism of the carcinogen benzo(*a*)pyrene[56,62] suggest that ozone exposures would also provide a useful model for investigating how altered lung pathology might contribute to oncogenic processes associated with polycyclic aromatic hydrocarbon inhalation.

REFERENCES

1. **Stokinger, H. E.,** Ozone toxicology, *Arch. Environ. Health,* 10, 719, 1965.
2. **Gordon, T., Venugopalan, C. S., Amdur, M. O., and Drazen, J. M.,** Ozone-induced airway hyperreactivity in the guinea pig, *J. Appl. Physiol.,* 57, 1034, 1984.
3. **Murlas, C. G. and Roum, J. H.,** Sequence of pathologic changes in the airway mucosa of guinea pigs during ozone-induced bronchial hyperreactivity, *Am. Rev. Respir. Dis.,* 131, 314, 1985.
4. **Stephens, R. J., Sloan, M. F., Evans, M. J., and Freeman, G.,** Alveolar type 1 cell response to exposure to 0.5 ppm O_3 for short periods, *Exp. Mol. Pathol.,* 20, 11, 1974.
5. **Evans, M. J., Johnson, L. V., Stephens, R. J., and Freeman, G.,** Cell renewal in the lungs of rats exposed to low levels of ozone, *Exp. Mol. Pathol.,* 24, 70, 1976.
6. **Evans, M. J., Johnson, L. V., Stephens, R. J., and Freeman, G.,** Renewal of the terminal bronchiolar epithelium in the rat following exposure to NO_2 or O_3, *Lab. Invest.,* 35, 246, 1976.
7. **Lum, H., Schwartz, L. W., Dungworth, D. L., and Tyler, W. S.,** A comparative study of cell renewal after exposure to ozone or oxygen, *Am. Rev. Respir. Dis.,* 118, 335, 1978.
8. **Evans, M. J., Dekker, N. P., Cabral-Anderson, L. J., and Shami, S. G.,** Morphological basis of tolerance to ozone, *Exp. Mol. Pathol.,* 42, 366, 1985.
9. **Plopper, C. G., Chow, C. K., Dungworth, D. L., Brummer, M., and Nemeth, T. J.,** Effect of low level ozone on rat lungs. II. Morphological responses during recovery and reexposure, *Exp. Mol. Pathol.,* 29, 400, 1978.
10. **Boorman, G. A., Schwartz, L. W., and Dungworth, D. L.,** Pulmonary effects of prolonged ozone insult in rats, *Lab. Invest.,* 43, 108, 1980.
11. **Last, J. A. and Greenberg, D. B.,** Ozone-induced alterations in collagen metabolism of rat lungs. II. Long-term exposures, *Toxicol. Appl. Pharmacol.,* 55, 108, 1980.
12. **Gardner, D. E.,** Use of experimental airborne infections for monitoring altered host defenses, *Environ. Health Perspect.,* 43, 99, 1982.
13. **Mustafa, M. G., Elsayed, N. M., Quinn, C. L., Postlethwait, E. M., Gardner, D. E., and Graham, J. A.,** Comparison of pulmonary biochemical effects of low-level ozone exposure on mice and rats, *J. Toxicol. Environ. Health,* 9, 857, 1982.
14. **Mustafa, M. G., Elsayed, N. M., Graham, J. A., and Gardiner, D. E.,** Effects of ozone exposure on lung metabolism: influence of animal age, species, and exposure conditions, in *Biomedical Effects of O_3 and Related Photochemical Oxidants,* Lee, S. D., Mustafa, M. G., and Melhman, M. A., Eds., Princeton Scientific, Princeton, NJ, 1983, 57.
15. **Elsayed, N. M., Mustafa, M. G., and Postlethwait, E. M.,** Age-dependent pulmonary response of rats to ozone exposure, *J. Toxicol. Environ. Health,* 9, 835, 1982.

16. **Hamm, T. E., Jr., Raynor, T. H., and Sherrill, J. M.,** Procurement and management of rodents to minimize disease problems: the CIIT animal management program, in *Complications of Viral and Mycoplasmal Infections in Rodents to Toxicology Research and Testing,* Hamm, T. E., Jr., Ed., Hemisphere, Washington, D. C., 1986, chap. 9.
17. **Phalen, R. F.,** *Inhalation Studies: Foundations and Techniques,* CRC Press, Boca Raton, FL, 1984, chap. 3.
18. **Flamm, D. L.,** Analysis of ozone at low concentrations with boric acid buffered KI, *Environ. Sci. Technol.,* 11(10), 978, 1977.
19. **Henderson, R. F., Damon, E. G., and Henderson, T. R.,** Early damage indicators in the lung. I. Lactate dehydrogenase activity in the airways, *Toxicol. Appl. Pharmacol.,* 44, 291, 1978.
20. **Guth, D. J., Warren, D. L., and Last, J. A.,** Comparative sensitivity of measurements of lung damage made by bronchoalveolar lavage after short-term exposure of rats to ozone, *Toxicology,* 40, 131, 1986.
21. **Roth, R. A.,** Effect of pneumotoxicants on lactate dehydrogenase activity in airways of rats, *Toxicol. Appl. Pharmacol.,* 57, 69, 1981.
22. **Warr, G. A. and Jakab, G. J.,** Pulmonary inflammatory responses during viral pneumonia and secondary bacterial infection, *Inflammation,* 7, 93, 1983.
23. **Mauderly, J. L.,** Bronchopulmonary lavage of small laboratory animals, *Lab. Anim. Sci.,* 27, 255, 1977.
24. **Bergmeyer, H. U.,** Lactate dehydrogenase, in *Methods of Enzymatic Analysis,* Vol. 3, Verlag Chemie, Deerfield Beach, FL, 1983, 118.
25. **Lowry, O. H., Rosebrough, N. J., Farr, A. L., and Randall, R. J.,** Protein measurement with the folin phenol reagent, *J. Biol. Chem.,* 193, 265, 1951.
26. **Rodkey, F. L.,** Direct spectrophotometric determination of albumin in human serum, *Clin. Chem.,* 11, 478, 1965.
27. **Bassett, D. J. P. and Fisher, A. B.,** Metabolic responses to carbon monoxide by isolated rat lungs, *Am. J. Physiol.,* 230, 658, 1976.
28. **Dungworth, D. L., Schwartz, L. W., Tyler, W. S., and Phalen, R. F.,** Morphological methods for evaluation of pulmonary toxicity in animals, *Annu. Rev. Pharmacol. Toxicol.,* 16, 381, 1976.
29. **Crapo, J. D., Barry, B. E., Chang, L. Y., and Mercer, R. R.,** Alterations in lung structure caused by inhalation of oxidants, *J. Toxicol. Environ. Health,* 13, 301, 1984.
30. **Bennett, H. S., Wyrick, A. D., Lee, S. W., and McNeil, J. H.,** Science and art in preparing tissues embedded in plastic for light microscopy, with special reference to glycol methacrylate, glass knives and simple stains, *Stain Technol.,* 51, 71, 1976.
31. **Namab, M., Dannenberg, A. M., and Tanaka, F.,** Improvement in the histochemical demonstration of acid phosphatase, beta-galactosidase and non-specific esterase in glycol methacrylate tissue sections by cold temperature embedment, *Stain Technol.,* 58, 207, 1983.
32. **Sannes, P. L.,** The staining of connective tissue in glycol methacrylate sections for light microscopy, *Stain Technol.,* 62, 280, 1987.
33. **Weibel, E. R.,** Principles and methods for the morphometric study of the lung and other organs, *Lab. Invest.,* 12, 131, 1963.
34. **Lum, H., Tyler, W. S., Hyde, D. M., and Plopper, C. G.,** Morphometry of in situ and lavaged pulmonary alveolar macrophages from control and ozone-exposed rats, *Exp. Lung Res.,* 5, 61, 1983.
35. **Stegemann, H. and Stalder, K.,** Determination of hydroxyproline, *Clin. Chim. Acta,* 18, 207, 1967.
36. **Berg, R. A.,** Determination of 3- and 4-hydroxyproline, in *Methods in Enzymology,* Vol. 82, Academic Press, New York, 1982, 372.
37. **Fisher, A. B., Scarpa, A., LaNoue, K. F., Bassett, D., and Williamson, J. R.,** Respiration of rat lung mitochondria and the influence of Ca^{++} on substrate utilization, *Biochemistry,* 12, 1438, 1973.
38. **Munro, H. N. and Fleck, A.,** The determination of nucleic acids, *Methods Biochem. Anal.,* 14, 113, 1966.
39. **Burton, K.,** A study of the conditions and mechanisms of the diphenylamine reaction for the colorimetric estimation of deoxyribonucleic acid, *Biochem. J.,* 62, 315, 1956.
40. **Mustafa, M. G., DeLucia, A. J., York, G. K., Arth, C., and Cross, C. E.,** Ozone interaction with rodent lung. II. Effects on oxygen consumption of mitochondria, *J. Lab. Clin. Med.,* 82, 357, 1973.
41. **Chow, C. K., Hussain, M. Z., Cross, C. E., Dungworth, D. L., and Mustafa, M. G.,** Effect of low levels of ozone on rat lungs. I. Biochemical responses during recovery and reexposure, *Exp. Mol. Pathol.,* 25, 182, 1976.
42. **Jackson, R. M. and Frank, L.,** Ozone-induced tolerance to hyperoxia in rats, *Am. Rev. Respir. Dis.,* 129, 425, 1984.
43. **Douglas, J. S., Curry, G., and Geffkin, S. A.,** Superoxide dismutase and pulmonary ozone toxicity, *Life Sci.,* 20, 1187, 1977.
44. **Bassett, D. J. P. and Bowen-Kelly, E.,** Rat lung metabolism after 3 days of continuous exposure to 0.6 ppm ozone, *Am. J. Physiol.,* 13, E131, 1986.
45. **Bassett, D. J. P. and Fisher, A. B.,** Pentose cycle activity of the isolated perfused rat lung, *Am. J. Physiol.,* 231, 1527, 1976.

46. **Bassett, D. J. P., Hamosh, M., Hamosh, P., and Rabinowitz, J. L.**, Pathway of palmitate metabolism in the isolated rat lung, *Exp. Lung Res.*, 2, 37, 1981.
47. **Rannels, D. E., Low, R. B., Youdale, T., Volkin, E., and Longmore, W. J.**, The use of radioisotopes in quantitative studies of lung metabolism, *Fed. Proc.*, 41, 2833, 1982.
48. **Hesterberg, T. W. and Last, J. A.**, Ozone-induced acute pulmonary fibrosis in rats — prevention of increased rates of collagen synthesis by methylprednisolone, *Am. Rev. Respir. Dis.*, 123, 47, 1981.
49. **Last, J. A., Greenberg, D. B., and Castleman, W. L.**, Ozone-induced alterations in collagen metabolism of rat lungs, *Toxicol. Appl. Pharmacol.*, 51, 247, 1979.
50. **Reiser, K. M. and Last, J. A.**, Pulmonary fibrosis in experimental acute respiratory disease, *Am. Rev. Respir. Dis.*, 123, 58, 1981.
51. **Juva, K. and Prockop, D. J.**, Modified procedure for the assay of H^3- or C^{14}-labeled hydroxyproline, *Anal. Biochem.*, 15, 77, 1966.
52. **Laurent, G. J.**, Dynamic state of collagen: pathways of collagen degradation in vivo and their possible role in regulation of collagen mass, *Am. J. Physiol.*, 252, C1, 1987.
53. **Bassett, D. J. P., Bowen-Kelly, E., Brewster, E. L., Elbon, C. L., Reichenbaugh, S. S., and Kerr, J. S.**, A reversible model of acute lung injury based on ozone exposure, *Lung*, 166, 355, 1988.
54. **Hu, P. C., Miller, F. J., Daniels, M. J., Hatch, G. E., Graham, J. A., Gardner, D. E., and Selgrade, M. K.**, Protein accumulation in lung lavage fluid following ozone exposure, *Environ, Res.*, 29, 377, 1982.
55. **Mustafa, M. G. and Tierney, D. F.**, Biochemical and metabolic changes in the lung with oxygen, ozone, and nitrogen dioxide toxicity, *Am. Rev. Resp. Dis.*, 118, 1061, 1978.
56. **Takahashi, Y., Miura, T., and Kubota, M. K.**, In vivo effect of ozone inhalation on xenobiotic metabolism of lung and liver of rats, *J. Toxicol. Environ. Health*, 15, 855, 1985.
57. **Bassett, D. J. P., Bowen-Kelly, E., Elbon, C. L., and Reichenbaugh, S. S.**, Rat lung recovery from 3 days continous exposure to 0.75 ppm ozone, *J. Toxicol. Environ. Health*, 25, 329, 1988.
58. **Nambu, Z. and Yokoyama, E.**, Antioxidant sytem and ozone tolerance, *Environ. Res.*, 32, 111, 1983.
59. **Reiser, K. M. and Last, J. A.**, Early cellular events in pulmonary fibrosis, *Exp. Lung Res.*, 10, 331, 1986.
60. **Goldstein, R. H. and Fine, A.**, Fibrotic reactions in the lung: the activation of the lung fibroblast, *Exp. Lung Res.*, 11, 245, 1986.
61. **Haschek, W. M. and Witschi, H.**, Pulmonary fibrosis — a possible mechanism, *Toxicol. Appl. Pharmacol.*, 51, 475, 1979.
62. **Bassett, D. J. P., Bowen-Kelly, E., and Seed, J. L.**, Rat lung benzo(a)pyrene metabolism following three days continous exposure to 0.6 ppm ozone, *Res. Commun. Pathol. Pharmacol.*, 60, 291, 1988.

SHEEP AS AN ANIMAL MODEL OF ENDOTOXEMIA-INDUCED DIFFUSE LUNG INJURY

Richard E. Parker

INTRODUCTION

The adult respiratory distress syndrome (ARDS) is a clinical manifestation of diffuse lung injury that is associated with a high mortality rate. One of the leading causes of ARDS is Gram-negative bacterial septicemia. The agent released during Gram-negative septicemia that induces diffuse lung injury appears to be endotoxin or, more accurately, the lipopolysaccharide moiety of endotoxin. Despite several years of intensive investigation, the precise mechanisms of endotoxin-induced diffuse lung injury are unclear. Currently, patient care during ARDS is limited to only supportive therapy.

Several animal models of endotoxemia have been developed, many of which utilize relatively high doses of endotoxin and appear to be models of endotoxin-induced systemic shock. The ability of some of these animal models to develop endotoxin-induced diffuse lung injury is questionable. One of the most commonly used animal models of endotoxin-induced diffuse lung injury is the chronically instrumented unanesthetized sheep preparation. Advantages of the sheep model include the high sensitivity of sheep to endotoxin, the ability to conduct experiments without possibly complicating effects of anesthesia, the ability to perform multiple experiments on the same animal (thereby serving as its own control), the ability of sheep to tolerate large amounts of instrumentation and pharmacological interventions, and the docile nature of sheep.

There are two major criticisms of the sheep lung lymph model of endotoxemia. The first is that some systemic lymph contamination of the collected lung lymph remains despite efforts to remove such contamination.[2] The amount of contamination undoubtedly varies among sheep and the significance of this contamination is unresolved. The second criticism involves possible alteration of lymph as it traverses the caudal mediastinal lymph node.[3] The significance of such possible nodal alterations is also under debate.

One of the earliest investigations dealing with the effects of endotoxemia on sheep is that by Halmagyi et al.[4] However, this investigation was limited to hemodynamic and airway measurements following high-dose bolus infusions of endotoxin. Other complicating factors of this investigation were that most of the observations were limited to a few minutes after infusing endotoxin and the experiments were conducted on anesthetized sheep. The development of the sheep lung lymph model by Staub and associates[5] has allowed investigators to study in greater detail the effects of various interventions on lung fluid balance. The investigation by Brigham et al.[6] dealing with the infusion of live *Pseudomonas* into the sheep lung lymph model conclusively demonstrated an increase in lung microvascular permeability during septicemia. However, the use of live bacterial infusions to cause diffuse lung injury was less than ideal due to difficulties in obtaining consistent reproducible responses to the bacterial infusions. In a later publication, Brigham et al.[7] demonstrated that sublethal *Escherichia coli* endotoxin infusions caused reproducible responses in sheep that were qualitatively similar to *Pseudomonas* infusions. Several investigators have since conducted studies of endotoxin-induced diffuse lung injury using the sheep lung lymph model.

METHODS

Surgical Preparation of Sheep Lung Lymph Model

The following description of the surgical procedures and postsurgical care of sheep, as

Table 1
INSTRUMENTS AND MATERIALS USED IN SHEEP SURGERY

Surgical instruments
- 7.5-in. Jansen bayonets (Adson) with dissecting ends (2 each)
- 8-in. Russian tissue forceps (2 each)
- 5-in. Halstead mosquito forceps, straight (2 each)
- 5-in. Halstead mosquito forceps, curved (2 each)
- 6.25-in. Rochester-Pean Forceps (2 each)
- Mayo-Hegar needle holder (2 each)
- 9.5-in. Kantroqitz thoracic clamp, angular (1 each)
- 7.5-in. Varco thoracic clamp, full curve (1 each)
- Kapp-Beck blood vessel clamp, small curve (1 each)
- Kapp-Beck blood vessel clamp, large curve (1 each)
- 6-in. Metzenbaum scissors, curved (1 each)
- 7-in. Metzenbaum scissors, curved (1 each)
- 5-in. Operating scissors, blunt blade (1 each)
- 4.25-in. Suture wire scissors, angular blade and serrated edges (1 each)
- 4.5-in. Iris scissors, straight and sharp pointed (1 each)
- Finochietto rib spreader, 1.75-in. deep, 2.625-in. wide blades, total opening 8.5 in.
- Zalkind malleable retractor, 2 × 13 in., without handle (1 each)

Materials
- Sterile normal saline solution
- Sterile heparin solution (1000 U/ml)
- Sterile cardio-green dye
- 6-in. cotton-tipped applicators
- Gauze sponges, 4 × 4 in.
- Teflon felt, 1 × 1 in.
- #4 Black silk suture
- 30-in. 5-0 Cardiovascular silk suture
- 18-in. 2-0 Cuticular silk suture
- 27-in. 0 Chromic gut suture
- 30-in. 2-0 Gastrointestinal silk suture
- Monofilament steel wire
- 1-cc Syringes, 25-gauge needle
- 3-cc Syringes, 22-gauge needle
- Luer stub adapters, 15-gauge
- 16-in. Silicone lymphatic cannula (0.025-in. I.D., 0.047-in. O.D.)
- 16-in. Silicone pulmonary arterial and left atrial catheters (0.062-in. I.D., 0.125-in. O.D.
- 24-in. Jugular vein and carotid artery catheters
- 8-French Cordis sheath introducer
- Sterile stopcocks and injection caps
- Silicone rubber sealant (glue)

utilized in the laboratory of Vanderbilt University School of Medicine, is a modification of the procedures described by Staub et al.[5] These procedures are for a standard chronically instrumented lung lymph sheep model in which catheters are placed in the efferent duct of the caudal mediastinal lymph node, main pulmonary artery, left atrium, carotid artery, and jugular vein. Also detailed are the procedures to minimize systemic lymph contamination of the collected lymph. All surgical procedures are done under sterile conditions on the same day in the order described below.

The instruments, materials, and equipment needed in the surgical preparation of sheep are listed in Table 1. The choice of anesthesia during surgery is optional. However, gaseous anesthesia (e.g., halothane with nitrous oxide) is recommended because sheep recover rapidly from such anesthesia and the level of anesthesia is easily adjusted.

Presurgical Procedures

Healthy yearling sheep should be used from which all food has been withheld for at least

24 h prior to surgery. The wool should be clipped from all sites at which incisions are to be made. Although shaving of the incision sites is recommended, from a practical standpoint this is possible only after the sheep has been anesthetized. Initial anesthesia is by an intravenous infusion of a short-acting barbiturate (thiopental) to allow intubation of the sheep. A wooden "bite block" is recommended to prevent damage to the intubation tube and to secure the placement of the tube. The animal is placed on its left side on a surgical table and secured in position with rope. The surgical table elevator is raised approximately 6 in. to allow increased access to the caudal mediastinal lymph node (CMLN) and its efferent lymphatic. (If the surgical table is not equipped with an elevator, a disinfected wooden board may be used.) The surgical sites are then shaved and disinfected with a germicidal agent (e.g., betadine solution) and the incision area carefully draped.

Surgical Procedures

Lymphatic Cannulation

An incision parallel to the ribs is made through the skin and fascia over the fifth intercostal space and the underlying muscle layers are separated by blunt dissection. A thoracotomy is performed by dissection along the upper margin of the rib, taking extreme care not to injure the lung. The intercostal muscles are torn for approximately 6 to 8 in. using finger dissection. Retractors are inserted into the wound and carefully opened to avoid fracture of the ribs, and a malleable retractor is used to retract the lungs in a ventral direction. The CMLN is located under the pleura between the aorta and esophagus.

The efferent lymph duct of the CMLN is located by injecting 0.3 to 0.4 ml of indocyanine green dye (Hynson, Westcott and Dunning, Baltimore) into the head portion of the lymph node using a 25-gauge needle. (The manuscript by Landolt et al.[8] is an excellent reference relative to many of the possible anatomical variations of the efferent CMLN lymph duct.) The pleura overlying the duct is carefully torn with bayonet forceps and the lymph duct is ligated with 5-0 silk suture at a site as distal from the node as possible. It is suggested that some surrounding tissue be included when ligating to add support to the fragile lymph duct. Care should be taken to prevent ligation of any venous or arterial blood vessels that may supply the CMLN. Any extra efferent ducts should also be ligated whether from the head or dorsal portions of the node. Gentle yet firm countertraction is made with the ligating suture, which can be secured by wedging the suture between the retractors and ribs. A second 5-0 silk suture is loosely placed around the duct approximately 1 cm proximal of the site of ligation. The lymphatic is cleared of any fascia and fat tissue over an area 1 to 4 mm proximal of the ligation by gently rubbing with a cotton-tipped applicator. Only the surface portion of the duct should be stripped of fat and fascia because the medial, ventral, and dorsal fat and fascia offer support to the lymphatic duct and thus aid in cannulation. A silicone (silastic) lymph cannula is beveled approximately 2 cm from the silicone glue bead and filled with heparin using a syringe and 22-gauge needle; the syringe should remain attached to the cannula to prevent emptying of the heparin.

A transverse cut, one third of the duct diameter, is made with iridectomy scissors at the previously stripped area. After the incision is made, it should be continually observed for 10 to 15 s because the previously dilated duct will collapse and may distort the site of the cut. Using bayonet forceps, the lymphatic cannula is inserted (bevel down) as far as possible into the lymph duct. The previously placed untied 5-0 silk suture is then tightened to provide a firm seal without causing any occlusion of the cannula. The needle and syringe are then removed from the lymph cannula and lymph is allowed to passively drain. The position of the cannula should be adjusted by rotating and/or advancing and withdrawing the cannula to obtain the least resistance to lymph flow. The suture used to first ligate the duct is then tied around the cannula just behind the silicone glue bead. The cannula is tunneled under the pleura, exteriorized via a stab wound one interspace caudal to the thoracotomy, and secured to the skin with silicone glue. The surgical procedures thus far described are schematically illustrated in Figure 1.

The surgical table elevator is lowered and double stainless-steel wires are loosely looped around the separated ribs, avoiding damage to the intercostal arteries. The lungs are hyperinflated three times to 50 cm H_2O, allowing three normal respiratory cycles between each hyperinflation, to reverse any atelectasis. The ribs are reapproximated with the steel wires, the muscle layers closed with 0 chromic gut sutures, and the skin closed with 2-0 cuticular silk sutures. At each step of closure the lymphatic cannula should be checked to insure that no kinking or occlusion has occurred.

Systemic Lymph Contamination Removal

A second right thoracotomy is made at the ninth intercostal space in a manner similar to that described above. The CMLN is located and ligated with 4 silk sutures at the level of the inferior pulmonary ligament. A second ligation of the CMLN is made approximately 1.5 cm posterior of the first ligation. The lymph node can then be severed and a portion removed. All observable lymphatics deemed not to be of lung origin that enter the CMLN above the inferior ligament should be ligated and/or cauterized. These lymphatics are usually small and located on the esophagus and diaphragm. The lungs are hyperinflated as previously described and maintained on 10 to 15 cm H_2O positive end-expiratory pressure during closure of the wound.

Normally, a much larger amount of systemic contamination from nonlung lymphatics that enter the CMLN above the nodal ligation occurs on the left side.[2] Therefore, a third thoracotomy is made at the level of the left ninth intercostal space and the procedures as previously described are repeated. In addition, the left side pleura should be ligated from the esophagus to the aorta at the level of the left inferior pulmonary ligament.

Placement of Vascular Catheters

The sheep is taken off positive end-expiratory pressure and a fourth thoracotomy is made at the left fifth intercostal space. The lungs are retracted dorsally with a malleable retractor to allow access to the heart. The pericardium is punctured and cut to expose the main pulmonary artery. Care must be taken to prevent damage to the phrenic nerve. A small section of the pulmonary artery is lifted with Russian tissue forceps and clamped with small curved Kapp-Beck blood vessel clamps. A 2-0 gastrointestinal silk suture and two Teflon felt pledgets are placed in a purse-string configuration around the clamped pulmonary arterial segment, as illustrated in Figure 2. A transverse incision is made between the suture loop and a saline-filled silicone catheter (0.062 in. i.d., 0.125 in. o.d.) is inserted into the incision. The Kapp-Beck clamp is carefully opened to allow the catheter to be inserted 4 to 5 cm antegrade relative to blood flow with Russian forceps. The Kapp-Beck clamp is closed to prevent blood loss and the suture is tightened with a friction knot. The clamps are removed to test for a good seal and catheter patency. A second knot is tied and the catheter is "boot-laced" with the suture. Catheter patency is again checked to determine if occlusion has occurred. A catheter is inserted into the left atrium in a manner identical to the pulmonary artery, except the catheter is inserted only 2 cm to prevent it from entering the left ventricle. Any blood within the thoracic cavity should be removed and both catheters exteriorized through the sixth intercostal space and secured to the skin with 2-0 silk boot-laces.

During closure of the fourth thoracotomy, the lungs are again hyperinflated and maintained on 10 to 15 cm H_2O positive end-expiratory pressure. At each step of closure, catheter patency should be tested. Catheters of a known volume are then placed in the left jugular vein and carotid artery. In addition, a Cordis introducer sheath may be placed in the right jugular vein to allow subsequent introduction of a Swan-Ganz thermistor-tipped catheter for the measurement of body temperature and cardiac output. Each vascular catheter should be

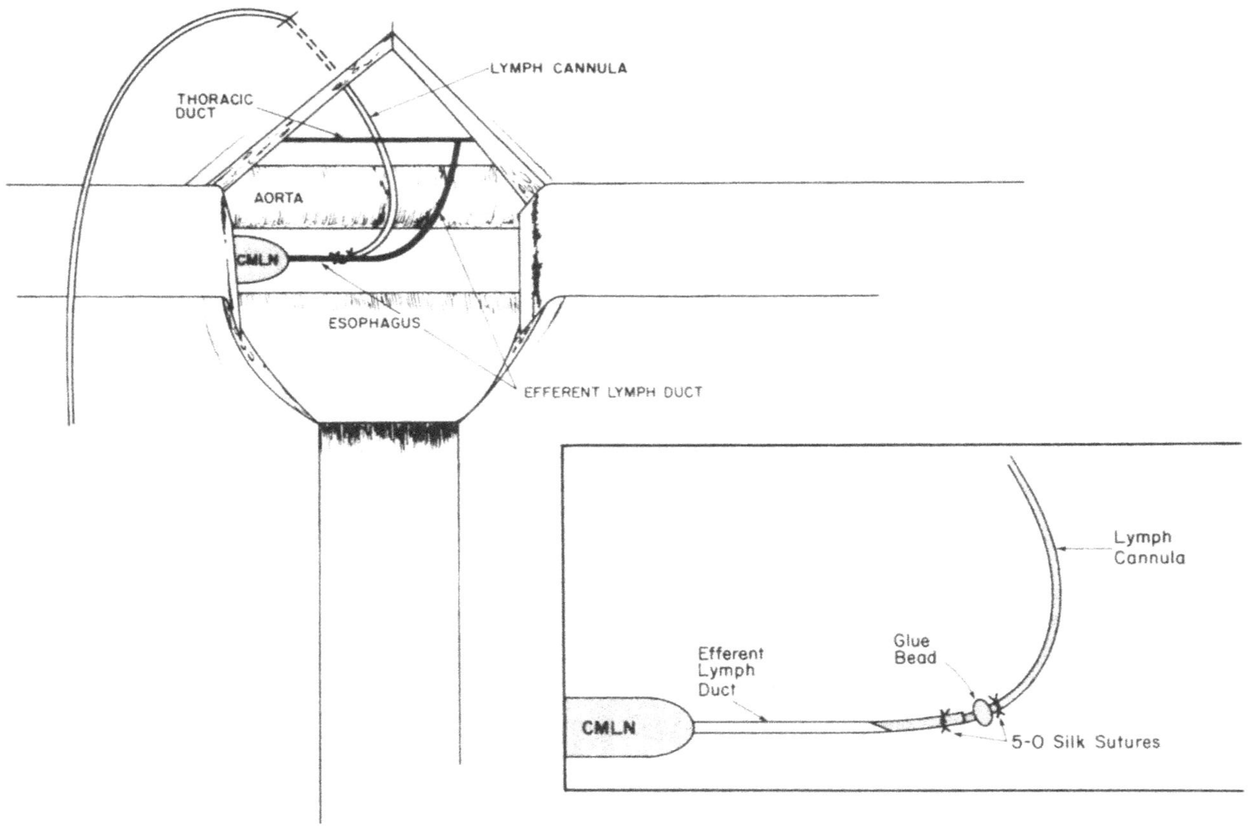

FIGURE 1. Schematic illustration of the anatomical features of the surgical cannulation of an efferent lymph duct of the caudal mediastinal lymph node. The insert figure is a more detailed schematic of the cannulated efferent lymph duct.

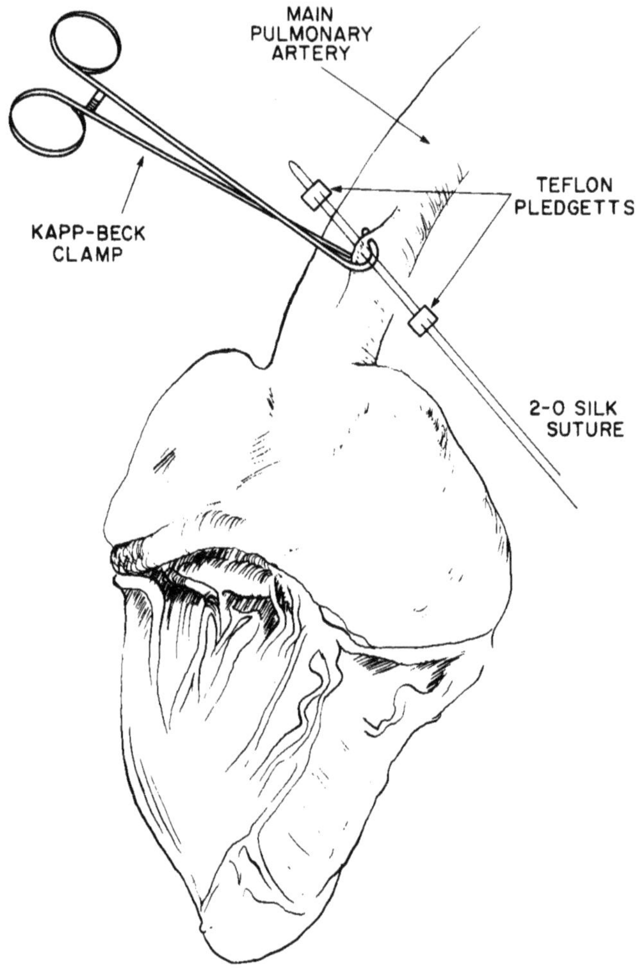

FIGURE 2. Schematic illustration of the placement of 2-0 gastrointestinal silk suture used in the placement of the pulmonary arterial catheter.

flushed with saline and filled with a heparin (1000 U/ml) and antibiotic (0.05 ml/10 ml) solution. The use of antibiotics (e.g., tetracycline) in the catheters is optional. However, it has been found that this often prevents septic episodes when the catheters are flushed on ensuing days.

The sheep is allowed to recover from anesthesia until it is alert and struggles against the restraining ropes. During this time a slow saline infusion (500 ml over 1 h) may be given to rehydrate the animal.

Postsurgical Sheep Care

A minimum of 7 d should be allowed for the sheep to recover from the surgical procedures prior to experimentation. Experience has shown that investigators should refrain from initiating experimental protocols until each sheep meets the following criteria: the general health is good (e.g., afebrile and eating well), lung lymph flow is stable, and lung lymph is clear (contains no red cells and the supernatant is not yellowish). During this time, proper sheep care is necessary to maintain patency of the lymph and vascular catheters. The vascular catheters should be aspirated at least once each day, flushed with saline, and refilled with a heparin/antibiotic solution. The lymphatic catheter should be stripped twice daily with the thumb and ball of the index finger to remove any small fibrin clots that may have formed.

Table 2
FREQUENTLY ENCOUNTERED PROBLEMS DURING SHEEP CARE

Problem	Cause	Remedy
Labored breathing	Hemothorax or pneumothorax	Evacuation of thorax via chest tube
Unable to aspirate vascular catheter	Catheter clotted	Forcefully inject 0.2 ml heparin and immediately aspirate[a]
Unable to aspirate vacular catheter and hard to inject	Catheter compressed or kinked	Remove securing sutures, withdraw catheter 2—4 cm and resecure
Unable to aspirate vascular catheter yet easy to inject	Catheter against vessel wall	Withdraw catheter 2—4 cm
Sheep sometimes becomes septic after flushing of vascular catheters	Contaminated vascular catheters	Aspirate catheters, flush vigorously with 20—30 ml sterile saline, fill with heparin/antibiotic solution 3 times daily over 3 d; give intramuscular antibiotics daily for 2 d

[a] To be used for jugular vein and carotid artery catheters only.

Lung lymph frequently ceases to flow for a number of reasons. The lymph cannula may have pulled out of the lymphatic, the cannula has become kinked or compressed, or a clot has formed in the cannula. If stripping the cannula does not meet with success, a small amount of heparin (0.2 ml) may be injected with a 22-gauge needle. At Vanderbilt University School of Medicine, the most successful method found to reestablish lymph flow[9] is to slowly inject 0.5 ml of a 1:1 (volume to volume) heparin/human plasma solution containing streptokinase and occlude the lymphatic cannula for 60 min.

If the lymphatic cannula can be easily injected yet remains nonfunctional after two consecutive streptokinase treatments, the cannula has most likely pulled out of the lymph vessel. This usually can be confirmed by injecting the cannula with heparin and elevating the tip of the cannula to cause a small amount of the heparin to flow retrograde along the cannula. If the fluid level in the cannula fluctuates noticeably with the respiratory cycles, it usually indicates that the cannula is no longer in the lymphatic. Unless the lymphatic cannula has pulled out during surgical procedures, little can be done to reestablish lymph flow. (Recannulation of the lymphatic after initial surgery is discouraged for two reasons. First, recannulating the lymphatic is much more difficult and meets with very little success. Second, sheep often do not tolerate the stress of a second cannulation procedure and thus from a humane perspective may be inappropriate.) Lymphatic cannula kinking and compression should not occur if preventive care is taken during surgery. However, if kinking should occur, the silicone glue securing the cannula to the skin of the sheep may be removed and the cannula withdrawn 2 to 3 cm before it is resecured with silicone glue.

Some of the problems most often encountered during sheep care and possible remedies are listed in Table 2.

Endotoxemia Experimental Protocols

General Protocol Considerations

Several factors should be considered prior to initiating specific experimental protocols. The type of endotoxin to be used in the experiments can be an important factor. It would be prudent to obtain a small quantity of endotoxin and conduct one or two preliminary experiments to determine the potency relative to the variables to be investigated. A quantity of suitable endotoxin from a single lot (i.e., batch) that will allow completion of all interrelated studies should then be obtained. Normally, a stock endotoxin solution (100 μg/ml saline) is made that is used for all experiments in a single study. However, if this is impractical, the stock endotoxin solution should be discarded after 10 weeks and another stock solution made from the same vial of endotoxin.

Table 3
VARIABLES MEASURED IN THE STANDARD SHEEP ENDOTOXEMIA MODEL

- Continuous measurements
  - Pulmonary arterial pressure
  - Left atrial pressure
  - Central venous pressure
  - Systemic arterial (aortic) pressure
- Measurements at 15 min intervals
  - Lung lymph flow rate
  - Lung lymph total protein concentration
  - Lung lymph total leukocyte levels
- Measurements at 30 and/or 60 min intervals
  - Plasma total protein concentration
  - Circulating blood hematocrit
  - Aortic and venous blood gases
  - Circulating blood leukocyte levels (total and differentials)
  - Body (blood) temperature
  - Heart rate
  - Cardiac output
- Discrete measurements
  - Blood and lung lymph eicosanoid concentrations
    - Thromboxane B_2
    - 6-Keto $PGF_{1\alpha}$ (prostacyclin)
    - PGE_2
    - LTB_4
    - LTC_4/LTD_4
  - Blood and lung lymph protein fraction concentrations (e.g., via polyacrylamide gradient gel electrophoresis)
  - Multiple indicator dilution measurements
    - Extravascular lung water
    - Lung intravascular blood volume
    - Lung microvascular permeability-surface area product (e.g., urea)

Careful consideration should be given to the experimental design. If a paired study is to be done (e.g., comparing control endotoxin responses to the responses after a drug intervention), the order should be randomized such that half of the control studies are conducted first. In addition, no more than three endotoxin experiments should be conducted on a single sheep due to possible development of endotoxin tolerance.

Specific Experimental Protocols

Measurements

A vast number of variables can be measured in the chronically instrumented unanesthetized lung lymph sheep model, some of which are listed in Table 3. Although methods for measuring most of the variables listed are self-evident, a few suggestions for some are described below.

On the day of the experiment, all vascular lines are aspirated, flushed with saline, and filled with heparin solution. Vascular catheters are connected to pressure transducers and the height of the transducers relative to the sheep (i.e., zero point) is marked on the side of the sheep for future reference. The zero point is arbitrary but should be consistent with some anatomical feature for each animal (e.g., olecranon, bottom of the chest, middle of the chest, etc.).

Lung lymph is collected in a graduated tube that contains a small amount of anticoagulant (heparin, EDTA, etc.) and the rate of flow is calculated from the volume collected over time. The lymph cannula should be positioned such that the tip of the cannula is never allowed to become immersed in the collected lymph, which may cause artifacts in the lymph flow rate via increased outlflow pressure. Lymph samples that are used to measure eicosanoid concentrations should be collected in ice with an inhibitor to stop the production of the arachidonate metabolite to be measured. These samples should be immediately centrifuged and the supernatant frozen at the appropriate temperature.

The method by which lung lymph is normally collected has recently been criticized by Drake et al.[10] These investigators have argued that since the lymphatics drain into the venous system, the pressure within the cannulated lymphatic should be maintained at or close to central venous pressure to simulate the effects on lung lymph flow if the lymphatic was not cannulated. This method requires the investigator to measure the resistance of the lymphatic cannula to lymph flow at various lymph protein concentrations and to calculate the appropriate height at which the distal tip of the cannula should be elevated to cause the proximal tip of the cannula to equal mean central venous pressure.

Blood samples should be obtained from the appropriate catheter(s) after first aspirating at least three times the catheter volume in a sterile syringe. After obtaining the blood sample(s), the aspirated blood may then be reinfused and the catheter flushed with a slightly heparinized solution. Those blood samples in which eicosanoid products are to be measured should be treated in a manner similar to the lymph described previously. When measuring blood gases and pH, the body temperature should also be recorded to allow correction of the measured values. This becomes more important during endotoxemia because the sheep body temperature usually increases substantially.

If cardiac output is measured by thermal dilution, at least two baseline determinations should be made at a minimum of 30 min apart. Each determination should be the average of three or four individual measurements. Prior to the first cardiac output measurement, the Swan-Ganz catheter should be flushed with 3 ml of the iced solution used in the measurements to prevent artifacts induced by the temperature of the fluid in the catheter.

Endotoxin Infusion Protocol

The amount of endotoxin and the time interval over which it is infused are arbitrary but should be consistent for all experiments. The precise method to infuse endotoxin is best described by a hypothetical experiment. Assuming that 0.5 μg/kg of endotoxin is to be infused intravenously in a 30-kg sheep over 15 min and the syringe pump rate of infusion is 1.6 ml/min, the total amount of endotoxin to be infused is 15 μg at a concentration of 0.625 g/ml. Although the total volume of the diluted endotoxin solution to be infused is 24 ml, a larger volume (e.g., 40 ml) is required to allow the filling of the catheter and extension tubing. If the stock endotoxin solution concentration is 100 μg/ml, then 0.25 ml of the stock solution is diluted with saline to a final volume of 40 ml. The diluted solution is mixed and aseptically filtered with a Millipore® filter (0.45-μm pore size; Millipore Corporation, Bedford, MA). The syringe pump is allowed to fill the tubing and venous catheter with the endotoxin solution approximately 5 min prior to beginning endotoxin infusion. (A small air bubble may be introduced to separate the endotoxin solution from the fluid within the venous catheter.) At the appropriate time the syringe pump is started and allowed to infuse for exactly 15 min, after which the pump is stopped and the venous catheter is aspirated to remove the remaining endotoxin solution and then flushed.

RESPONSES OF SHEEP TO ENDOTOXEMIA

Typical responses to a 15-min infusion of endotoxin (0.5 μg/kg) for a single sheep experiment are shown in Figures 3 to 6. The pulmonary hemodynamic response is depicted in Figure 3, the lung lymph response is shown in Figure 4, the lung lymph and circulating blood leukocyte responses are illustrated in Figure 5, and the 6-keto-$PGF_{1\alpha}$, thromboxane B_2 (the stable metabolites of prostacyclin and thromboxane A_2 respectively), and PGE_2 production as measured in lung lymph are shown in Figure 6.

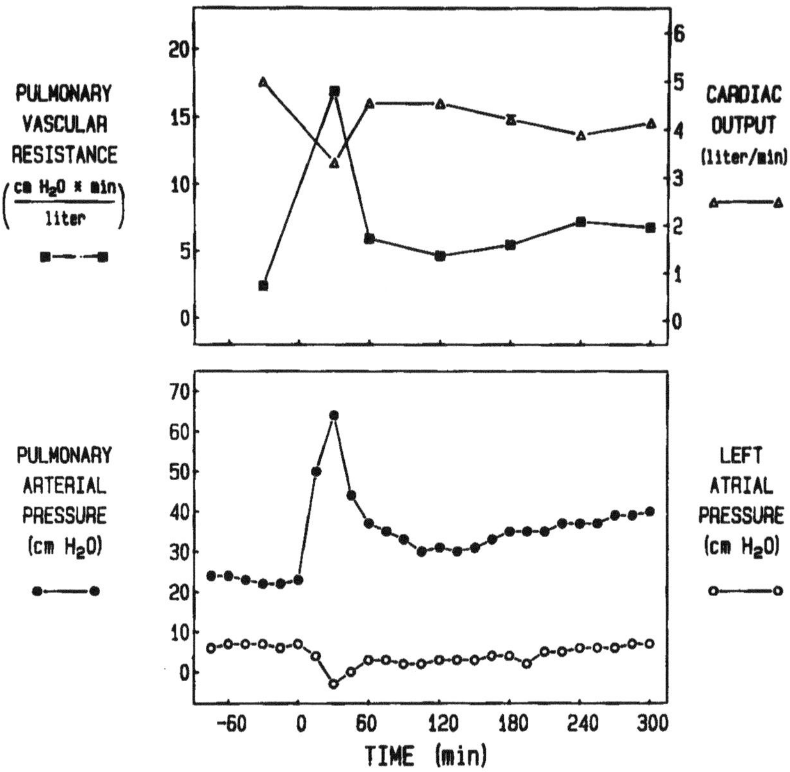

FIGURE 3. Pulmonary hemodynamic responses of one sheep to a 15 min infusion of *E. coli* endotoxin (0.5 μg/kg).

Phase 1 Responses

As can be seen from Figures 3 to 6, the responses to endotoxemia can be separated into two phases. Phase 1 occurs during the first hour after beginning the infusion of endotoxin and is characterized by pulmonary arterial hypertension, increased pulmonary vascular resistance, decreased cardiac output, increased lung lymph flow rate, decreased lung lymph to plasma total protein concentration ratio, hypoxemia, fever, and increased circulating hematocrit.

Thromboxane A_2 (TXA_2) is thought to be the direct mediator of endotoxemia-induced pulmonary arterial hypertension and hypoxemia.[11-12] It appears the TXA_2 causes an increase in pulmonary microvascular pressure via pulmonary venoconstriction, as evidenced by the observed increase in lung lymph flow rate and decrease in lung lymph to plasma total protein concentration ratio. However, the source of TXA_2 during endotoxemia is unclear, in that TXB_2 has been reported to be released by a number of cell types, including granulocytes, lymphocytes, alveolar macrophages, platelets, and endothelial cells.

Phase 2 Responses

Phase 2 occurs 2 to 5 h after endotoxin infusion, during which time pulmonary arterial pressure remains significantly elevated from baseline values but is decreased from the Phase 1 values, cardiac output increases toward normal, hypoxemia abates, body temperature remains increased, lung lymph flow rate remains elevated, and the lung lymph to plasma total protein concentration ratio returns to near baseline values. Phase 2 characteristics are considered to be indicative of an increase in lung microvascular permeability.

The time course of the increase in pulmonary microvascular permeability is difficult to

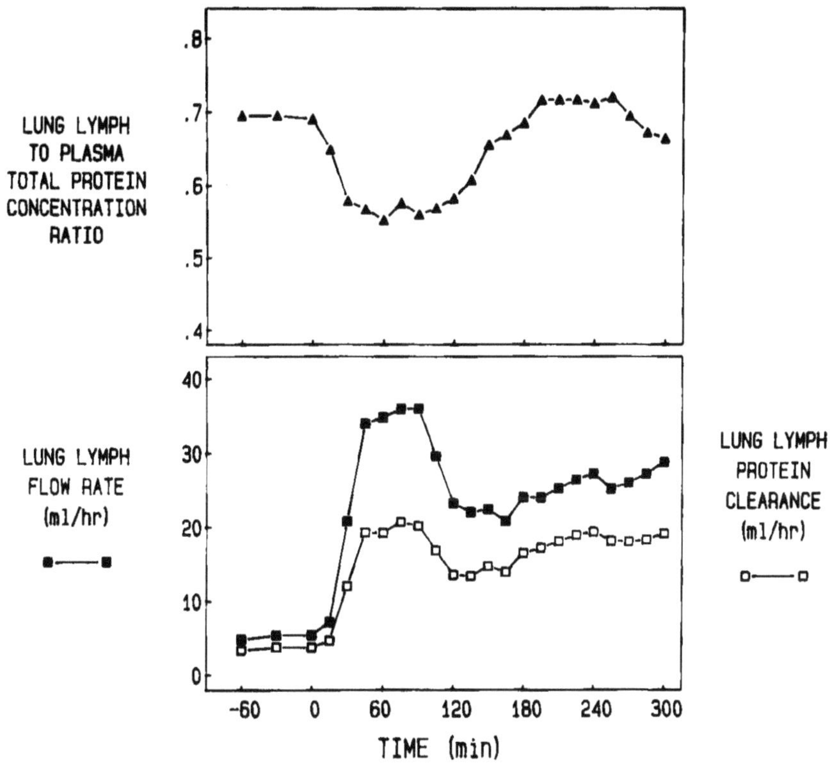

FIGURE 4. Lung lymph responses to endotoxin for the experiment of Figure 3.

assess with the unanesthetized preparation. However, based on histological data obtained from anesthetized sheep,[13] it is possible that an increase in permeability actually begins 15 to 30 min after endotoxin infusion, as shown in Table 4. The cause of the increase in permeability may be due to toxic oxygen metabolites,[14-16] protease release,[17] and/or direct cytotoxicity of endotoxin on the microvascular endothelial cells.[18,19] Pretreatment with several pharmacological agents (e.g., methylprednisolone,[20] antihistamines,[21] aminophylline + isoproterenol,[22] and verapamil[23]) has been reported to attenuate the increase in lung microvascular permeability. In addition, hydroxyurea-induced granulocytopenia has been reported to attenuate the permeability phase of endotoxemia in sheep,[24] whereas nitrogen-mustard-induced granulocytopenia did not attenuate the response.[25]

VARIATIONS OF THE SHEEP ENDOTOXEMIA MODEL

The previous sections have described the chronically instrumented unanesthetized sheep lung lymph model of endotoxemia. However, there are variations in endotoxin infusion and the sheep preparation that may be of interest to an investigator.

Variations in Endotoxin Infusion

Bolus and constant intravenous infusions of endotoxins are two additional methods to induce endotoxemia that warrant further investigation. Apparently, only the early study of Halmagyi et al.[4] reported the effects of a bolus infusion of endotoxin (0.3 mg/kg) in sheep. These authors reported many of the same endotoxin-induced effects on pulmonary hemodynamics in anesthetized sheep as were described above for the unanesthetized sheep. Interestingly, these authors reported that peak pulmonary arterial hypertension occurred

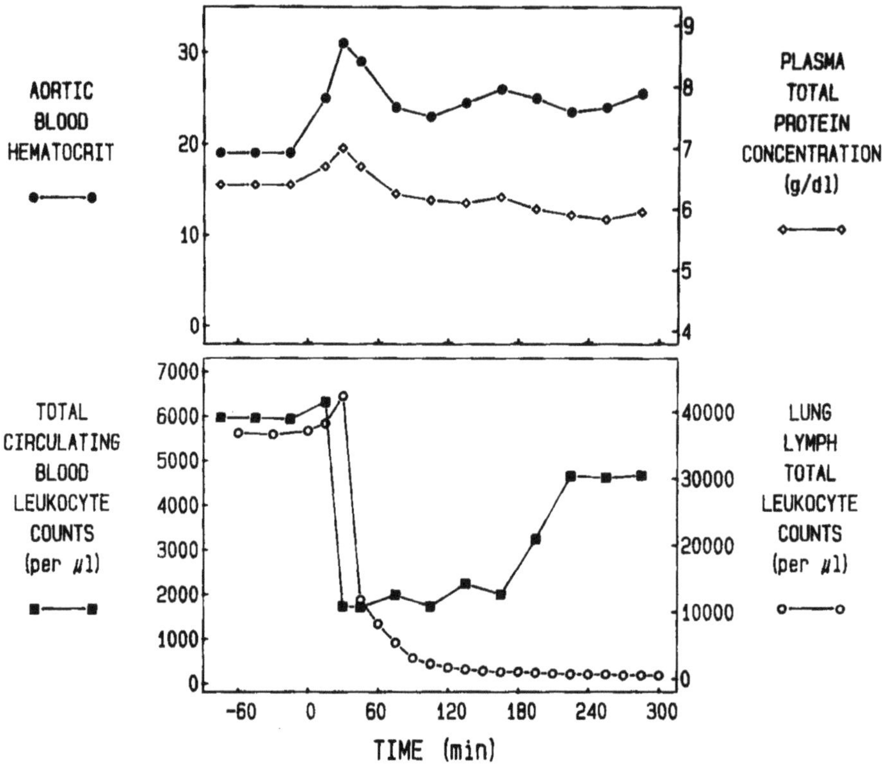

FIGURE 5. Effects of endotoxin on total leukocyte counts in circulating blood and lung lymph (lower panel) and aortic blood hematocrit and plasma total protein concentration (upper panel) in one sheep.

within 1 to 1.5 min after endotoxin infusion, rather than the 25 to 35 min for experiments described previously.

More recently, Traber and Traber[26] compared the effects of a constant infusion of endotoxin (24 ng/kg/h) to a 30-min infusion of endotoxin (1.5 μg/kg) in unanesthetized sheep. These investigators observed a modified Phase 1 response with a less intense pulmonary arterial hypertension, no decrease in cardiac output, and no decrease in lung lymph to plasma total protein concentration ratio. However, the Phase 2 responses between the two groups were qualitatively similar. An experiment similar to that of Traber and Traber[26] was conducted at Vanderbilt University School of Medicine in which a constant infusion of 50 ng/kg/h was given for 8.5 h. The results of this experiment are show in Figure 7. If the responses of this single experiment during constant endotoxin infusion are found to be consistent for other sheep, this method to induce endotoxemia may prove to be a more appropriate model to study the effects of pharmacological interventions during increased lung microvascular permeability.

Variations of the Sheep Preparation

Anesthetized Sheep Preparation

Experimental circumstances, such as painful procedures or the need to obtain multiple lung tissue samples, may preclude the use of unanesthetized sheep. It is suggested that, when possible, both lymphatic cannulation and removal of systemic lymph contamination be done at least 7 d in advance of experimental procedures. The rationale for doing as much of the surgical preparation beforehand is based on both the time involved in the preparation of the animal and possible deterioration of the animal during extended anesthesia. Moreover,

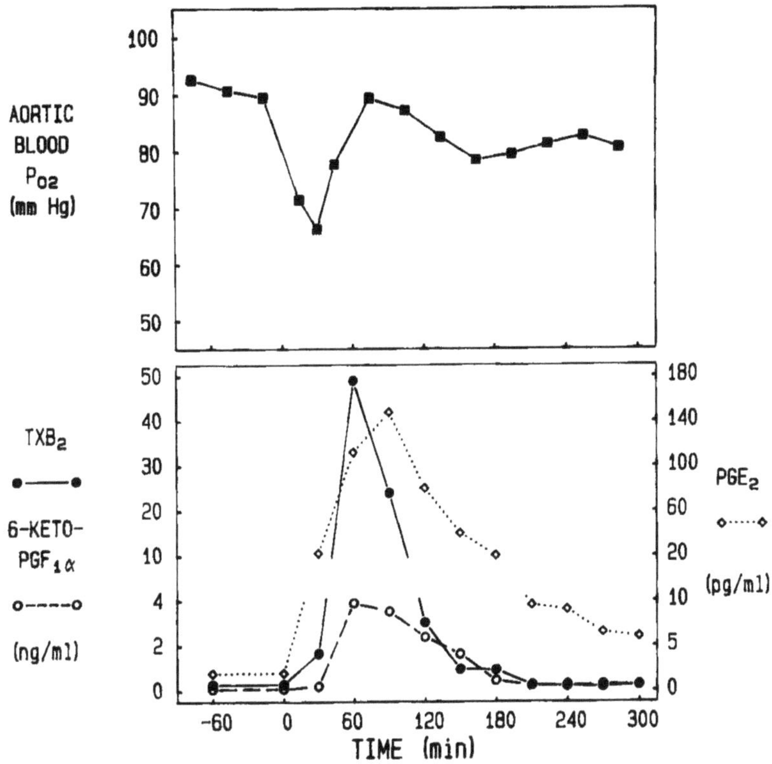

FIGURE 6. Effects of endotoxemia on aortic blood P_{O_2} (upper panel) and various eicosanoid production (lower panel) in one sheep.

Table 4
HISTOLOGICAL AND PHYSIOLOGICAL ALTERATIONS FOLLOWING *ESCHERICHIA COLI* ENDOTOXEMIA IN ANESTHETIZED SHEEP

Time after starting endotoxin infusions (min)	Structural alterations	Physiological alterations
15	Granulocyte accumulation, margination, degranulation, and fragmentation; accumulation of activated lymphocytes in pulmonary microvessels	Increased pulmonary arterial pressure, increased flow of protein poor lung lymph, hypoxemia, leukopenia, alterations of lung mechanics
30	Migration of leukocytes into lung interstitium, some interstitial edema	Peak pulmonary arterial hypertension and lung mechanics alterations
60	Perivascular edema, damaged type I and interstitial cells, endothelial and vascular wall damage	Pulmonary arterial pressure and lung mechanics return normal
≥120	Endothelial layer disruption	Increased lung microvascular permeability, pulmonary arterial pressure and lung mechanics stable

From Meyrick, B. and Brigham, K. L., *Lab. Invest.*, 48, 458, 1983. With permission.

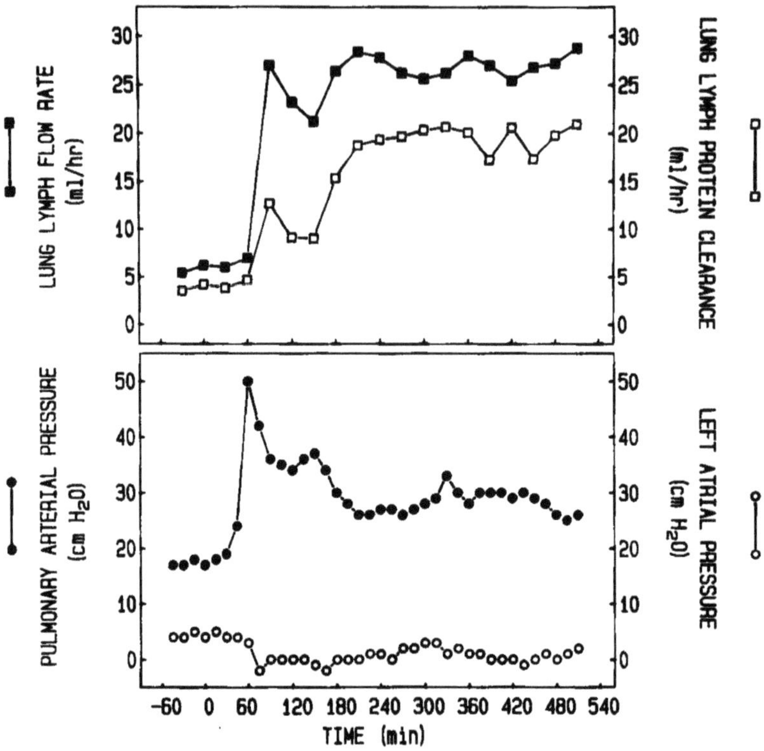

FIGURE 7. Effects of a constant endotoxin infusion (50 ng/kg/h) on pulmonary arterial and left atrial pressures (lower panel) and lung lymph flow rate and lung lymph total protein clearance (upper panel) for one sheep.

any damage caused by the surgical procedures will also be allowed to heal prior to the experiment. The disadvantage of utilizing an anesthetized preparation is the possible effects of anesthesia on the endotoxin-induced responses to be studied.

In Situ Sheep Preparation

Another potentially useful sheep preparation in the study of endotoxemia is the pump perfused *in situ* lung lymph preparation. The major advantage of this preparation is the ability to control many of the variables important in determining lung fluid balance (e.g., pulmonary blood flow, outflow pressure, perfusate composition, perfusate temperature, ventilation rate, etc.). In addition, the site of action for endotoxin is limited to the lungs (thereby permitting investigation of mediators and reactions devoid of systemic contamination) and the elimination of systemic contamination and nodal modification of the collected lung lymph. However, there are major disadvantages of the *in situ* preparation. First, extreme care must be taken to insure that all portions of the perfusion circuit are free of bacterial and pyrogen contamination. Second, normally only a single experiment can be conducted on each preparation. As was suggested for the anesthetized preparation, as much of the surgical procedures should be done 7 d in advance of the experiment.

REFERENCES

1. **Gabel, J. C., Drake, R. E., Arens, J. F., and Taylor, A. E.,** Unchanged pulmonary capillary filtration coefficients after *Escherichia coli* endotoxin infusion, *J. Surg. Res.*, 25, 97, 1978.
2. **Drake, R., Adair, T., Traber, D., and Gabel, J.,** Contamination of caudal mediastinal node efferent lymph in sheep, *Am. J. Physiol.*, 241, H354, 1981.
3. **Adair, T. H., Montani, J.-P., and Guyton, A. C.,** Modification of lymph by sheep caudal mediastinal node. Effect of intranodal endotoxin, *J. Appl. Physiol.*, 57, 1597, 1984.
4. **Halmagyi, D. F. J., Starzecki, B., and Horner, G. J.,** Mechanism and pharmacology of endotoxin shock in sheep, *J. Appl. Physiol.*, 18, 544, 1963.
5. **Staub, N. C., Bland, R. D., Brigham, K. L., Demling, R., Erdmann, A. J., and Woolverton, W. C.,** Preparation of chronic lung lymph fistulas in sheep, *J. Surg. Res.*, 19, 315, 1975.
6. **Brigham, K. L., Woolverton, W. C., Blake, L. H., and Staub, N. C.,** Increased sheep lung vascular permeability caused by *Pseudomonas* bacteremia, *J. Clin. Invest.*, 54, 792, 1974.
7. **Brigham, K. L., Bowers, R. E., and Haynes, J.,** Increased sheep lung vascular permeability caused by *Escherichia coli* endotoxin, *Circ. Res.*, 45, 292, 1979.
8. **Landolt, C. C., Matthay, M. A., and Staub, N. C.,** Anatomic variations of efferent duct from caudal mediastinal lymph node in sheep, *J. Appl. Physiol.*, 50, 1372, 1981.
9. **Traber, D. L., Adams, T., Jr., Henriksen, N., Traber, L. D., and Thomson, P. D.,** Reproducibility of cardiopulmonary effects of different endotoxins in the same sheep, *J. Appl. Physiol.*, 54, 1167, 1983.
10. **Drake, R., Giesler, M., Laine, G., Gabel, J., and Hansen, T.,** Effect of outflow pressure on lung lymph flow in unanesthetized sheep, *J. Appl. Physiol.*, 58, 70, 1985.
11. **Ogletree, M. L. and Brigham, K. L.,** Imidazole, a selective inhibitor of thromboxane synthesis, inhibits pulmonary vascular responses to endotoxin in awake sheep, *Am. Rev. Respir. Dis.*, 123, 247, 1981.
12. **Kubo, K. and Kobayashi, T.,** Effects of OKY-046, a selective thromboxane synthetase inhibitor, on endotoxin-induced lung injury in unanesthetized sheep, *Am. Rev. Respir. Dis.*, 132, 494, 1985.
13. **Meyrick, B. and Brigham, K. L.,** Acute effects of *Escherichia coli* endotoxin on the pulmonary microcirculation of anesthetized sheep: structure-function relationships, *Lab. Invest.*, 48, 458, 1983.
14. **Milligan, S. A., Hoeffel, S. A., and Flick, M. R.,** Endotoxin-induced acute lung injury in unanesthetized sheep is prevented by catalase, *Am. Rev. Respir. Dis.*, 131, A422, 1985.
15. **Demling, R. H., LaLonde, C., Jin, L.-J., Ryan, P., and Fox, R.,** Endotoxemia causes increased lung tissue lipid peroxidation in unanesthetized sheep, *J. Appl. Physiol.*, 60, 2094, 1986.
16. **Bernard, G., Lucht, W., Niedermeyer, M., Snapper, J., Ogletree, M., and Brigham, K. L.,** Effect of n-acetylcysteine on the pulmonary response to endotoxin in awake sheep and upon *in vitro* granulocyte function, *J. Clin. Invest.*, 73, 1772, 1984.
17. **Demling, R. H., Proctor, R., and Starling, J.,** Lung injury and lung lysosomal enzyme release during endotoxemia, *J. Surg. Res.*, 30, 135, 1981.
18. **Harlan, J. M., Harker, L. A., Reidy, M. A., Gajdusek, C. M., Schwartz, S. M., and Striker, G. E.,** Lipopolysaccharide-mediated bovine endothelial cell injury *in vitro, Lab. Invest.*, 48, 269, 1983.
19. **Meyrick, B. O.,** Endotoxin-mediated pulmonary endothelial cell injury, *Fed. Proc.*, 45, 19, 1986.
20. **Brigham, K. L., Bowers, R. E., and McKeen, C. R.,** Methylprednisolone prevention of increased lung vascular permeability following endotoxemia in sheep, *J. Clin. Invest.*, 67, 1103, 1981.
21. **Brigham, K. L., Padove, S. J., Bryant, K., McKeen, C. R., and Bowers, R. E.,** Diphenhydramine reduces endotoxin effects on lung vascular permeability in sheep, *J. Appl. Physiol.*, 49, 516, 1980.
22. **Foy, T., Marion, J., Brigham, K. L., and Harris, T. R.,** Isoproterenol and aminophylline reduce lung capillary filtration during high permeability, *J. Appl. Physiol.*, 46, 146, 1979.
23. **Parker, R. E. and Brigham, K. L.,** Verapamil attenuates increased sheep lung microvascular permeability during endotoxemia, *Fed. Proc.*, 45, 524, 1986.
24. **Heflin, A. C., Jr. and Brigham, K. L.,** Prevention by granulocyte depletion of increased vascular permeability of sheep lung following endotoxemia, *J. Clin. Invest.*, 68, 1253, 1981.
25. **Winn, R., Maunder, R., Chi, E., and Harlan, J.,** Neutrophil depletion does not prevent lung edema after endotoxin infusion in goats, *J. Appl. Physiol.*, 62, 116, 1987.
26. **Traber, D. and Traber, L.,** Continuous infusion of endotoxin (LPS) increases lung lymph flow and cardiac output, *Fed. Proc.*, 45, 525, 1986.

INJURY TO THE ISOLATED, PERFUSED LUNG INDUCED BY PHORBOL MYRISTATE ACETATE

Laurie J. Carpenter-Deyo and Robert A. Roth

INTRODUCTION

The adult respiratory distress syndrome (ARDS) is an acute respiratory illness which affects approximately 150,000 people per year in the United States.[1] Risk factors for development of ARDS include hemorrhagic, cardiogenic, septic, or anaphylactoid shock; gastric aspiration; smoke inhalation; pancreatitis; drug overdose; near drowning, or trauma.[2,3] Once fully developed, the syndrome is characterized by pulmonary edema and respiratory failure which is often unresponsive to ventilative therapy. Despite advances in intensive care, the mortality rate is greater than 50%.[1]

Although the exact mechanism(s) whereby ARDS develops is unknown, evidence exists which suggests that neutrophils participate in its pathogenesis. A correlation is seen between the magnitude of neutrophil influx into the lungs and the severity of pulmonary edema.[4] Increased activities of neutrophil-derived enzymes which are capable of injuring tissue have also been detected in bronchoalveolar lavage (BAL) fluid of ARDS patients.[4-7]

Since many clinical studies have been performed in patients with the fully developed syndrome, it has been difficult to determine whether neutrophils play a causal or a secondary role in ARDS. Because of this and other problems inherent in studying the disease process in humans, various animal models of ARDS have been developed. One such model employs phorbol myristate acetate (PMA).

When administered intravenously to rabbits, PMA produces changes in lung morphology which are similar in many ways to those seen in ARDS patients. Within 5 h of injection, PMA causes diffuse alveolar damage with intra-alveolar hemorrhage, fibrin deposits, and pulmonary edema.[8-10] By this time, neutrophil migration into interstitial and alveolar spaces and damage to epithelial and endothelial cells have occurred.[10] A hypercellular stage ensues 2 to 4 d after injection which is characterized by hyperplasia of type II cells and interstitial inflammation.[9,10] A similar proliferative stage occurs in ARDS patients who survive the acute stage. Both in humans afflicted with ARDS and in rabbits given daily injections of PMA, pneumonitis eventually progresses to fibrosis.[9,10]

In vitro, PMA stimulates neutrophils (PMN) to adhere, aggregate, degranulate, and release toxic oxygen metabolites.[8,11] Because of this property, it has been hypothesized that PMA pneumotoxicity is mediated by products from activated neutrophils. In the past few years, several studies have been performed *in vivo* and in the isolated lung to elucidate the role of the PMN in PMA-induced lung injury. It is hoped that a better understanding of the role of the neutrophil in ARDS will arise from the results of these studies.

This chapter is devoted predominantly to the use of the isolated, perfused lung (IPL) preparation in evaluating the PMA model of acute lung injury. Several investigators have used the IPL to study how PMA injures the lung because this preparation affords certain advantages for this purpose.[12-19] Some of these are listed in Table 1. Unlike other preparations *in vitro,* cells in the IPL are maintained in their normal anatomical and physiological associations. Thus, potential interactions among the many cell types which can occur *in vivo* are preserved. At the same time, use of the isolated lung eliminates extrapulmonary factors which may influence the toxicity of PMA *in vivo.* Unlike *in vivo,* test substances can be added to or removed from the pulmonary vasculature of the IPL very easily. Both respiratory and nonrespiratory functions may be evaluated with relative ease in this preparation. These are sometimes more cumbersome to study in other systems *in vitro* or *in vivo.*

Table 1
ADVANTAGES AND DISADVANTAGES OF ISOLATED LUNG PREPARATIONS

Advantages	
	Normal pulmonary architecture is maintained
	Extrapulmonary influences are eliminated or can be controlled
	Easy to add substances directly into the pulmonary circulation
	Samples of medium perfusing the lungs can be collected with ease
	Respiratory and nonrespiratory functions are readily evaluated
	Composition of perfusion medium can be determined by the investigator
Limitations	
	Alterations in specific cell types are not easily identified
	Period of viability is limited
	Results obtained may not be representative of what occurs in the intact animal

In the isolated lung preparation, the composition of the medium perfusing the lungs can be determined by the investigator. For example, by perfusing the vasculature with medium containing PMA and various blood components, one can determine how each of these components, separately or collectively, participates in the pathogenesis of PMA-induced lung injury. This is more difficult to do *in vivo*.

Although there are many advantages associated with using the IPL in pneumotoxicity studies, there are also some limitations (Table 1). Because the intact lung contains many different cell types, it is often difficult to identify which cell type(s) may be involved in a toxic response. Because the IPL, like any isolated organ, must be considered a deteriorating preparation from the outset, its usefulness is usually limited to, at most, a few hours of perfusion. In addition, because extrapulmonary influences which are present *in vivo* are eliminated from the IPL, results obtained in the IPL may not be representative of what occurs in the intact animal. Clearly, the advantages and disadvantages of using a particular prepartion should be considered before choosing one to use in addressing a specific toxicologic question. Most often, questions regarding mechanisms of action of toxic chemicals are most effectively addressed using several different but complementary biological preparations. Thus, although use of the isolated lung is emphasized in this chapter, the sole use of this preparation in evaluating the pneumotoxicity of PMA or other chemicals is not advocated.

IMPLEMENTING THE MODEL

Because several reviews and articles have been published which describe how to set up and perform isolated lung experiments,[20-24] these procedures are not discussed at length in this section. Rather, important details which should be considered when using this preparation to study PMA pneumotoxicity are addressed.

Surgical Procedure for Lung Isolation

Isolated lungs from rabbits, dogs, and rats have been used to study the mechanism of PMA-induced lung injury.[12-19] Because rats are routinely used in studies by the authors of this chapter, surgical techniques used to isolate lungs from this species are described. For the most part, similar surgical procedures are used to isolate lungs from other species.

The surgical steps required for successful isolation of rat lungs are diagrammed in Figure 1. Prior to surgery, rats are anesthetized with sodium pentobarbital (50 mg/kg, intraperitoneally). Xylazine (25 mg/kg, intramuscularly) and ketamine (2-5 mg/kg, intravenously) have been used to anesthetize rabbits before lung isolation.[12-14] After the animal is anesthetized, the trachea is cannulated. Some investigators connect a ventilator to this cannula to

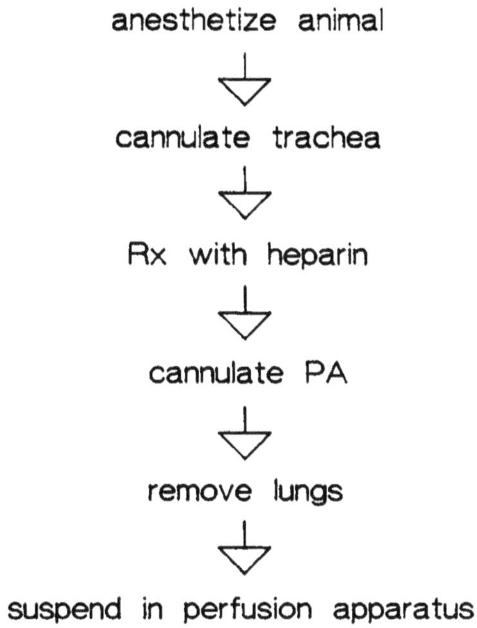

FIGURE 1. Flowchart of surgical procedure for isolated lung preparation.

prevent anoxia during the remainder of the surgical procedure.[12-14] However, short periods of anoxia do not appear to be detrimental to lung tissue.[24]

Next, a laparotomy is performed and heparin (500 U) is injected into the vena cava. Treatment with an anticoagulant is important to prevent thrombosis of the pulmonary vasculature. Next, the diaphragm and rib cage are cut away to expose the lungs and trachea. The pulmonary artery (PA) is then cannulated with a polyethylene cannula via a cut made in the right ventricle. In our studies, a stoppered, fluid-filled cannula is used to prevent the introduction of air into the pulmonary artery. If desired, medium may be pumped through this cannula *in situ* to clear blood from the vasculature.[12-14] After the PA is cannulated, the heart is cut away from the lungs. If venous outflow cannulation is required, a cannula is tied into the left atrium and the heart is now cut away.[12-14]

After removing the lungs from the thoracic cavity, they are placed in a dish containing normal saline and inflated and deflated repeatedly until the surface is smooth. This procedure tends to minimize atelectasis during perfusion. The pump to the perfusion apparatus is then turned on and the PA and tracheal cannulae are inserted into the proper sleeves. Care must be taken to avoid introduction of air into the PA cannula. Inappropriate twisting of either of the cannulae will increase perfusion pressure and may induce edema formation.

Before performing experimental manipulations, lungs are perfused with buffer in a single-pass manner to clear the vasculature of blood and to stabilize perfusion pressure. If blood has been cleared from the lungs during surgery, lungs should also be preperfused with buffer for a short period once in the perfusion apparatus until perfusion pressure has stabilized and lungs have equilibrated with this environment. Inflow pressure, which is monitored with a pressure transducer and recorded on a polygraph, should be approximately 4 to 6 mmHg for rat lungs perfused with buffer at a flow of 10 ml/min. Perfusion at higher flows or with different media may result in higher inflow pressure. If perfusion pressure remains abnormally high or tends to increase with time during the equilibration period, or if normal lungs are not visibly cleared of blood by 10 to 15 min of this preperfusion, the preparation should not be used for an experiment.

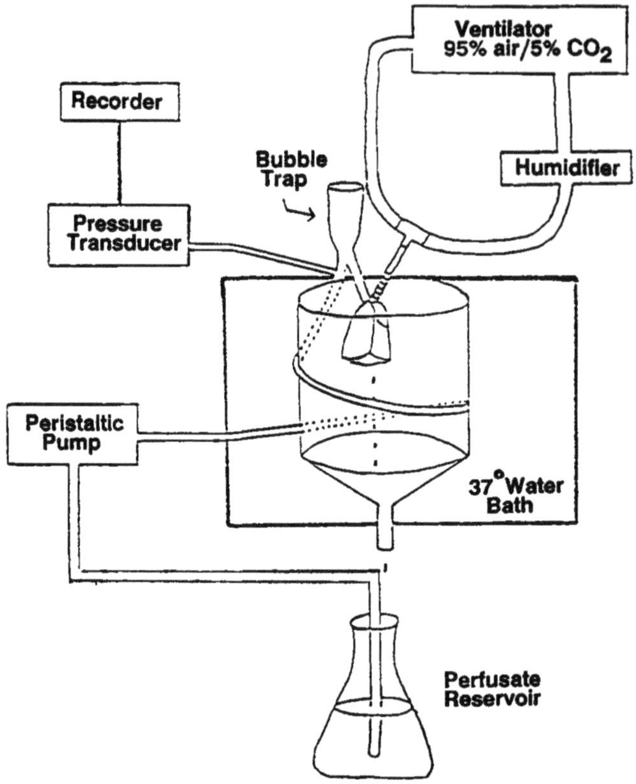

FIGURE 2. Components of lung perfusion apparatus.

Apparatus and Conditions of Perfusion

To investigate the mechanism of PMA toxicity to the isolated lung, several investigators have used systems which are designed to perform perfusions at a constant flow.[12-16] The system used by the authors (Figure 2) also delivers medium to the lungs at a constant flow. Perfusion medium is pumped through polyethylene tubing and into a siliconized glass coil which surrounds the plastic cylinder in which the lungs are suspended. The plastic cylinder is anchored into a tank which is filled with water maintained at body temperature. A glass bubble trap is placed immediately proximal to the pulmonary arterial cannula to prevent bubbles in the medium from entering the pulmonary circulation.

After perfusing the lungs, medium flows through the chamber housing the lungs and back into the reservoir for recirculating perfusions, as shown in Figure 2, or into a separate container for single-pass perfusions. The choice of either a single-pass or recirculating system will be dictated by the aims of the experiment and by cost. For example, single-pass perfusions are sometimes useful for studying the release, uptake, or metabolism of compounds by the lung. Recirculating perfusions are commonly used for toxicity studies.

When being perfused, lungs should be kept in a warm, moist environment. This is accomplished by surrounding the perfusion chamber with a heated waterbath (Figure 1) or a hood saturated with water vapor or by perfusing the medium through a heat exchanger.[12-19] Whichever method is used is not important; however, the investigator should be certain that the system maintains the perfusion medium and the lungs at the desired temperature.

In addition to constant flow perfusions, constant pressure perfusions have been used to study the mechanism of PMA toxicity to the isolated lung.[19] In this type of system, perfusion medium from an inflow reservoir placed above the level of the lungs flows into the lungs by gravity. Constant pressure perfusions are useful in determining whether a pneumotoxicant produces edema by a hydrostatic or permeability mechanism, since the contribution that elevated inflow pressure makes to the development of edema is eliminated.[19]

Before performing an isolated lung experiment, conditions of flow or perfusion pressure must be selected. If a system employing constant inflow pressure is used, it should be borne in mind that perfusion of the lung is dependent on arterial, venous, and alveolar pressures. To maximize uniformity of perfusion distribution, these pressures ideally should be adjusted to maintain zone 3 conditions. In zone 3, pulmonary arterial pressure exceeds pulmonary venous pressure and venous pressure exceeds alveolar pressure. In the isolated dog lobe, zone 3 conditions are achieved by adjusting the heights of the arterial and venous reservoirs to 15 to 20 cm H_2O and 4 to 5 cm H_2O, respectively, and inflating the lungs to an alveolar pressure of about 3 cm H_2O.[19]

In a constant flow system, generally the higher the flow, the greater the tendency toward edema formation. A commonly used flow in buffer-perfused rat lungs is about 10 ml/min, which is less than one fourth of pulmonary flow (i.e., cardiac output) *in vivo* in the rat. At higher flows, the addition of red blood cells (RBCs) to the perfusion medium seems to make for a more viable preparation.[25,26] If animals are to be used which vary substantially in body weight, flow can be adjusted according to lung weight or body weight. A commonly used flow in PMA experiments employing buffer-perfused lungs from rabbits weighing 1.8 to 3.8 kg is 40 to 45 ml/kg/min.[12-14]

Ventilation

To maintain the paO_2 and $paCO_2$ at physiological levels, lungs can be ventilated with a warmed, humidified gas mixture containing O_2 and CO_2. Alternatively, oxygenation may be achieved by bubbling the perfusion medium or statically inflating the lungs with these gases. Although ventilation is the more popular and more physiological means of maintaining adequate oxygenation of isolated lungs, the use of other methods is appropriate under some circumstances (i.e., when constant airway pressure is desired or when lungs are perfused with buffers devoid of plasma expanders).

Ventilation of lungs from animals such as rats and rabbits is commonly implemented with a small-animal respirator. With this type of respirator, one can adjust inspiratory and expiratory profiles to achieve the desired inspiratory pressure, end-expiratory pressure, and respiratory rate. For our studies in isolated rat lungs, maximal inspiratory pressure of 12 to 14 cm H_2O results in adequate inflation, and respiratory frequency is set at a rate which approximates that of the resting rat (90 to 110 c/min). In the isolated rabbit lung, a respiratory rate of 20 to 25 c/min has been used in studies with PMA.[12-14] In an isolated lung preparation, positive end-expiratory pressure should be maintained at about 2 to 3 cm H_2O to minimize atelectasis and edema formation.

Commonly used gas mixtures for ventilation include 95% O_2/5% CO_2 and 95% air/5% CO_2. Usually, the air/CO_2 mixture is used because the pO_2 of this gas more closely approximates the pO_2 of inspired gas *in vivo*. While the higher inspiratory pO_2 afforded by the mixture containing 95% O_2 may insure adequate oxygenation of the medium, it may also favor formation of tissue-damaging oxygen radicals and lipid peroxidation. Thus, the disadvantages of using this mixture in pneumotoxicity studies in the IPL may outweigh the advantages.

Perfusion Medium

To investigate the role of the neutrophil in PMA toxicity, several investigators have perfused lungs with a buffer containing a plasma expander and neutrophils.[12-18] Plasma expanders are added to the perfusion medium to increase osmolarity; consequently, they prevent fluid extravasation from the capillaries. Several different buffers have been employed (e.g., Krebs-Henseleit, Krebs-Ringer, or Greenberg-Bohr). Most investigators use bovine serum albumin (2 to 5%) as the plasma expander. Other substances which have been used in buffer-perfused lungs include Ficoll and dextran.[15,16] The choice of which expander to use should be made carefully, for some of these agents may influence the outcome of certain experiments. Dextran, for example, may itself produce edema in lungs of some species (unpublished observations). The use of albumin may be contraindicated in some studies because it has the ability to bind to and thereby influence the toxicokinetics of small molecular weight compounds in the isolated lung.[27,28] Its ability to scavenge oxygen radicals may also preclude its use in some studies.[29]

Experiments with PMA have also been performed in lungs perfused with blood or blood plasma supplemented with dextran.[19] The large quantity of anticoagulated blood which is needed to perfuse a lung can be obtained from a convenient vessel of an animal after intravenous injection with heparin or sodium citrate.[19,30] Because the amount of plasma which is obtained from one animal may not be sufficient to perfuse a lung, it may be necessary to combine plasma from several animals or to dilute the plasma with buffer supplemented with plasma expander.

Perfusion medium pH should be adjusted to approximately 7.35 to 7.45, before being used in the isolated lung, and the pH should be maintained throughout the duration of the perfusion. Ventilation of the lungs with a gas mixture containing 5% CO_2 helps to prevent excessive loss of CO_2 and resultant alkalosis. Gas should not be bubbled directly into medium containing bovine serum albumin as this causes uncontrollable foaming. Weak bases (e.g., sodium bicarbonate) or weak acids (e.g., ammonium chloride) may be added to the medium to maintain pH.

Isolation of Neutrophils

Several scientists investigating the role of the neutrophil in PMA toxicity perfuse lungs with buffer containing neutrophils isolated from human blood.[12-16] Procedures used by these investigators to isolate and purify neutrophils from human blood are variations of the methods published by Boyum[31] or by Hjorth and co-workers.[32] Before use, venous blood from human donors is anticoagulated with heparin (20 to 40 U/ml) or sodium citrate (0.38%). Erythrocytes are removed from the blood by sedimentation with dextran (Sigma Chemical Company, St. Louis or Pharmacia Fine Chemicals, Piscataway, NJ) or Hespan (American Hospital Supply, McGraw Parks, IL). Some investigators remove platelets from blood by centrifugation before removing erythrocytes.[12,13] Monocytes are then separated from the leukocyte-rich supernatant by centrifugation (275 to 400 $\times$ *g*) with Ficoll-Hypaque (Pharmacia Fine Chemicals or Winthrop Laboratories, New York) or Percoll (Pharmacia Fine Chemicals). The pellet containing neutrophils is resuspended in 0.15 *M* NH_4Cl to lyse remaining erythrocytes. After centrifugation, the neutrophils are suspended in the desired buffer. Preparations obtained by this method contain greater than 95% neutrophils.[12,14]

PMN have been isolated from the blood of animals using similar methods or from the peritoneum of rats by glycogen elicitation (Figure 3). The large number of neutrophils required can be obtained from the peritoneal cavities of adult male Sprague-Dawley rats (400 to 600 g) 4 h after the intraperitoneal injection of 1.0% glycogen in saline.[17,18,33] Contaminating erythrocytes (RBCs) are removed from this preparation by lysis with NH_4Cl (0.15 *M*). The pellet obtained after centrifugation at 100 $\times$ *g* is then washed with phosphate-buffered saline (PBS) and resuspended in the desired buffer. Suspensions obtained by this method contain approximately 95% neutrophils.

Currently, PMA (12-*O*-tetradecanoyl-phorbol-13 acetate) can be purchased from either Sigma Chemical Company (St. Louis) or L. C. Services Corporation (Woburn, MA). Use of PMA from Consolidated Midland Corporation (Brewster, NY) is also described in the literature; however, it is no longer available from this company. Upon receipt, it is diluted

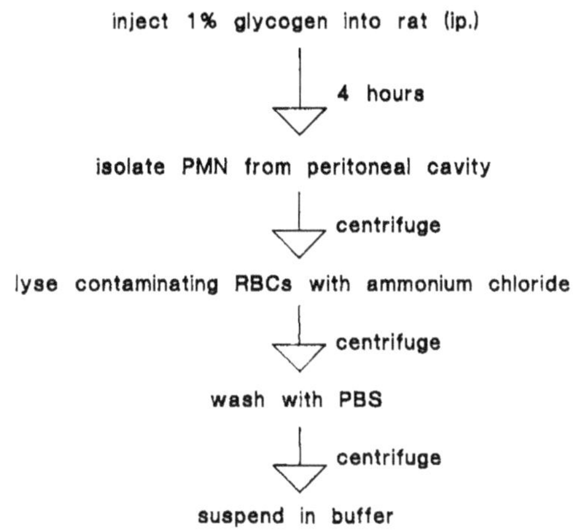

FIGURE 3. Procedure for isolation of neutrophils (PMN) from the peritoneum of the rat.

with dimethylsulfoxide (DMSO) to a concentration of 2 to 5 mg/ml, divided into aliquots of 50 to 100 μl, and stored at −70°C.[12-14,17-19] It is diluted further to the desired concentration with buffer just prior to use. Additional dilution with DMSO should be avoided, as DMSO may influence the degree of lung injury.

Experimental Design

Before initiating experiments with PMA in isolated lungs, several decisions regarding experimental design must be made. For example, the investigator must decide whether medium should be perfused through lungs in a single-pass or a recirculating manner. To date, experiments with PMA have been performed in lungs perfused with medium which is recirculating. Perhaps more information about PMA toxicity can be gained from this type of system, for toxic mediators which may be produced when PMA is perfused through the lungs are being supplied continuously to the vasculature. These mediators may be washed out of the lungs if a single-pass system is used.

The duration of the perfusion must also be determined. This choice should not be made arbitrarily, but should be based on the results of preliminary experiments. Lungs should be perfused with medium containing the vehicle for PMA to determine how long control preparations can be perfused without showing signs of deterioration. The length of time chosen for subsequent experiments should be less than that which leads to an increase in pressure or edema in this preparation. Although the authors have perfused isolated lungs for up to 2 h in the system described above without observing weight changes, shorter periods of perfusion are generally used. Duration of perfusion in published studies using PMA ranges anywhere from 40 to 90 min.[12-19]

Consideration must also be given to the route of administration. When choosing whether to administer PMA intrabronchially or intravenously, one should be aware that the mechanism of toxicity of PMA given intravenously may be different from that of PMA given intrabronchially.[12,34,35] When administering PMA to the isolated lung, most investigators have added PMA directly to the perfusion medium.[14-19]

Before performing experiments to determine if a particular blood element (e.g., neutrophils) mediates edema induced by PMA, lungs should be perfused with various concentrations of PMA to determine if PMA produces edema in the absence of the blood element. Subsequent experiments should be performed with a concentration of PMA which does not produce injury in lungs perfused only with buffer. In contrast, when investigating whether a particular blood element may protect lungs against PMA toxicity, one may want to use a concentration of PMA which produces edema in the absence of the blood element being investigated. The threshold concentration for toxicity may differ among species and may depend upon the rate of perfusion, binding of the toxicant to perfusion medium constituents, or other factors. Toxicity does not develop in rabbit lungs perfused at 40 ml/kg body weight per minute with buffer containing 120 ng/ml PMA.[14] In the isolated, buffer-perfused rat lung preparation used by the authors, concentrations of approximately 50 ng/ml and above result in rapid fluid accumulation.[17,18]

Similar experiments should also be performed with buffer containing only the blood element of interest to determine the number or concentration of cells which should be used in subsequent experiments. The concentration or number of cells used in experiments with PMA should be physiologically relevant and should not by itself produce toxicity. In experiments designed to investigate the role of the neutrophil in PMA toxicity, concentrations of PMN in the perfusion medium have ranged from 0.7 to 3 $\times$ 10^6/ml.[12-18]

In our experiments in the isolated rat lung, edema is induced by perfusion with 14 to 28 ng/ml PMA and 1 $\times$ 10^8 PMN for 30 min.[17,18] A range of concentrations of PMA is given because the concentration required to produce PMN-dependent lung injury may vary depending on the lot of PMA used, on the duration of storage, or on other factors.

EVALUATING THE MODEL

After insult with a pneumotoxicant, isolated lungs may undergo changes in physiology, morphology, biochemistry, or metabolic function. Lung injury may be assessed by evaluating such changes. In the following section, markers of lung injury which have been used to evaluate PMA toxicity to the isolated lung are described. Although other methods and markers exist, the discussion is limited to those which have been used in studies involving PMA.

Physiological Changes

Alterations in vascular integrity or in resistance to flow are common responses to pneumotoxicants. In lungs perfused at a constant flow, an increase in vascular resistance is manifest as an increase in perfusion pressure. In lungs perfused at constant pressure, an increase in resistance retards the flow. An increase in inflow pressure can be monitored with a pressure transducer placed between the pump and the lungs and recorded on a polygraph.[12-18] Venous pressure can also be recorded if the left ventricle has been cannulated. Flow can be monitored by a flowmeter placed distal to the arterial reservoir.[19]

Injury to the vasculature often results in edema formation in the isolated lung preparation. Development of edema can be assessed by weighing the lungs at the end of the perfusion[16-18] or by monitoring the weight change during perfusion with a force displacement transducer.[12-15] Lung weight can be expressed as an absolute number or as a fraction of either the body weight or the dry lung weight.

In isolated lungs perfused at constant flow, edema may be induced by an increase in vascular permeability and/or by an increase in hydrostatic pressure. To determine the extent to which each of these factors contributes to the development of edema, lungs can be perfused with a vasodilator such as papaverine or nitroglycerine to prevent an increase in inflow pressure and then be subjected to a venous pressure challenge.[14,36] Venous pressure can be increased (by about 10 mmHg) by elevating and/or by partially occluding the outflow perfusion line. After approximately 10 min, outflow pressure is returned to baseline. An increase in lung weight that occurs during this challenge is due to edema, and differences between experimental conditions are attributed to differences in vascular permeability.

An increased concentration of albumin in BAL fluid from lungs perfused with medium containing bovine serum albumin has also been used as an index of increased vascular permeability.[12-14] BAL fluid is obtained by instilling a predetermined amount of isotonic salt solution into the trachea and then withdrawing it. The fluid should then be spun in a centrifuge to remove cells. Concentrations of albumin in supernatant fluids are determined by measuring the change in absorbance at 630 nm after a short incubation (<1 min) with Bromcresol green (Sigma Chemical Company, St. Louis) and comparing this change with those observed with prepared albumin standards.

An increase in vascular leak can be detected by perfusing the lungs with medium containing ^{125}I-radiolabeled albumin.[16] After a period of time, radioactivity is removed from the vascular space by single-pass perfusion with radioactivity-free medium. Radioactivity in the lung tissue, BAL fluid, and a 1-ml aliquot of perfusion medium is then measured by a gamma counter. The lung albumin leak index is calculated by dividing the counts in the lung plus BAL fluid by the counts in 1 ml of perfusion medium.

Permeability changes in lungs perfused at constant pressure may also be assessed by determining the capillary filtration coefficient (K_f), isogravimetric capillary pressure (P_{ci}), and in blood-perfused lungs, the reflection coefficient (αd). K_f, which is a measure of water flux, increases when vascular permeability increases. P_{ci}, which is the pressure at which the lung does not gain or lose weight, and αd, which is a measure of the selectivity of the wall to retarding passage of plasma proteins, decrease when vascular permeability increases. K_f is determined by increasing both arterial and venous pressures by equal amounts and observing the weight gain for a short period of time.[19] It is calculated by dividing the initial rate of weight gain by the induced pressure change.[37] The P_{ci} is determined by decreasing arterial and increasing venous pressure in isogravimetric steps, which results in altered perfusion medium flow. The pressure at which these latter pressures converge at zero flow is the P_{ci}. The reflection coefficient can be estimated from changes in hematocrit and plasma protein concentration of a blood perfusion medium.[19,38]

Evaluation of respiratory functions may also provide valuable information about the nature of PMA toxicity in the isolated lung. However, while measurements of compliance, airflow resistance, functional residual capacity, and specific conductance have been determined in intact animals given PMA, they have not been reported in isolated lungs perfused with PMA. However, with the proper equipment, pulmonary mechanical functions commonly measured *in vivo* can also be measured in the isolated lung.[39-41]

Morphologic Analysis

After being perfused, lungs may be fixed for subsequent morphologic evaluation by light (LM) or electron microscopy (EM). Lungs may be fixed in a manner suitable for examination by LM or EM by intratracheal and/or intravenous infusion of 2 to 4% glutaraldehyde in 0.1 *M* cacodylate buffer at constant pressure.[9,10,16,17] Sections of lung can be prepared for examination by LM by postfixation in 10% buffered formalin and embedding in paraffin.[9,10] Samples to be examined by EM can be postfixed in osmium tetroxide, dehydrated, and embedded in epon.[10,17] Also, 1-μm sections suitable for examination by LM can be prepared by this process.[17] Because sections for EM examination can be prepared from these sections, use of this method eliminates the need to prepare separate sections for examination by both LM and EM. The conditions for fixation, embedding, etc., should be chosen with the objectives of a particular experiment in mind.

Alterations in Biochemistry or Metabolic Functions

When integrity of lung cells is compromised, enzymes which are normally intracellular, such as lactate dehydrogenase (LDH), may be released into the airways. Thus, increased activities of LDH may be detected in cell-free BAL fluid after perfusion with certain pneumotoxicants. LDH activity is commonly measured using the method of Bergmeyer and Bernt.[42] This method measures LDH activity by quantifying spectrophotometrically the disappearance of the cofactor NADH using pyruvate as the substrate.

Alterations in metabolic functions of lungs have also served as indicators of toxicity. For example, certain circulating biogenic amines such as serotonin (5HT) and norepinephrine (NE) are actively transported from perfusion medium into lung tissue by a carrier which is apparently on the endothelial cell surface. Certain injuries to vascular endothelium impair the ability of the lung to remove these amines from the circulation. Removal of 5HT or NE can be determined by perfusing the lungs in a single-pass manner with ^{14}C-5HT or ^{14}C-NE and measuring the amount of radiolabeled amine in samples of effluent medium collected shortly after (1 to 20 min) the addition of the amine to the perfusion medium.[43] Before being counted in a scintillation counter, each sample is passed through a cation exchange column to separate amines from acid metabolites. Removal of a particular amine is defined as the difference between the concentration of unchanged amine in the inflow perfusion medium and that in the collected effluent and is usually expressed as a fraction of the inflow concentration.[17,43,44]

Activities of enzymes such as angiotensin-converting enzyme (ACE) and 5′-nucleotidase, which are bound to the luminal surface of endothelial cells, may also be altered by toxicants which injure these cells. Activity of ACE can be determined by measuring the conversion of the synthetic substrate ^{3}H-benzoyl-phe-ala-pro (^{3}HBPAP) to ^{3}H-benzoylphenylalanine (^{3}HBPA) or the conversion of [glycine-1-^{14}C]hippuryl-histidyl-L-leucine ([^{14}C]HHL) to glycine-1-^{14}C-hippuric acid.[45,46] Before radiochemical analysis, ^{3}HBPA in effluent samples must be separated from unmetabolized ^{3}HBPAP by acidification with HCl and extraction with toluene. When using the [^{14}C]HHL method, samples are extracted with ethyl acetate prior to analysis to remove radioactivity associated only with the metabolite. To assess 5′-nucleotidase activity, lungs are perfused in a single pass manner with ^{14}C-2-adenosine-5′-monophosphate (5′-[^{14}C]AMP) and the amount of ^{14}C-adenosine which appears in the effluent is quantified.[47] Unmetabolized 5′[^{14}C]AMP is removed from these samples via passage through an anion exchange column.

When perfused with certain toxicants, isolated lungs may synthesize and release eicosanoids which, because of their vasoconstrictor or permeability-inducing properties, may contribute to lung injury. Thus, measurement of arachidonic acid metabolites may provide information about the mechanism of toxicity of a particular compound. Samples of cell-free perfusion medium can be analyzed for content of specific eicosanoids by high performance liquid chromatography (HPLC) and/or radioimmunoassay (RIA). Choice of assay should be dictated by assay specificity and sensitivity. At present, fairly specific antibodies are available to several prostanoids. However, since several currently available antibodies to specific leukotrienes exhibit undesired cross-reactivities, a useful method of analysis of these metabolites is reversed-phase HPLC with ultraviolet (UV) detection.[48,49] Analysis by HPLC is also advocated if concentrations of several different prostanoids in a given sample are to be determined, but sensitivity can be problematic. Fractions eluting from the HPLC with retention times corresponding to those of prostanoid standards may be collected and subsequently analyzed by the appropriate RIA.[49,50]

Detection of Neutrophil-Derived Products

In the past few years, it has become evident that inflammatory cells such as neutrophils and macrophages may participate in lung injury caused by certain toxicants. When activated by specific stimuli, these cells undergo a respiratory burst and release active oxygen species which are capable of degrading tissues.[51] These include the superoxide anion ($O\cdot-$), hydrogen peroxide (H_2O_2) and the hydroxyl radical ($OH^{\cdot}$). Presently, several methods exist for the detection of each of these species.[52,53] Rather than describing each of these assays at length, those which have been more commonly used when studying the mechanism of PMA pneumotoxicity are mentioned. For details on these and other procedures, the references above should be consulted.

Generation of superoxide by phagocytic cells is commonly measured spectrophotometrically by monitoring at 550 nm the superoxide dismutase-inhibitable reduction of ferricytochrome C.[54] This method has been used to detect superoxide release in isolated lungs as well as from isolated cells.[16] Release of H_2O_2 by neutrophils may be quantified by measuring the oxidation of reduced cytochrome C at 550 nm,[55] by monitoring the absorbance at 480 nm of ferrithiocyanate formed from the reaction of ferrous ammonium sulfate with potassium thiocyanate,[56] or by measuring the decrease in scopoletin fluorescence in the presence of horseradish peroxidase.[56] Hydroxyl radical release from cells *in vitro* has been detected using electron paramagnetic resonance coupled with spin trapping techniques.[57,58]

Although the release of oxidants *in vitro* is readily detected by the aforementioned methods, it has been difficult to quantify the release of specific oxygen metabolites *in vivo* with the majority of these methods. However, methods which are based on the ability of these metabolites to oxidize or to inactivate endogenous substances can be used to detect production of these metabolites *in situ*. The generation of oxidants *in vivo* has been estimated by measuring the specific activity of alpha$_1$-protease inhibitor (α_1PI) or catalase (after the intraperitoneal injection of 3-amino-1,2,4-triazole) in BAL fluid or by quantifying the glutathione content in lung tissue or cells in BAL fluid.[35,59-62] Measurement of the extent of lipid peroxidation of membranes has also been used as a marker of lung injury induced by chemicals which may act by an oxygen radical-dependent mechanism (i.e., paraquat).[63]

When stimulated by certain agents, neutrophils may also release lysosomal enzymes, some of which are capable of injuring cells and tissues. These include myeloperoxidase, lysozyme, collagenase, elastase, cathepsins, and cationic proteins.[64] To simplify this discussion, there is no elaboration on methods used to detect these enzymes. Rather, methods used to detect the enzymes that have been found in BAL fluid from animals treated with PMA or in medium from lungs perfused with PMA are briefly discussed. Detailed methods on how to perform assays for these and other neutrophil-derived enzymes are described elsewhere.[65,66]

The activities of some neutrophil-derived enzymes can be measured by spectrophotometric or immunologic methods. Phenophthalein glucuronide is hydrolyzed by β-glucuronidase to a species which absorbs light at 550 nm.[67] In the presence of H_2O_2 and myeloperoxidase, *O'*-dianisidine is oxidized to a compound which absorbs light at 560 nm.[68] Lysozyme can be quantified by measuring the turbidity of a solution containing *Micrococcus lysodeikticus*.[69] Elastolytic activity may be assessed using the synthetic substrate succinyl-tri-L-alanyl-p-nitroanilide.[70] Enzyme-linked immunoassays can be used to measure lactoferrin[71] or elastase complexed with α_1PL.[59] Activity of acidic cathepsins can be assessed by measuring cleavage fragments formed via the action of these enzymes on denatured hemoglobin.[35]

In summary, isolated lungs perform several nonrespiratory as well as respiratory functions. Given the diversity of these functions, one must choose which will be useful as markers of injury for a particular experiment. The choice should be dictated by the nature of the experimental protocol, by the hypothesis to be tested, and by the sensitivity of the marker in detecting lung injury.

EXPERIMENTAL FINDINGS

As indicated in Table 2, perfusion of lungs with PMA and buffer containing neutrophils produces vascular injury. In isolated rat or rabbit lungs perfused at a constant flow with medium containing PMA and neutrophils, increases in perfusion pressure and lung weight are observed.[12-18] Because the increase in pressure occurs prior to development of edema, it is possible that edema is produced by a hydrostatic rather than a permeability mechanism. However, results from several studies indicate that vascular permeability is increased in these lungs. In rat lungs perfused with buffer containing neutrophils, PMA, and ^{125}I-BSA, accumulation of radioactivity in lung tissue and BAL fluid is greater than in controls.[16] The

Table 2
CHANGES SEEN IN LUNGS PERFUSED WITH PMA AND PMN

Change	Ref.
Early	
↑ Perfusion pressure	12—18
↑ Synthesis of thromboxane	18
$O_2^{\cdot-}$ generation	16
Release of lysosomal enzymes	16
Neutrophil sequestration	16
Later	
↑ Lung weight	12—18
↑ Synthesis of prostacyclin	18
↑ Albumin in lavage fluid	12—14
Accumulation of ^{125}I-BSA in lavage fluid	16

concentration of albumin in BAL fluid from rabbit lungs perfused with neutrophils and PMA is also increased relative to lungs perfused with only neutrophils or with PMA.[12-14] The pressor response does not appear to be necessary for manifesting edema, since lungs which are perfused with a concentration of nitroglycerine that blunts the pressor response develop edema when subjected to a venous pressure challenge.[14] In isolated dog lobes perfused at constant pressure with blood containing PMA, permeability is increased, as evidenced by a decrease in the isogravimetric capillary pressure and reflection coefficient and an increase in the capillary filtration coefficient.[19]

Examination by electron microscopy of tissue from lungs perfused with PMN and PMA reveals extensive endothelial cell blebbing and exposure of basement membranes.[16,17] Because endothelial cell integrity is altered in rat lungs perfused with PMA and PMN, metabolic functions of endothelium in these lungs have been examined. In isolated rat lungs perfused with PMN and PMA, disposition of perfused 5HT does not differ from controls.[17] In contrast, recent results obtained in isolated, buffer-perfused rabbit lungs indicate that PMA (in the absence of neutrophils) decreases the V_{max} for ACE and decreases the V_{max} and increases the K_m for 5HT uptake.[72] Alterations in kinetics of ACE, 5′-nucleotidase, and 5HT uptake also occur in rabbits treated with PMA *in vivo*.[73,74] Furthermore, effects on ACE and 5′-nucleotidase kinetics are produced by a dose of PMA which does not produce morphologic evidence of lung injury or an alteration in blood flow.[73] Thus, in some but not all instances, measurements of the activity of these metabolic functions may provide sensitive markers of endothelial cell injury caused by PMA.

In vitro, PMA stimulates PMN to release toxic oxygen species and potentially harmful lysosomal enzymes (Figure 4).[8,11,75,76] It is also a weak stimulus for the release of arachidonic acid metabolites from PMN.[77] Results from several studies suggest that PMA produces injury to the isolated lung via a PMN-dependent mechanism. In isolated rat lungs perfused with PMN and PMA, degranulated PMN appear to be in close approximation or are adherent to areas of endothelium that are damaged.[17] Increased concentrations of neutrophil-derived oxidants and enzymes are detected in the perfusion medium of rat lungs perfused with PMA and PMN before the development of edema.[16] In addition, results from many studies indicate that lung injury induced by PMA does not occur in lungs which are not coperfused with neutrophils.[12-16]

Active oxygen species from PMN appear to mediate edema, since lungs which are coperfused with PMA and PMN that cannot undergo a respiratory burst do not exhibit an increase in lung weight or enhanced permeability to protein.[12] In addition, lungs are protected against development of edema by perfusion with oxygen radical scavengers such as dime-

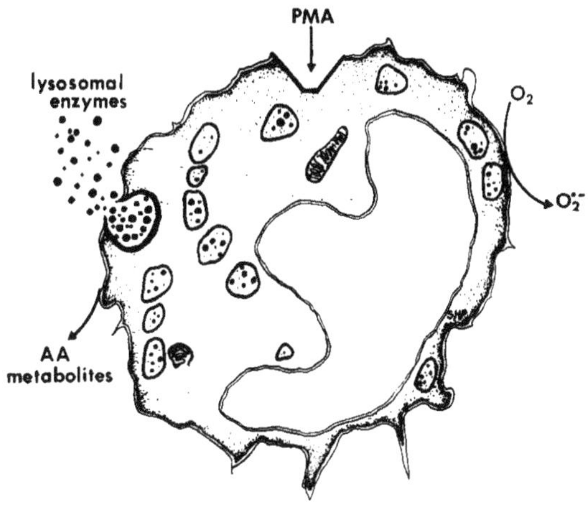

FIGURE 4. Products released from neutrophils stimulated with PMA.

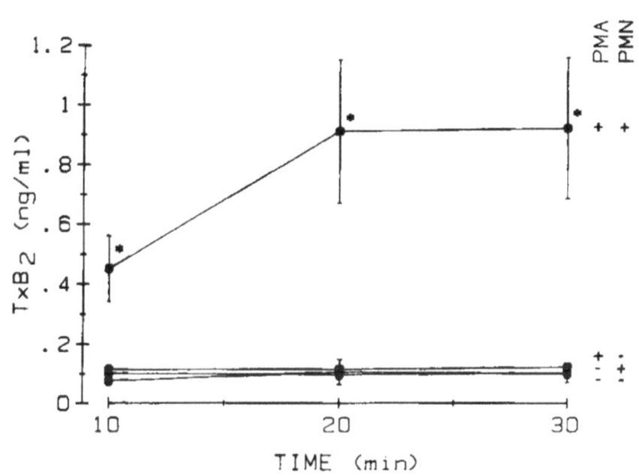

FIGURE 5. Effect of PMA and neutrophils (PMN) on thromboxane (TxB_2) production in isolated, perfused rat lungs. Lungs were perfused for 30 min with PMA (14 ng/ml) or DMSO vehicle and with PMN (1 × 10^8) or PMN vehicle. Samples of effluent were collected at various times and were analyzed for TxB_2 by radioimmunoassay. In some cases, error bars are obscured by points. *Significantly different from all other treatment groups. (From Carpenter, L. J. and Roth, R. A., *Toxicol. Appl. Pharmacol.*, 91, 33, 1987. With permission.)

thylthiourea or superoxide dismutase and catalase.[14-17] Neutrophil adherence appears to be obligatory for the manifestation of edema, since manipulations which inhibit neutrophil adherence (cytochalasin B or anti-Mo1 monoclonal antibody) but do not inhibit oxygen radical production or degranulation provide protection.[13,16]

As indicated in Figure 5, synthesis of thromboxane in isolated rat lungs is stimulated by perfusion with PMN and PMA.[15,18] The observation that an increase in thromboxane synthesis precedes edema development suggests that thromboxane may contribute to the pathogenesis of edema in this preparation. The fact that the thromboxane synthetase inhibitor, Dazmegrel, provides virtually complete protection from edema produced in rat lungs perfused with PMA and PMN supports this hypothesis.[18] Thromboxane appears to increase vascular permeability in isolated dog lobes perfused with blood containing PMA, since alterations in K_f, P_{ci}, and the reflection coefficient are not observed when the lobes are coperfused with the thromboxane synthetase inhibitor OKY-046.[19]

In summary, perfusion with PMA and PMN produces acute injury and edema in isolated lungs. Although increases in pressure are observed in these lungs, the bulk of evidence suggests that an increase in vascular permeability is responsible for edema development. Although the exact mechanism of edema development is not known, results suggest that active oxygen species and thromboxane contribute to its pathogenesis.

DISCUSSION

For the most part, the pathological and biochemical changes that are observed in isolated lungs perfused with PMA are similar to those seen acutely in animals treated with PMA. Both *in vivo* and in the isolated lung, a single administration of PMA produces an increase in pulmonary arterial pressure and vascular injury.[8-10,12-19,34,35,59,78-81] As in the isolated lung, results from many studies *in vivo* suggest that neutrophils are involved in the toxicity of PMA. Rabbits given PMA intravenously are protected against lung injury by neutrophil depletion.[10,12,35] The generation of oxidants, as assessed by a decrease in the activities of α_1-PI and catalase in BAL fluid and a decrease in glutathione in lung tissue, is observed *in vivo* after PMA.[35,59] Active oxygen species appear to mediate edema in rabbits treated intravenously with PMA, for edema does not occur in rabbits pretreated with a hydroxyl radical scavenger or a chemical which inhibits the respiratory burst of neutrophils.[14,78]

Results from studies *in vivo* also suggest that a cyclooxygenase metabolite(s) of arachidonic acid is involved in the pathogenesis of PMA toxicity. Increased concentrations of TxB_2 and 6-keto $PGF_{1\alpha}$ are detected in lung lymph in sheep treated with PMA.[79,80] Treatment with a cyclooxygenase inhibitor attentuates pulmonary hypertension which occurs in these sheep. However, in contrast to what is seen in the isolated rat lung perfused with PMA and PMN or in the isolated, blood-perfused dog lobe, inhibition of thromboxane synthesis by meclofenamate does not protect sheep from the PMA-induced increase in vascular leak.[80] One possible explanation for this difference in results is that some extrapulmonary factor(s) which is present *in vivo* but absent in the isolated lung may override the protective effect of cyclooxygenase inhibition.

In vitro, PMA stimulates PMN to release enzymes from specific, but not from azurophilic, granules.[76] Release of enzymes from specific granules, i.e., lactoferrin and lysozyme, is also observed in rat lungs perfused with PMA and human PMN.[16] However, in the intact animal treated with PMA, increased concentrations of azurophilic granule constituents, such as myeloperoxidase, acid protease, β-glucuronidase, and elastase, have been detected in BAL fluid.[35,59] Presently, it is unknown if release of enzymes from azurophilic granules occurs in isolated lungs perfused with PMA and PMN, or if release of specific granule constituents occurs *in vivo* after PMA. In addition, it is also unknown if lysosomal enzymes play a role in the pathogenesis of PMA toxicity *in vivo*. Because edema does not occur when lungs are perfused with granulocytes which can release lysosomal enzymes but cannot release oxidants, it has been suggested that lysosomal enzymes are not important mediators of PMA toxicity. However, since experiments with specific inhibitors of degranulation or lysosomal enzyme activity have not yet been performed, the possibility remains that PMN-derived enzymes may play a role in PMA toxicity. Indeed, the observation that oxygen radicals can inactivate certain enzymes such as alpha$_1$-protease inhibitor,[59] which regulates the activity of some lysosomal enzymes,[82] raises the possibility of interaction between these PMN products.

Although the evidence which implicates PMN involvement in PMA toxicity is quite strong, results from several studies indicate that PMA can produce neutrophil-independent injury in some instances. Neutrophil depletion does not protect rats or rabbits from pulmonary edema which occurs after the intrabronchial administration of PMA.[34,35] Intravenous PMA also produces lung injury in neutrophil-depleted sheep and in isolated lungs which are not coperfused with PMN.[17,18,81] Results from our studies indicate that the ability of PMA to produce lung injury in the absence of neutrophils is dependent on the PMA concentration used. At high concentrations, PMA produces increased perfusion pressure and edema which are independent of perfused PMN and intravascularly-generated oxygen metabolites. At low concentrations, however, edema is mediated by active oxygen species derived from neutrophils. Currently, the mechanism by which PMA produces edema in the absence of PMN is not known. Studies with the vasodilator papaverine suggest that a factor(s) which increases hydrostatic pressure may be responsible.[83] Indeed, PMA may cause smooth muscle contraction directly by activating protein kinase C.[84] Alternatively, thromboxane released by action of PMA on lung cells may contribute to the increase in pressure and the formation of edema.[18]

In rabbits given daily intravenous injections of PMA for several weeks, acute edematous lung injury is followed by the development of interstitial pneumonitis and pulmonary fibrosis.[9,10] These changes are also seen in patients who survive the acute phase of ARDS.[85] Although neutrophil-depleted rabbits are protected against edema produced by chronic PMA treatment, they still develop fibrosis. This suggests that fibrosis which develops in PMA-treated rabbits is not precipitated by acute lung injury. Other than this, relatively little is known about how PMA produces pulmonary fibrosis. Unfortunately, since the period of viability in the IPL preparation is limited, it cannot be used for chronic toxicity studies. Further studies *in vivo* are needed to investigate this aspect of PMA pneumotoxicity.

Although the role of the neutrophil in the development of pulmonary edema by PMA has been extensively investigated, little attention has been given to the role which other lung cells play in this process. *In vitro,* PMA elicits the release of toxic oxygen species from eosinophils, macrophages, and endothelial cells.[86-88] In the presence of platelets, release of oxidants by activated neutrophils is enhanced.[89] Because PMA-induced injury is oxygen radical-dependent, these cells may participate in the pathogenesis of lung injury by increasing the overall oxidative burden. In addition, these cells may also release other mediators of lung injury upon stimulation by PMA or by substances which are formed from PMA-activated cells. Thus, further studies designed to investigate the role which cells other than PMN play in PMA toxicity are necessary to determine how this chemical produces lung injury *in vivo.*

In summary, the PMA model of pneumotoxicity has and will likely continue to provide important information about the biology of lung injury. Effects of PMA on the lung have been fairly well characterized, but much remains unknown about the mechanisms by which PMA causes pulmonary vascular injury. Results from experiments in the isolated lung coincide largely with those obtained from studies *in vivo,* and this preparation affords advantages in elucidating how PMA inflicts lung injury. The use of this and other models should contribute to our understanding of the pathophysiology of ARDS and other acute lung disorders in humans.

ACKNOWLEDGMENTS

This work was supported by USPHS Research Grant HL32244. We thank Diane Hummel for typing this manuscript and Susan H. Parkinson for her artistic work (Figure 4).

REFERENCES

1. **Saunders, N. A.,** Adult respiratory distress syndrome: mechanisms of lung injury, *Aust. N.Z. J. Med.*, 14, 769, 1984.
2. **Edde, R. R. and Burtis, B. B.,** Lung injury in anaphylactoid shock, *Chest*, 63, 637, 1973.
3. **Petty, T. L.,** Adult respiratory distress syndrome (a perspective based on 18 years' personal experience), *Eur. J. Respir. Dis.*, 126, 3, 1983.
4. **Weiland, J. E., Davis, W. B., Holter, J. F., Mohammed, J. R., Dorinsky, P. M., and Gadek, J. E.,** Lung neutrophils in the adult respiratory distress syndrome: clinical and pathophysiologic significance, *Am. Rev. Respir. Dis.*, 133, 218, 1986.
5. **Lee, C. T., Fein, A. M., Lippman, M., Holtzman, H., Kimbel, P., and Weinbaum, G.,** Elastolytic activity in pulmonary lavage fluid from patients with adult respiratory distress syndrome, *N. Engl. J. Med.*, 304, 192, 1981.
6. **McGuire, W. W., Spragg, R. C., Cohen, A. B., and Cochrane, C. G.,** Studies on the pathogenesis of the adult respiratory distress syndrome, *J. Clin. Invest.*, 69, 543, 1982.
7. **Christner, R., Fein, A., Goldberg, S., Lippman, M., Abrams, W., and Weinbaum, G.,** Collagenase in the lower respiratory tract of patients with adult respiratory distress syndrome, *Am. Rev. Respir. Dis.*, 131, 690, 1985.
8. **O'Flaherty, J. T., Cousart, S., Lineberger, A. S., Bond, E., Bass, D. A., DeChatelet, L. R., Leake, E. S., and McCall, C. E.,** Phorbol myristate acetate: *in vivo* effects upon neutrophils, platelets and lung, *Am. J. Pathol.*, 101, 79, 1980.
9. **McCall, C. E., Taylor, R. G., Cousart, S. L., Woodruff, R. D., Lewis, J. C., and O'Flaherty, J. T.,** Pulmonary injury induced by phorbol myristate acetate following intravenous administration in rabbits: acute respiratory distress followed by interstitial pneumonitis and pulmonary fibrosis, *Am. J. Pathol.*, 111, 258, 1983.
10. **Taylor, R. G., McCall, C. E., Thrall, R. S., Woodruff, R. D., and O'Flaherty, J. T.,** Histopathologic features of phorbol myristate acetate-induced lung injury, *Lab. Invest.*, 52, 61, 1985.
11. **Repine, J. E., White, J. G., Clawson, C. C., and Holmes, B. M.,** The influence of phorbol myristate acetate on oxygen consumption by polymorphonuclear leukocytes, *J. Lab. Clin. Med.*, 83, 911, 1974.
12. **Shasby, D. M., Van Benthuysen, K. M., Tate, R. M., Shasby, S. S., McMurtry, I. F., and Repine, J. E.,** Granulocytes mediate acute edematous lung injury in rabbits and in isolated rabbit lungs perfused with phorbol myristate acetate: role of oxygen radicals, *Am. Rev. Respir. Dis.*, 125, 443, 1982.
13. **Shasby, D. M., Shasby, S. S., and Peach, M. J.,** Granulocytes and phorbol myristate acetate increase permeability to albumin of cultured endothelial monolayers and isolated perfused lungs: role of oxygen radicals and granulocyte adherence, *Am. Rev. Respir. Dis.*, 127, 72, 1983.
14. **Jackson, J. H., White, C. W., McMurtry, I. F., Berger, E. M., and Repine, J. E.,** Dimethylthiourea decreases acute lung edema in phorbol myristate acetate-treated rabbits, *J. Appl. Physiol.*, 61, 353, 1986.
15. **McDonald, R. J., Berger, E. M., and Repine, J. E.,** Neutrophil-derived oxygen metabolites stimulate thromboxane release, pulmonary artery pressure increases and weight gains in isolated perfused rat lungs, *Am. Rev. Respir. Dis.*, 135, 957, 1987.
16. **Ismail, G., Morganroth, M. L., Todd, R. F., and Boxer, L. A.,** Prevention of pulmonary injury in isolated perfused rat lungs by activated human neutrophils preincubated with anti-mol monoclonal antibody, *Blood*, 69, 1167, 1987.
17. **Carpenter, L. J., Johnson, K. J., Kunkel, R. G., and Roth, R. A.,** Phorbol myristate acetate produces injury to isolated rat lungs in the presence and absence of perfused neutrophils, *Toxicol. Appl. Pharmacol.*, 91, 22, 1987.
18. **Carpenter, L. J. and Roth, R. A.,** Involvement of thromboxane in injury to isolated rat lungs perfused with phorbol myristate acetate in the presence and absence of neutrophils, *Toxicol. Appl. Pharmacol.*, 91, 33, 1987.
19. **Allison, R. C., Marble, K. T., Hernandez, E. M., Townsley, M. I., and Taylor, A. E.,** Attenuation of permeability lung injury after phorbol myristate acetate by verapamil and OKY-046, *Am. Rev. Respir. Dis.*, 134, 93, 1986.
20. **Rosenbloom, P. M. and Bass, A. D.,** A lung perfusion preparation for the study of drug metabolism, *J. Appl. Physiol.*, 29, 138, 1970.
21. **Niemeier, R. W. and Bingham, E.,** An isolated perfused lung preparation for metabolic studies, *Life Sci.*, 11, 807, 1972.
22. **Mehendale, H. M., Angevine, L. S., and Ohmiya, Y.,** The isolated perfused lung: a critical evaluation, *Toxicology*, 21, 1, 1981.
23. **Smith, B. R. and Bend, J. R.,** Lung perfusion techniques for xenobiotic metabolism and toxicity studies, in *Methods in Enzymology*, Vol. 77, Jakoby, W. B., Ed., Academic Press, New York, 1981, 105.
24. **Niemeier, R. W.,** The isolated perfused lung, *Environ. Health Perspect.*, 56, 35, 1984.

25. **Wiersma, D. A. and Roth, R. A.**, Clearance of 5-hydroxytryptamine by rat lung and liver: the importance of relative perfusion and intrinsic clearance, *J. Pharmacol. Exp. Ther.*, 212, 97, 1980.
26. **Wiersma, D. A., Braselton, W. E., and Roth, R. A.**, The influence of flow on the metabolism of perfused benzo[a]pyrene by isolated rat lung, *Chem. Biol. Interact.*, 43, 1, 1982.
27. **Hofman, W. F. and Ehrhart, I. C.**, Albumin attenuation of oleic acid edema in dog lung depleted of blood components, *J. Appl. Physiol.*, 58, 1949, 1985.
28. **Nicolaysen, G.**, Perfusate qualities and spontaneous edema formation in an isolated perfused lung preparation, *Acta Physiol. Scand.*, 83, 563, 1971.
29. **Holt, M. E., Ryall, M. E. T., and Campbell, A. K.**, Albumin inhibits human-polymorphonuclear leukocyte luminol-dependent chemiluminescence: evidence for oxygen radical scavenging, *Br. J. Exp. Pathol.*, 65, 231, 1984.
30. **Ganey, P. E. and Roth, R. A.**, 6-Keto Prostaglandin $F_{1\alpha}$ and thromboxane in isolated, blood-perfused lungs from moncrotaline pyrrole-treated rats, *J. Toxicol. Environ.*, 23, 127, 1988.
31. **Boyum, A.**, Isolation of mononuclear cells and granulocytes from human blood. Isolation of mononuclear cells by one centrifugation, and of granulocytes by combining centrifugation and sedimentation at 1 g, *Scand. J. Clin. Lab. Invest.*, 21, 77, 1968.
32. **Hjorth, R., Jonsson, A. K., and Vretblad, P.**, A rapid method for purification of human granulocytes using Percoll. A comparison with dextran sedimentation, *J. Immunol. Methods*, 43, 95, 1981.
33. **Ward, P. A., Sulavik, M. C., and Johnson, K. J.**, Rat neutrophil activation and effects of lipoxygenase and cyclooxygenase inhibitors, *Am. J. Pathol.*, 116, 223, 1984.
34. **Johnson, K. J. and Ward, P. A.**, Acute and progressive lung injury after contact with phorbol myristate acetate, *Am. J. Pathol.*, 107, 29, 1982.
35. **Schraufstatter, I. U., Revak, S. D., and Cochrane, C. G.**, Proteases and oxidants in experimental pulmonary inflammatory injury, *J. Clin. Invest.*, 73, 1175, 1984.
36. **Tate, R. M., Van Benthuysen, K. M., Shasby, D. M., McMurtry, I. F., and Repine, J. E.**, Oxygen radical-mediated permeability edema and vasoconstriction in isolated perfused rabbit lungs, *Am. Rev. Respir. Dis.*, 126, 802, 1982.
37. **Drake, R., Gaar, K. A., and Taylor, A. E.**, Estimation of the filtration coefficient of pulmonary exchange vessels, *Am. J. Physiol.*, 234, 266, 1978.
38. **Maron, M. B.**, Differential effects of histamine on protein permeability in dog lung and forelimb, *Am. J. Physiol.*, 242, 565, 1982.
39. **Drazen, J. M.**, Physiologic basis and interpretation of common indices of respiratory mechanical function, *Environ. Health Perspect.*, 16, 11, 1976.
40. **Drazen, J. M.**, Physiological basis and interpretation of indices of pulmonary mechanics, *Environ. Health Perspect.*, 56, 3, 1984.
41. **Costa, D. L.**, Interpretation of new techniques used in the determination of pulmonary function in rodents, *Fundam. Appl. Toxicol.*, 5, 423, 1985.
42. **Bergmeyer, H. U. and Bernt, E.**, Lactate dehydrogenase UV assay with pyruvate and NADH, in *Methods of Enzymatic Analysis*, Vol. 2, Bergmeyer, H. U., Ed., Academic Press, New York, 1974, 574.
43. **Gillis, C. N., Huxtable, R. J., and Roth, R. A.**, Effects of monocrotaline pretreatment of rats on removal of 5-hydroxytryptamine and noradrenaline by perfused lung, *Br. J. Pharmacol.*, 63, 435, 1978.
44. **Block, E. R. and Schoen, F. J.**, Effect of alpha naphthylthiourea on uptake of 5-hydroxytryptamine from the pulmonary circulation, *Am. Rev. Respir. Dis.*, 123, 69, 1981.
45. **Rohrbach, M. S.**, [Glycine-1-^{14}C]hippuryl-L-histidyl-L-leucine: a substrate for the radiochemical assay of angiotensin converting enzyme, *Anal. Biochem.*, 84, 272, 1978.
46. **Catravas, J. D., Lazo, J. S., Dobuler, K. J., Mills, L. R., and Gillis, C. N.**, Pulmonary endothelial dysfunction in the presence or absence of interstitial injury induced by intratracheally injected bleomycin in rabbits, *Am. Rev. Respir. Dis.*, 128, 740, 1983.
47. **Catravas, J. D. and White, R. E.**, Kinetics of pulmonary angiotensin-converting enzyme and 5′-nucleotidase *in vivo*, *J. Appl. Physiol.*, 57, 1173, 1984.
48. **Sun, F. F. and McGuire, J. C.**, Metabolism of arachidonic acid by human neutrophils: characterization of the enzymatic reactions that lead to the synthesis of leukotriene B_4, *Biochim. Biophys. Acta*, 794, 56, 1984.
49. **Schulz, R. and Seeger, W.**, Release of leukotrienes into the perfusate of calcium-ionophore stimulated rabbit lungs: influence of 5-lipoxygenase inhibitors, *Biochem. Pharmacol.*, 35, 183, 1986.
50. **Salmon, J. A., Simmons, P. M., and Palmer, R. M. J.**, A radioimmunoassay for leukotriene B_4, *Prostaglandins*, 24, 225, 1982.
51. **Fantone, J. C. and Ward, P. A.**, Role of oxygen-derived free radicals and metabolites in leukocyte-dependent inflammatory reactions, *Am. J. Pathol.*, 107, 397, 1982.
52. **Parker, L.**, Oxygen radicals in biological systems, in *Methods in Enzymology*, Vol. 105, Parker, L., Ed., Academic Press, New York, 1984, 167.

53. **Greenwald, R. A., Ed.,** *Handbook of Methods for Oxygen Radical Research,* CRC Press, Boca Raton, FL, 1985, 117.
54. **Babior, B. M., Kipnes, R. S., and Curnutte, T.,** Biological defense mechanisms; the production by leukocytes of superoxide, a potential bactericidal agent, *J. Clin. Invest.,* 52, 741, 1973.
55. **Toth, K. M., Clifford, D. P., White, C. W., and Repine, J. E.,** Intact human erythrocytes prevent hydrogen peroxide mediated damage to isolated perfused rat lungs and cultured bovine pulmonary artery endothelial cells, *J. Clin. Invest.,* 74, 292, 1984.
56. **Thurman, R. G., Ley, H. G., and Scholz, R.,** Hepatic microsomal ethanol oxidation: hydrogen peroxide formation and role of catalase, *Eur. J. Biochem.,* 25, 420, 1972.
57. **Green, M. R., Hill, H. A. O., Okolow-Zubkowska, M. J., and Segal, A. W.,** The production of hydroxyl and superoxide radicals by stimulated neutrophils — measurement by EPR spectroscopy, *FEBS Lett.,* 100, 23, 1979.
58. **Britigan, B. E., Cohen, M. S., and Rosen, G. M.,** Detection of the production of oxygen-centered free radicals by human neutrophils using spin trapping techniques: a critical perspective, *J. Leuk. Biol.,* 41, 349, 1987.
59. **Revak, S. D., Rice, C. L., Schraufstatter, I. U., Halsey, W. A., Bohl, B. P., Clancy, R. M., and Cochrane, C. G.,** Experimental pulmonary inflammatory injury in the monkey, *J. Clin. Invest.,* 76, 1182, 1985.
60. **Aebi, H.,** Catalase, in *Methods of Enzymatic Analysis,* Vol. 2, Bergmeyer, H. U., Ed., Academic Press, New York, 1974, 673.
61. **Brehe, J. E. and Burch, H. B.,** Enzymatic assay for glutathione, *Anal. Biochem.,* 74, 189, 1976.
62. **Abrams, W. R., Weinbaum, G., Weissbach, L., Weissbach, H., and Brot, N.,** Enzymatic reduction of oxidized a-1-proteinase inhibitor restores biological activity, *Proc. Natl. Acad. Sci. U.S.A.,* 78, 7483, 1981.
63. **Aldrich, T. K., Fisher, A. B., Cadenas, E., and Chance, B.,** Evidence for lipid peroxidation by paraquat in the perfused rat lung, *J. Lab. Clin. Med.,* 101, 66, 1983.
64. **Falloon, J. and Gallin, J. L.,** Neutrophil granules in health and disease, *J. Allergy Clin. Immunol.,* 77, 653, 1986.
65. **Laszlo, C.,** Proteolytic enzymes, in *Methods in Enzymology,* Vol. 80, Laszlo, C., Ed., Academic Press, New York, 1981, 535.
66. **Barrett, A. J.,** Lysosomal enzymes, in *Lysosomes: A Laboratory Handbook,* 2nd ed., Dingle, J. T., Ed., North-Holland, Amsterdam, 1979, chap. 2.
67. **Fishman, W. H.,** β-Glucuronidase, in *Methods of Enzymatic Analysis,* Vol. 2, Bergmeyer, H. U., Ed., Academic Press, New York, 1974, 929.
68. **Bretz, U. and Baggiolini, M.,** Biochemical and morphological characterization of azurophil and specific granules of human neutrophilic polymorphonuclear leukocytes, *J. Cell Biol.,* 63, 251, 1974.
69. **Yuli, I., Tomonaga, A., and Snyderman, R.,** Chemoattractant receptor functions in human polymorphonuclear leukocytes are divergently altered by membrane fluidizers, *Proc. Natl. Acad. Sci. U.S.A.,* 79, 5906, 1982.
70. **Bieth, J., Spiess, B., and Wermuth, C. G.,** The synthesis and analytical use of a highly sensitive and convenient substrate of elastase, *Biochem. Med.,* 11, 350, 1974.
71. **Hetherington, S. V., Spitznagel, J. K., and Quie, P. G.,** An enzyme-linked immunoassay (ELISA) for measurement of lactoferrin, *J. Immunol. Methods,* 65, 183, 1983.
72. **Myers, C. L. and Pitt, B. R.,** Effect of phorbol ester (PMA) on kinetics of lung angiotensin converting enzyme (ACE) and serotonin (5HT) uptake, *Fed. Proc.,* 46, 665, 1987.
73. **McCormick, J. R., Chrzanowski, R., Andreani, J., and Catravas, J. D.,** Early pulmonary endothelial enzyme dysfunction after phorbol ester in conscious rabbits, *J. Appl. Physiol.,* 63, 1972, 1987.
74. **Havill, A. M., Riggs, D., Pitt, B. R., and Gillis, C. N.,** Pulmonary injury and altered metabolic function following intratracheal instillation of phorbol myristate acetate, *Am. Rev. Respir. Dis.,* 133, A19, 1986.
75. **Wright, D. G., Bralove, D. A., and Gallin, J. I.,** The differential mobilization of human neutrophil granules. Effects of phorbol myristate acetate and ionophore A23187, *Am. J. Pathol.,* 87, 237, 1977.
76. **Estensen, R. D., White, J. G., and Holmes, B.,** Specific degranulation of human polymorphonuclear leukocytes, *Nature,* 248, 347, 1974.
77. **Ward, P. A., Sulavik, M. C., and Johnson, K. J.,** Activated rat neutrophils: correlation of arachidonate products with enzyme secretion but not with O_2^- generation, *Am. J. Pathol.,* 120, 112, 1985.
78. **Canham, E. M., Shoemaker, S. A., Tate, R. M., Harada, R. N., McMurtry, I. F., and Repine, J. E.,** Mepacrine but not methylprednisolone decreases acute edematous lung injury after injection of phorbol myristate acetate in rabbits, *Am. Rev. Respir. Dis.,* 127, 594, 1983.
79. **Loyd, J. E., Newman, J. H., English, D., Ogletree, M. J., Meyrick, B. O., and Brigham, K. L.,** Lung vascular effects of phorbol myristate acetate in awake sheep, *J. Appl. Physiol.,* 54, 267, 1983.
80. **Newman, J. H., Loyd, J. E., Ogletree, M. L., Meyrick, B. O., and Brigham, K. L.,** Cyclooxygenase inhibition during phorbol-induced granulocyte stimulation in awake sheep, *J. Appl. Physiol.,* 56, 999, 1984.

81. **Dyer, E. L. and Snapper, J. R.,** Role of circulating granulocytes in sheep lung injury produced by phorbol myristate acetate, *J. Appl. Physiol.*, 60, 576, 1986.
82. **Lonky, S. A. and McCarren, J.,** Neutrophil enzymes in the lung: regulation of neutrophil elastase, *Am. Rev. Respir. Dis.*, 127, S9, 1983.
83. **Carpenter, L. J. and Roth, R. A.,** 12-Tetradecanoyl phorbol-13-acetate (PMA) produces injury to isolated rat lungs in the presence and absence of perfused neutrophils (abstract), *Fed. Proc.*, 45, 220, 1986.
84. **Rasmussen, H., Forder, J., Kojima, I., and Scriabine, A.,** TPA-induced contraction of isolated rabbit vascular smooth muscle, *Biochem. Biophys. Res. Commun.*, 122, 776, 1984.
85. **Zapol, W. M., Trelstad, R. L., Coffey, J. W., Tsai, L., and Salvador, R. A.,** Pulmonary fibrosis in severe acute respiratory failure, *Am. Rev. Respir. Dis.*, 119, 547, 1979.
86. **Hoidal, J. R., Repine, J. E., Beall, G. D., Rasp, R. L., and White, J. G.,** The effect of phorbol myristate acetate on the metabolism and ultrastructure of human alveolar macrophages, *Am. J. Pathol.*, 91, 469, 1978.
87. **Yamashita, T., Someya, A., and Hara, E.,** Response of superoxide anion production by guinea pig eosinophils to various soluble stimuli: comparison to neutrophils, *Arch. Biochem. Biophys.*, 241, 447, 1985.
88. **Matsubara, T. and Ziff, M.,** Superoxide anion release by human endothelial cells: synergism between phorbol ester and a calcium ionophore, *J. Cell Physiol.*, 127, 207, 1986.
89. **Ward, P. A., Cunningham, T. W., Till, G. O., and Johnson, K. J.,** Role of platelets in oxygen radical mediated responses, *Fed. Proc.*, 46, 444, 1987.

NITROSOURETHANE-INDUCED LUNG INJURY

Stephen F. Ryan, C. Redington Barrett, and Deng F. Liau

INTRODUCTION

Acute Alveolar Injury in the Human

Acute alveolar injury (AAI) is a type of injury to the lung with many causes or antecedent events. It is the lesion seen in the adult respiratory distress syndrome (ARDS) following surgical or traumatic shock, cutaneous burns, aspiration of gastric secretions, or drug overdose. It can also be caused by a variety of other agents, including immune complexes, antineoplastic and other drugs, hyperoxia, and viral infections of the lung. Despite the diversity of causes or antecedent events, the type of injury seen in the lung is generally stereotyped and follows a fairly predictable time course.

AAI is defined morphologically and the several descriptions of its histopathology are in general agreement on its essential features.[1-4] The dominant lesion in AAI is injury to alveolar epithelium with variable but usually inconspicuous injury to the capillary endothelium and variable pulmonary edema, the mechanism of which is to date unexplained. The earliest lesion recognizable as AAI usually occurs between 24 and 36 h after the onset of respiratory symptoms when small airspaces are lined by hyaline membranes which become thicker and more extensive during the ensuing days (Figures 1 and 2). Ultrastructurally, the hyaline membranes almost always rest on denuded basement membranes and are composed of the residua of necrotic epithelial cells admixed with granular proteinacious material and occasionally small amounts of fibrin (Figure 3). While type I alveolar epithelial cells are usually more severely injured, type II cells frequently undergo necrosis as well (Figures 4 and 5).

The acute phase with hyaline membranes is closely followed, usually beginning about 3 d after onset of symptoms, by regeneration of alveolar epithelial cells. The regenerating cells first appear in alveoli near small blood vessels and airways. These cells are large and round or elliptical and contain vacuolated cytoplasm and large nuclei with prominent nucleoli and occasional mitoses (Figure 6). The number of these regenerating cells increases until, by 10 to 12 d after onset of respiratory symptoms, they completely line many alveoli. Between 5 to 7 and 10 to 12 d after onset, the regenerating cells steadily become more uniform and cuboidal (Figure 7). Electron microscopically, many of the regenerating cells during the 3- to 7-d period closely resemble the epithelium of the fetal lung and are devoid of lamellar bodies (Figure 8). By 7 d, some of the cells are recognizable as mature type II cells by the presence of typical cytoplasmic lamellar bodies. These mature-appearing type II cells increase in number until, by 10 or 12 d, they compose the entire population of regenerating cells (Figure 9). During the first few days of injury, neutrophils can often be seen aggregating in capillary lumina and infiltrating interalveolar septa. Occasional capillary endothelial cells are lifted from the basement membranes and swollen, but generally this endothelial injury is inconspicuous. In milder cases the process may progress to recovery with little or no residual disturbance of lung architecture, but in the most severe cases architectural disorganization becomes marked and irreversible (Figure 10). This disorganization is caused by irreversible closure of alveoli, probably as a result of total loss of epithelium, and by organization of hyaline membranes in a process similar to that seen in organizing pneumonia. Following recovery from acute alveolar injury, some patients are left with residual abnormalities on the chest X-ray and impairment of gas exchange.

The hallmark of AAI is extensive necrosis of alveolar epithelium followed by regeneration of cells. These cells at first appear immature and resemble those of the fetal lung. They later mature to type II cells recognizable by their content of cytoplasmic lamellar bodies in a process which requires approximately 2 weeks to complete.

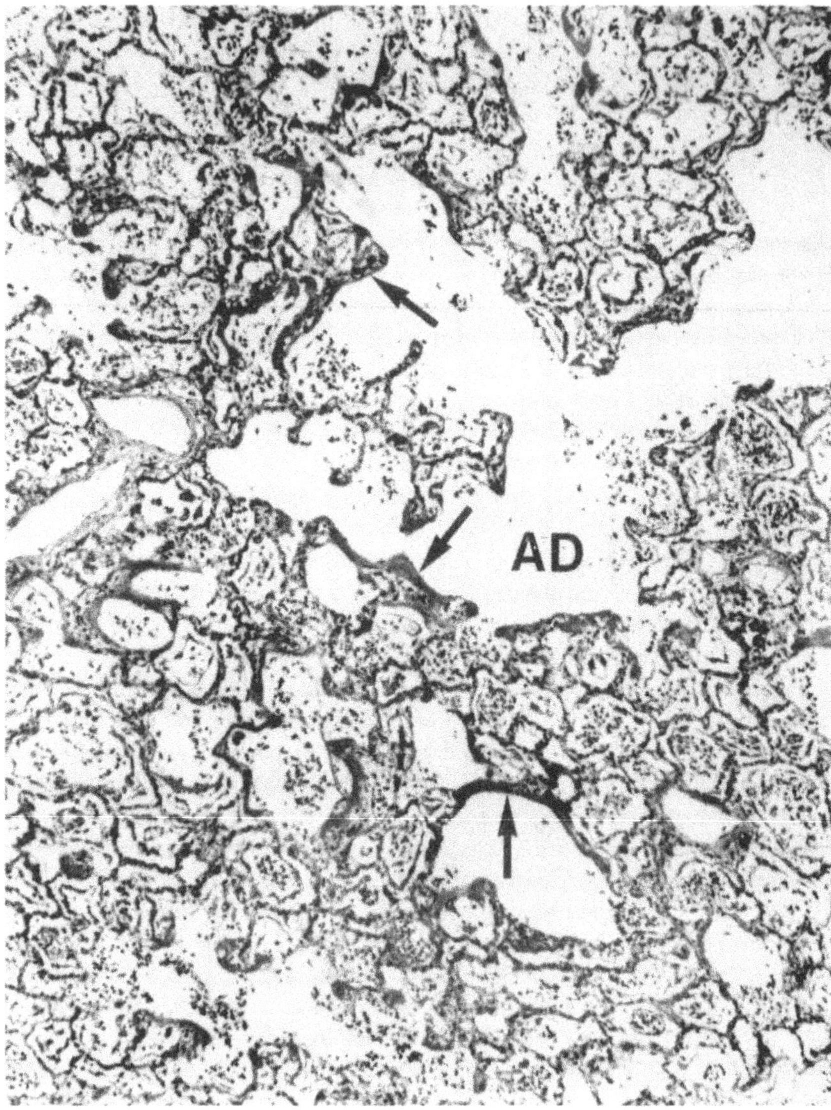

FIGURE 1. Early AAI. This section is from the lung of a patient who died 6 d after sustaining massive trauma with shock in a fall. Respiratory failure developed 20 h after the fall and continued until death. Lung architecture is essentially intact. Small air spaces, especially alveolar ducts (AD), are lined by hyaline membrane (arrows). (H&E × 87.)

Experimental Nitrosourethane Lung Injury

In 1967, Herrold[5] reported that subcutaneous injection of the carcinogen *N*-nitroso-*N*-methylurethane (nitrosourethane) in the hamster caused interstitial injury to the lung which resembled fibrosing alveolitis. Subsequent studies showed that after several weeks, repeated weekly injections of small doses led to histologic changes in the lungs which closely resembled those in human fibrosing alveolitis.[6] Larger doses led to more rapid onset of respiratory failure and to more florid lung injury, often with hyaline membranes. It has since been shown that a single subcutaneous injection of 5 to 8 mg/kg body weight of nitrosourethane in dogs causes AAI practically indistinguishable by light and electron microscopy from that seen in humans.[7] Both the evolving morphologic features and the time course are similar. Because of this structural similarity, because the severity of the physiologic and

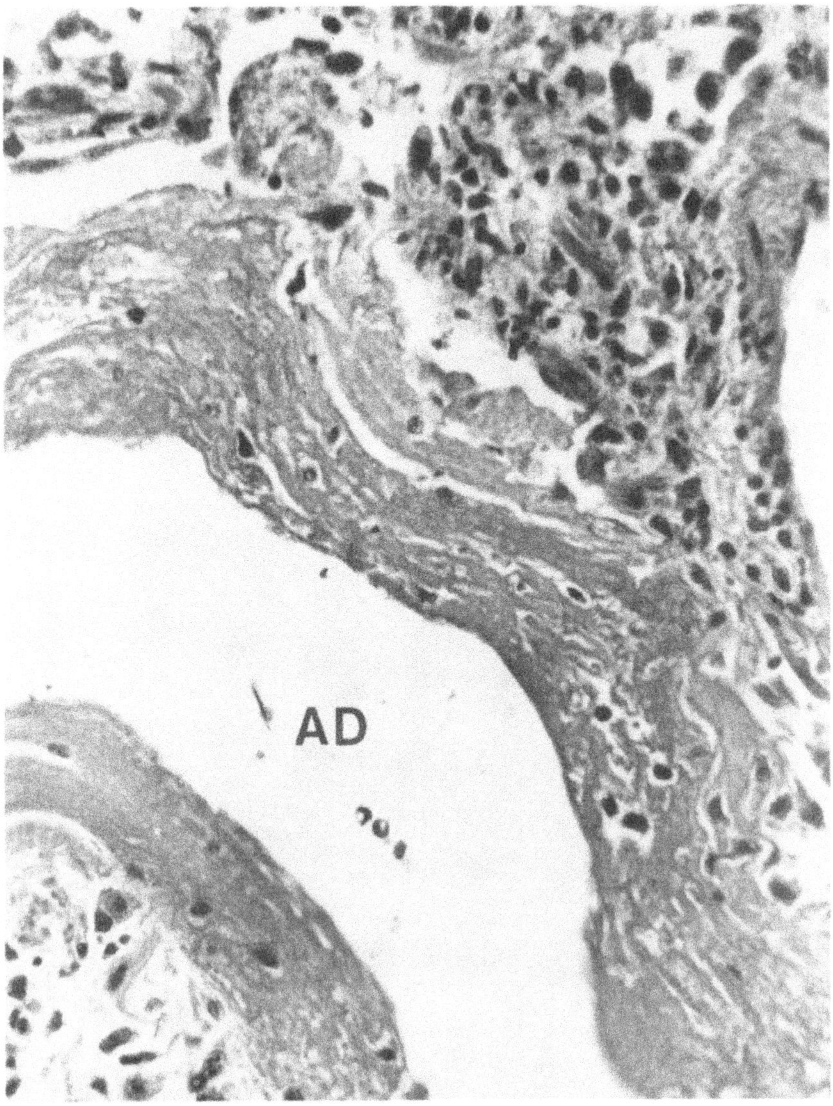

FIGURE 2. A thick hyaline membrane containing pyknotic and fragmented nuclei from a patient with AAI several days more advanced than that of the patient depicted in Figure 1. (H&E × 355.)

structural changes induced by nitrosourethane are dose related, and because the animal used is large, this empirical lung injury has provided an ideal model in which to study mechanical, physiologic, and metabolic consequences of AAI. On the other hand, because the mechanism of nitrosourethane-induced lung injury is not known, the model has no known etiologic relevance to the human injury.

EXPERIMENTAL METHODS AND FINDINGS

Induction of Injury

N-nitroso-*n*-methylurethane (NNNMU) (Kings' Labs Inc., Blythewood, SC) is stored in the laboratory in 5-ml sterile multiple-dose vials (Elkins Sinn Inc., Cherry Hills, NJ, product code 452700) at −20°C. When stored at 0 to 4°C, it loses potency over a period of 1 to 2

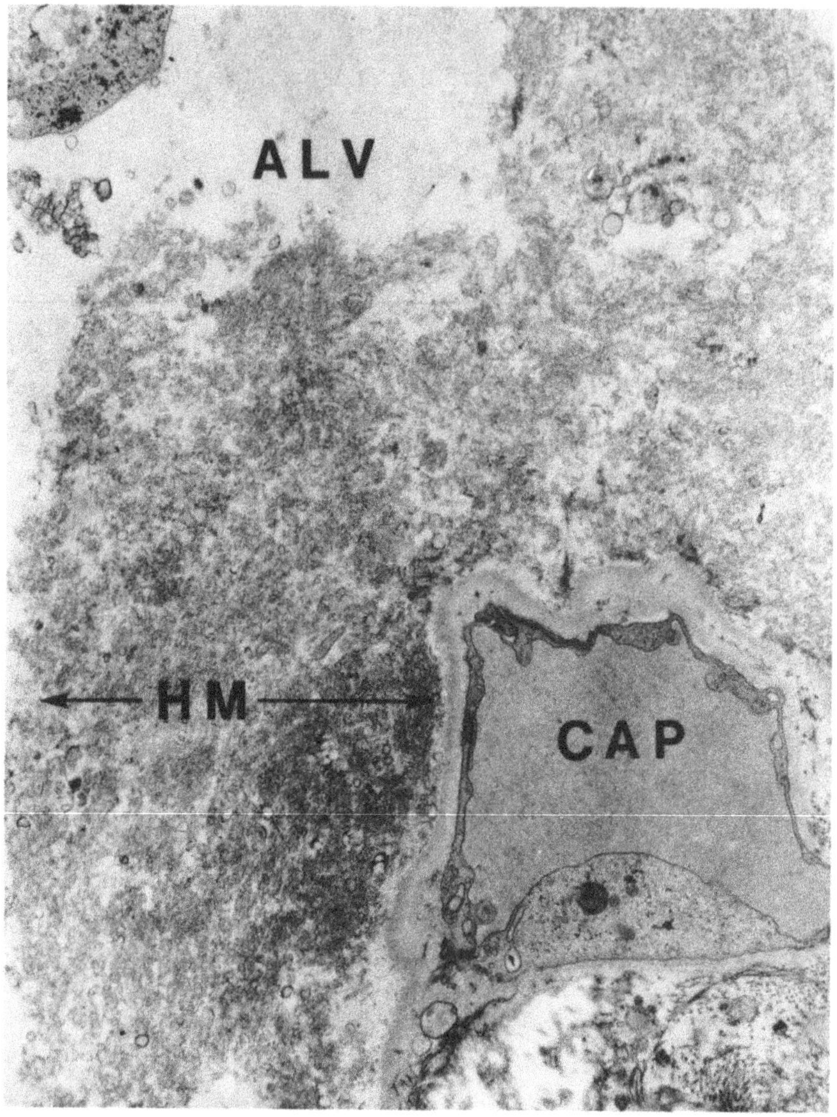

FIGURE 3. A hyaline membrane (HM), lying on a basement membrane denuded of epithelium, is composed of granular material admixed with fragments of cell membranes and various organelles. These features indicate that the hyaline membrane, at least in part, is composed of the remains of necrotic alveolar epithelial cells (ALV). The capillary endothelium (CAP) is intact. (Electron micrograph × 10,260.)

years. Nitrosourethane is a potent carcinogen and extreme caution should be exercised in handling it. Not only ingestion but skin contact and inhalation of fumes must be rigorously avoided.

For use in dogs, a 4% solution of nitrosourethane is prepared with 0.5% ethanol as solvent. After preparation, the solution must be shaken vigorously because nitrosourethane is not highly soluble at this concentration. The solution must also be shaken prior to withdrawal of the dose into the syringe and several times during withdrawal. With the needle in the vial it is shaken for 1 min, withdrawn approximtely 1 ml, shaken for a few seconds, withdrawn another milliliter, and so on until the entire dose is measured. The injection is immediately made subcutaneously in the scruff of the neck. The site of injection is massaged vigorously

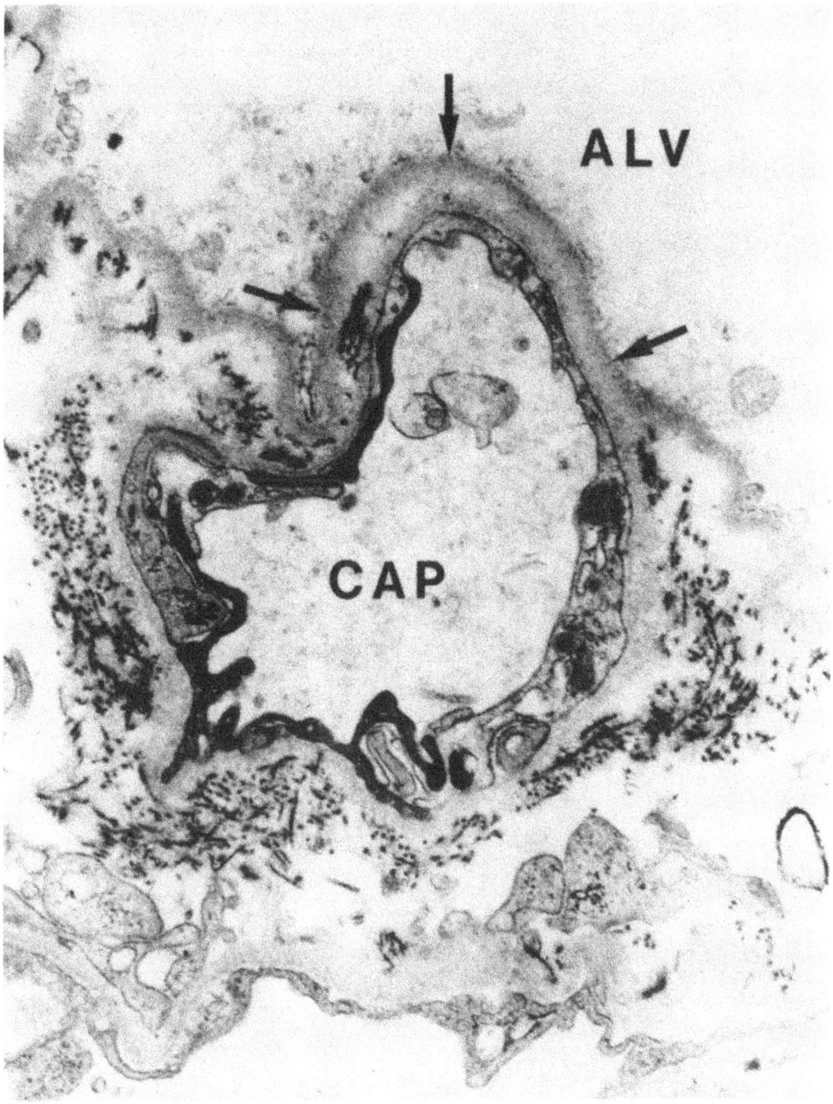

FIGURE 4. An interalveolar septum from the lung of a patient 5 d after the onset of respiratory failure following trauma. Histologic sections show typical AAI. The alveolar epithelium (ALV) is destroyed, leaving a denuded basement membrane (arrows) while the capillary endothelium (CAP) is intact. (Electron micrograph × 16,200.)

to disperse the injectate. Identical lung injury has been caused in hamsters, dogs, rabbits, and rats. Unused nitrosourethane solution must never be poured down the drain. Do not discard NNNMU; it is stored indefinitely.

Adult male mongrel dogs weighing between 9 and 40 kg have been used for these studies. The animals are housed and fed according to the *Guide for Care and Use of Laboratory Animals of the U.S. Public Health Service*. Smears of peripheral blood are examined for *d. immitis* and the animals are judged to be free of acute respiratory disease on physical examination prior to use. The animals are deprived of food and water for at least 16 h prior to a physiologic study.

After injection of 5 mg of nitrosourethane per kilogram of body weight, approximately one third of the animals die of respiratory failure, often complicated by pneumothorax. Most deaths occur between 5 and 18 d after injection.[8]

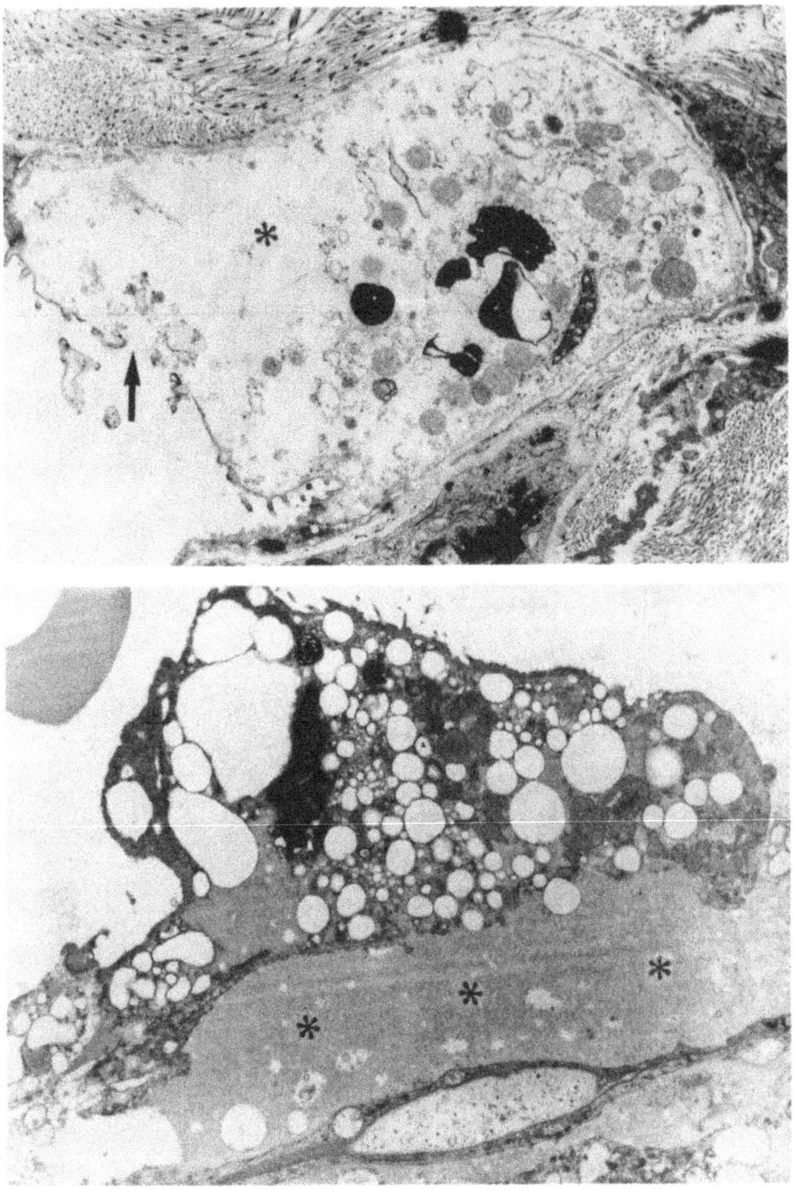

FIGURE 5. Injured type II alveolar epithelial cells from two cases of early AAI. In the upper micrograph, the cell membrane is focally disrupted (arrow) and the cytosol is rarified (asterisk). (Electron micrograph × 16,000.) In the lower micrograph, the cell, recognizable as a type II cell by its content of one or two lamellar bodies, is necrotic and vacuolated and lifted from the basement membrane by edema fluid. (Electron micrograph × 16,000.)

Morphologic Studies

Methods

For morphologic studies, the animals are killed with intravenous pentobarbital at various intervals following injection of nitrosourethane and the lungs quickly excised, weighed, and inflated through a tracheal cannula with cold 3% gluteraldehyde buffered with phosphate to pH 7.4. Both lungs are then immersed in a container of the same fixative and the tracheal cannula attached via a tube to a reservoir of fixative situated above the immersed lung to provide an inflation pressure of 25 cm of H_2O. A pump circulates fluid which accumulates

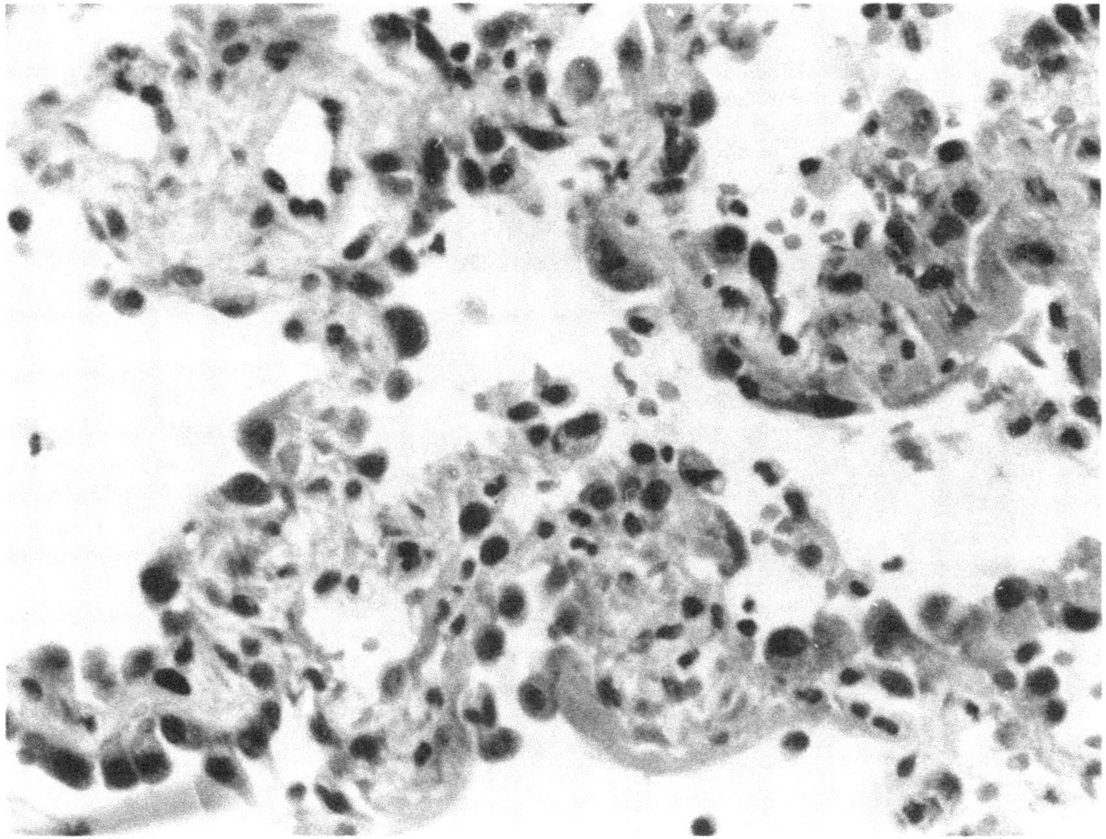

FIGURE 6. Early regeneration of alveolar epithelium. This section is from a biopsy of the lung of a patient 7 d after onset of respiratory symptoms. Typical acute alveolar injury was presumed to be due to viral pneumonia. Several alveoli are partly lined by large cuboidal or elliptical epithelial cells with large nuclei. (H&E × 250.)

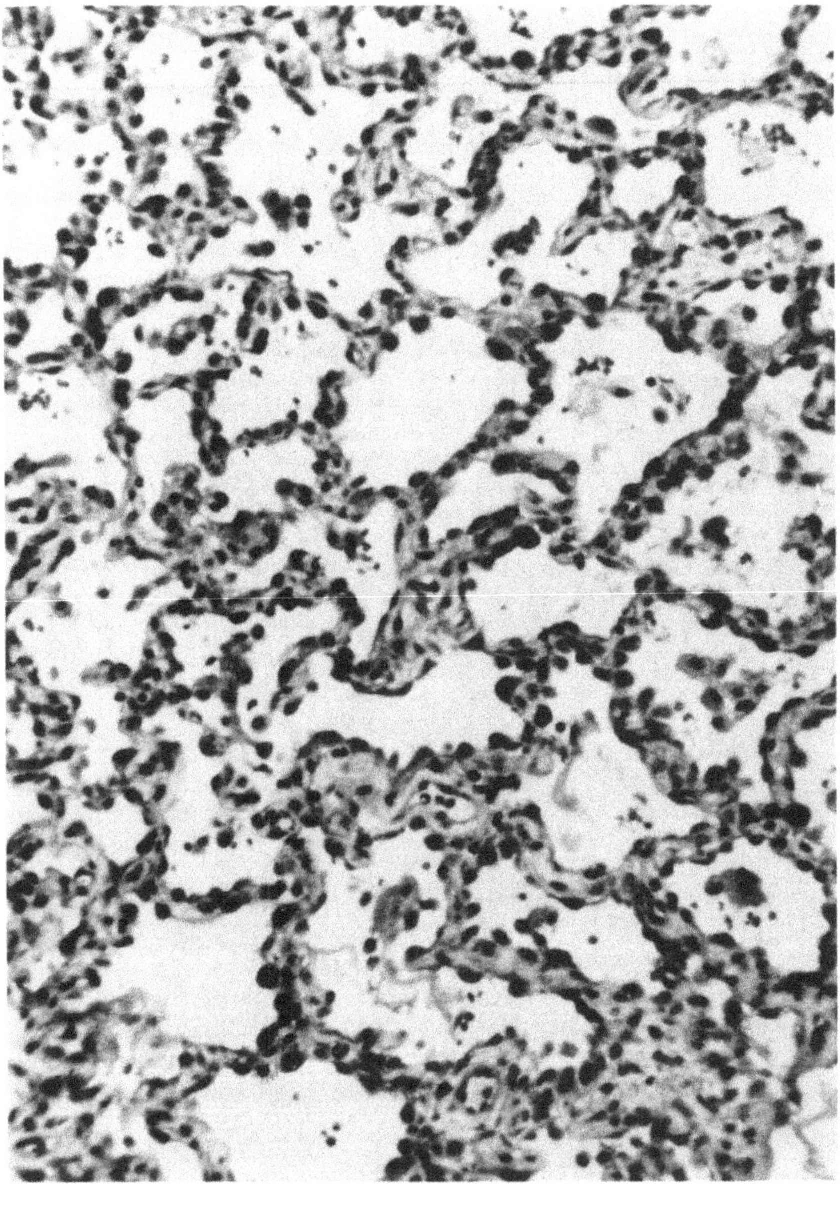

FIGURE 7. Advanced regeneration of alveolar epithelium. This section is from a biopsy of the lung of a patient 14 d after onset of symptoms. AAI was presumed to be due to viral pneumonia. Small airspaces are almost completely lined by fairly uniform cuboidal or elliptical cells. (H&E × 210.)

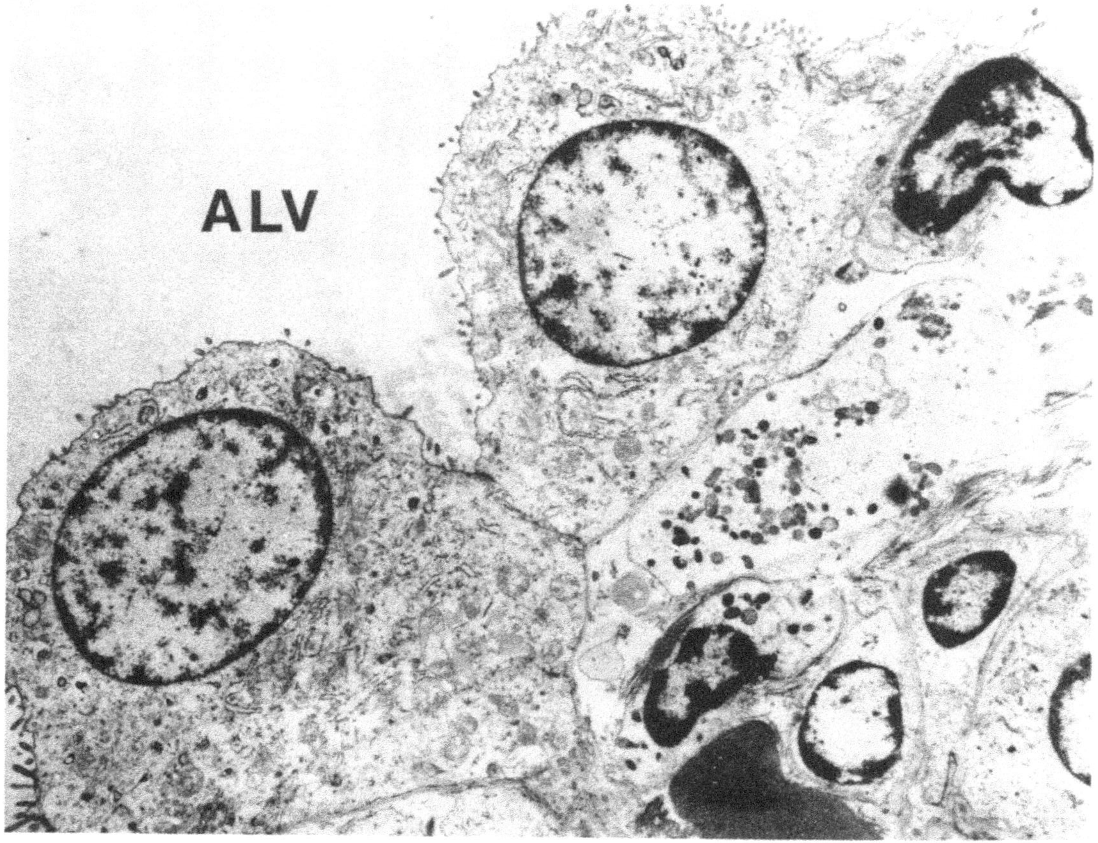

FIGURE 8. Alveolar epithelial cells (ALV) during early regeneration. The cells contain large nuclei and abundant endoplasmic reticulum but are devoid of lamellar bodies. At this stage, the cells resemble those of the fetal lung. (Electron micrograph × 4,300.)

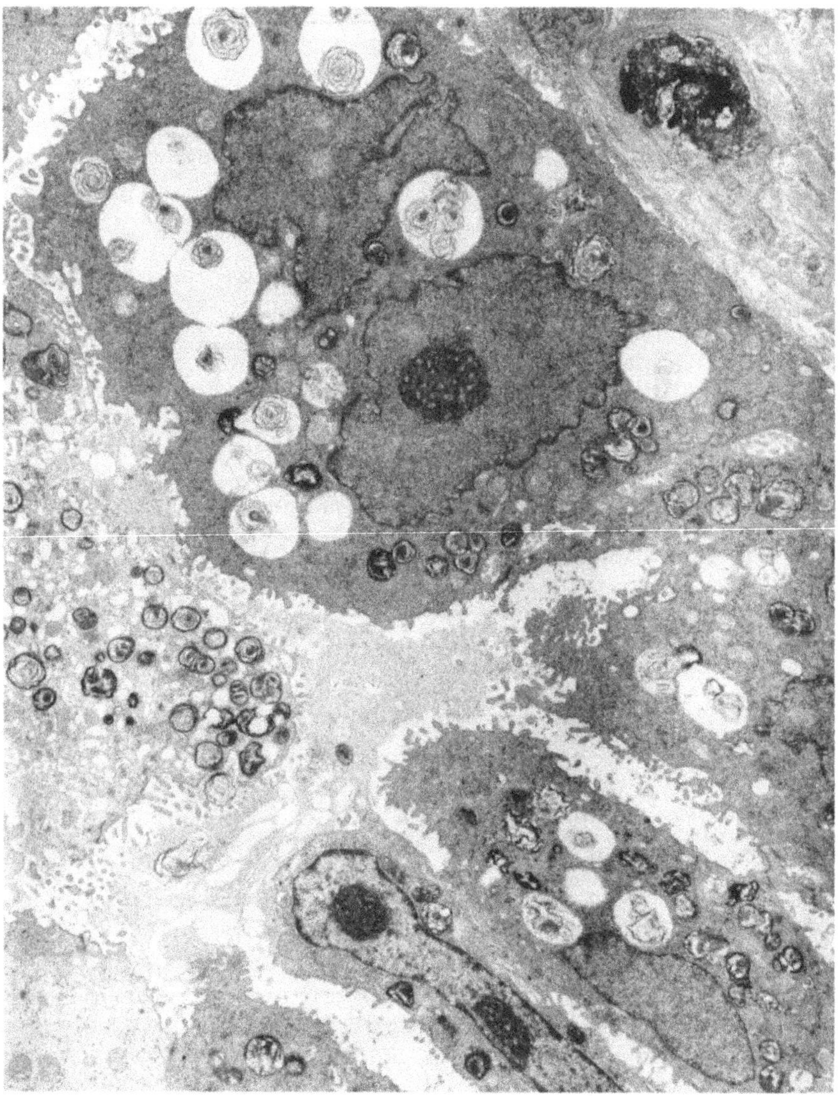

FIGURE 9. Alveolar epithelial cells in the phase of advanced regeneration. This alveolus is completely lined by mature-appearing type II cells with abundant cytoplasmic lamellar bodies. (Electron micrograph × 5,900.)

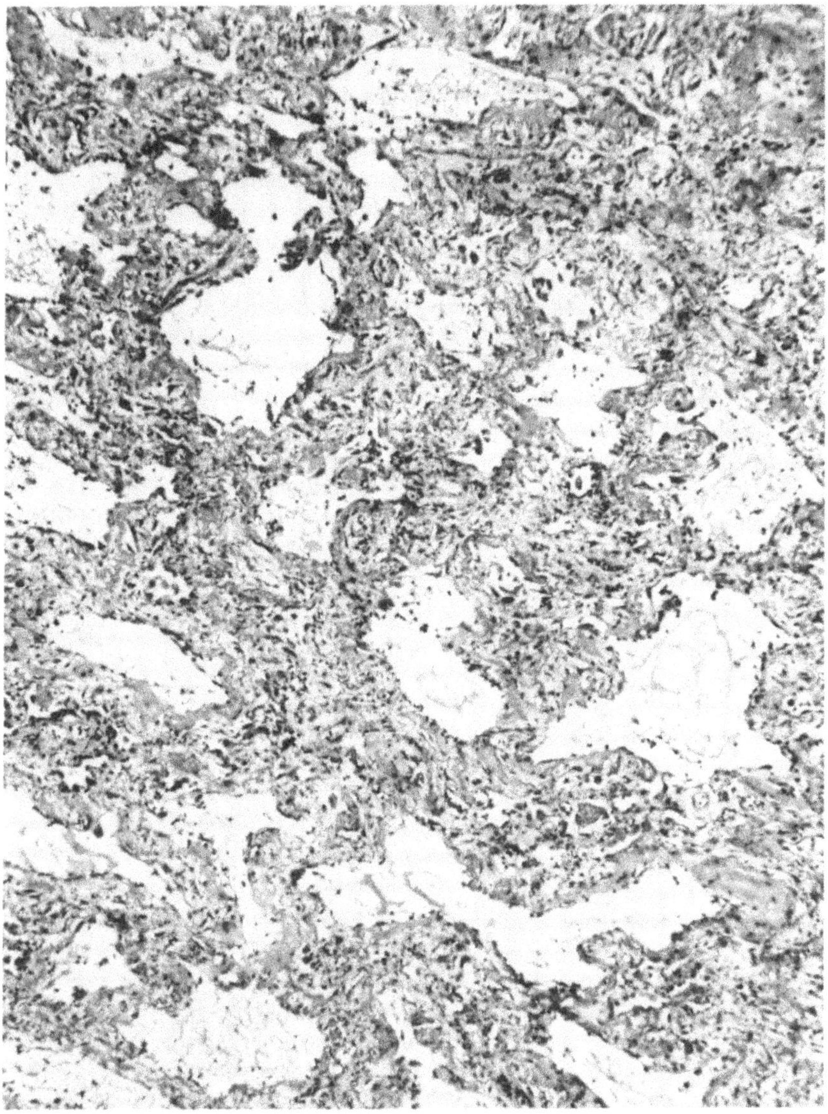

FIGURE 10. AAI with severe architectural revision. This section is from a biopsy of the lung of a patient 14 d after onset of symptoms. AAI was presumed to be due to viral pneumonia. Enlarged and contorted airspaces are lined by hyaline membranes and separated by clusters of closed alveoli. (H&E × 75.)

in the immersion container back into the upper reservoir. The entire apparatus is situated in a walk-in cold room which provides temperature of 4°C throughout the fixation period of 12 h. The lungs are then sliced in 1-cm thick frontal sections, examined grossly, and sampled for light and electron microscopic examination.

Results

The earliest morphologic changes, seen at 24 h, are perivascular edema and cytoplasmic vacuolation of type II alveolar epithelial cells.[7] By 2 d, scattered alveolar epithelial cells are necrotic and their numbers increase until 5 to 6 d after injection when many necrotic cells of both types are seen and large areas of basement membrane are denuded (Figures 11 and 12). Hyaline membranes are readily identified and rest on these denuded basement

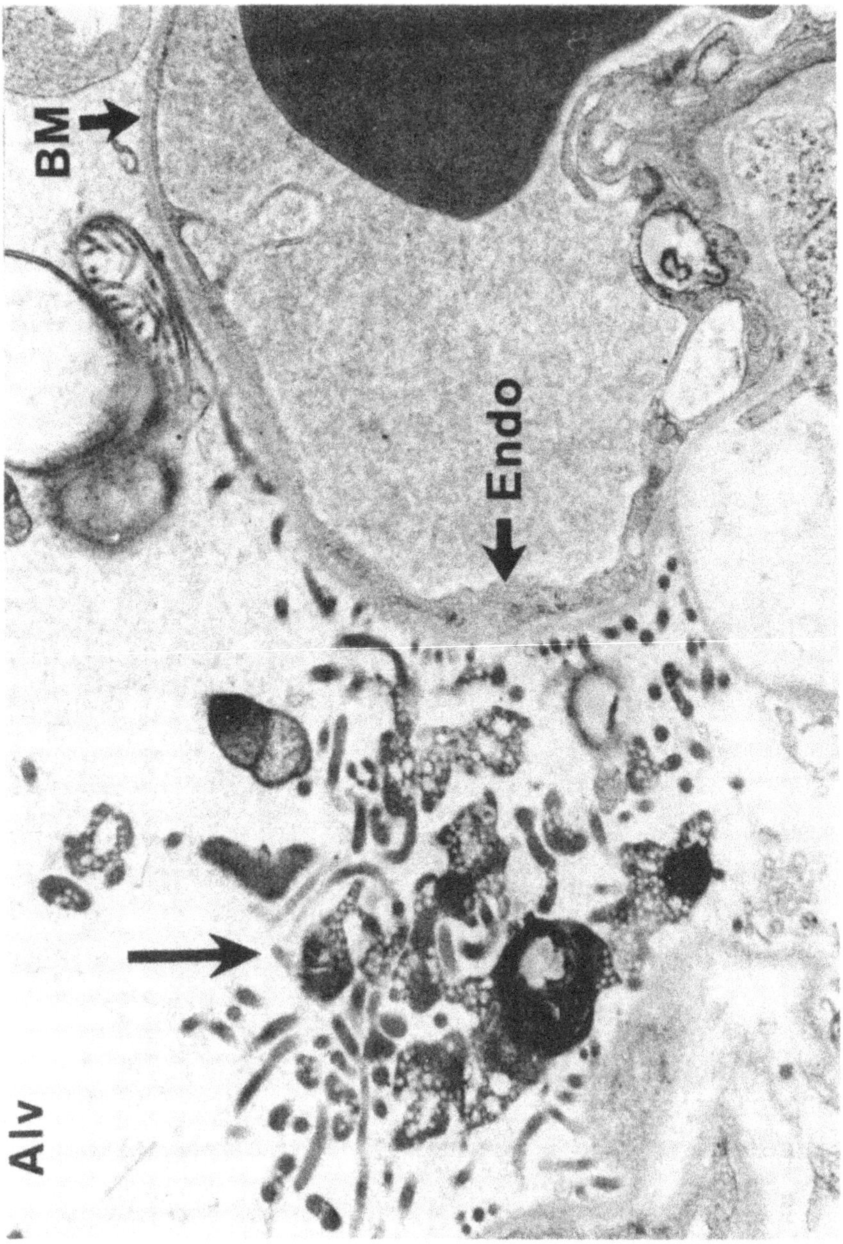

FIGURE 11. **Interalveolar septum form the lung of a dog 7 d after injection of nitrosourethane. An alveolar epithelial cell, probably type I** (arrow), is necrotic and sloughing, leaving part of the basement membrane denuded (BM). The capillary endothelium (Endo) is intact. (Electron micrograph × 17,500.)

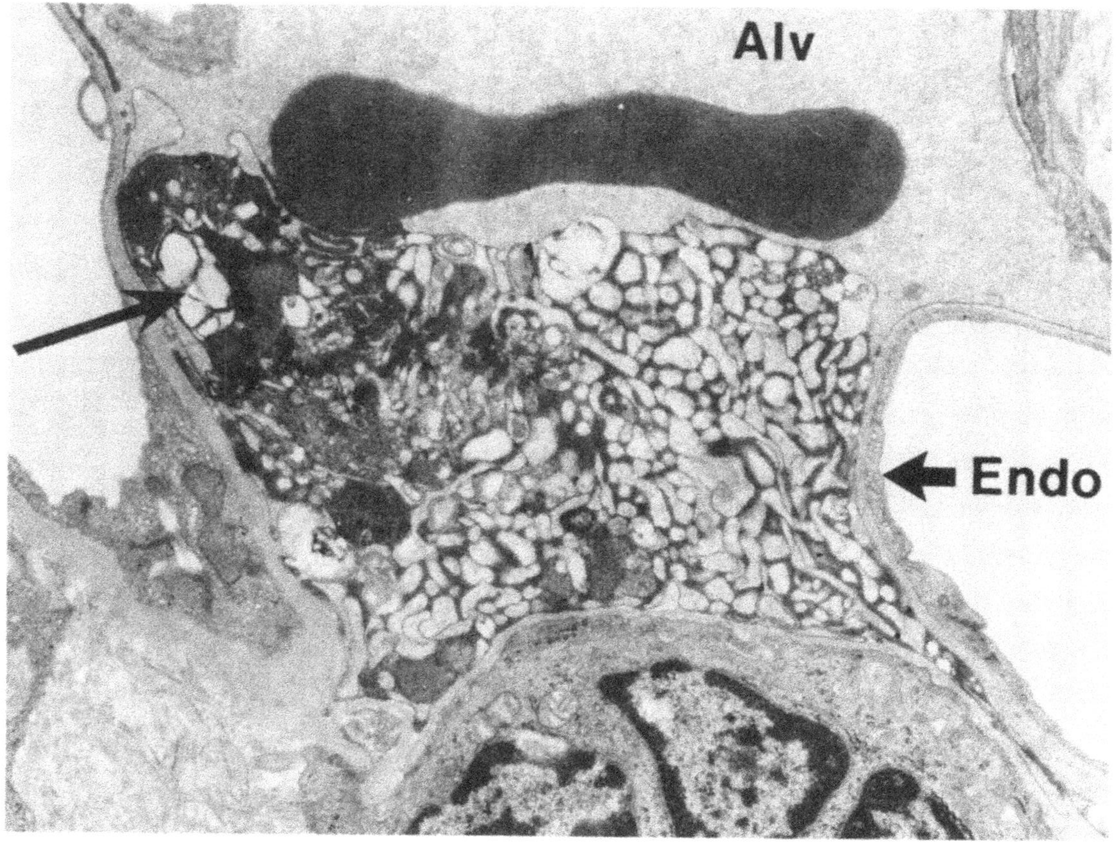

FIGURE 12. Interalveolar septum from the lung of a dog 7 d after injection of nitrosourethane. An epithelial cell, recognizable as type II by its position in an alveolar recess and its content of a cytoplasmic lamellar body, is necrotic while the capillary endothelium (Endo) remains intact. (Electron micrograph × 10,000.)

membranes. They are composed of granular material enmeshing fragmented membranes and disrupted organelles indicating that they are the residua of necrotic epithelial cells (Figure 13). During this early period, alveolar septa are infiltrated by neutrophils and atelectasis and patchy alveolar edema are marked (Figure 14). As early as 2 d postinjection, a few alveoli, almost always centriacinar, are partly lined by regenerating epithelial cells. These cells, which steadily increase in number, are at first irregular in shape with swollen mitochondria and only rare lamellar bodies. Nuclei and nucleoli are large. As their numbers increase, the new cells become more uniform and cuboidal. By 9 to 10 d postinjection, when they have come to occupy a large proportion of the alveolar surface, lamellar bodies are numerous (Figure 15). This process of maturation continues until, by 20 d, almost all the cells closely resemble mature type II cells although some contain masses of cytoplasmic glycogen. During the entire process, capillary endothelium remains essentially intact with only scattered subepithelial vesicles and a rare swollen cell. Because the lungs in these studies were examined after inflation fixation at a constantly maintained translung pressure, architectural revision was readily recognized. As early as 5 d postinjection, groups of alveoli show loss of volume or collapse. The extent of patchy alveolar closure increases with time and is accompanied by enlargement of intervening small airspaces, mostly alveolar ducts, which gives a coarse appearance to the parenchyma even on gross examination. This close similarity of the evolving morphologic picture with that of the human lesion, dominated by epithelial injury and repair with early edema and late irreversible architectural revision, is the basis of the utility of nitrosourethane injury as a model of human disease.

Physiologic Studies

Methods

Anesthetic procedures — The animals are premedicated with 0.5 mg/kg of xylazine (Rompun) or with 10 to 15 mg of acepromazine administered subcutaneously. A 20-gauge cannula is inserted into a forelimb vein 20 min after premedication and its position is secured by adhesive tape. For studies lasting less than 2 h, the animal is anesthetized with a 4% solution of sodium thiamylal (40 mg/ml), a short-acting barbiturate. For studies of longer duration, sodium pentobarbital (60 mg/ml) is used. With either preparation, the expected total dose is usually 0.15 to 0.5 ml/kg body weight (0.33 to 0.5 ml/5 lb body weight). In practice, 10 ml of the selected barbiturate is drawn into a syringe, the syringe is attached to the cannula, and a bolus equal to one half of the calculated total dose is administered intravenously. Supplemental doses of the barbiturate are administered intravenously every 30 s until the animal is unconscious and the corneal reflex abolished. During the experiment, supplemental doses (0.5 to 1 ml) of the barbiturate are administered when needed.

Mechanical ventilation — The animal is placed in the supine position on a table with a grooved top. Its legs are secured to the side of the table with wide gauze. A 9-mm cuffed endotracheal tube is inserted into the trachea and, after the cuff is inflated, its position is secured by a bite-block. For the duration of the experiment, the animal is ventilated with a Harvard® dog ventilator (Model 613) at a respiratory frequency of 20 per minute and a tidal volume of 200 ml. During the study, the cardiac rate and rhythm are monitored continuously by an electrocardiogram. The dog is paralyzed with 5 mg of intravenously administered succinylcholine. A continuous infusion of a 0.18% solution of succinylcholine at a rate of 0.19 ml/min is used to maintain paralysis. Occasionally, supplemental boluses of 5 mg of the drug are administered intravenously if the degree of paralysis is insufficient to abolish spontaneous respiratory muscle activity. At the conclusion of the study, 600,000 U of procaine penicillin is administered intramuscularly to dogs undergoing serial studies.

Lung mechanics — All measurements of lung volume and lung mechanics are made after the lungs have been inflated to a transpulmonary pressure of 30 cm of water to standardize the volume history of the respiratory system. Static volume-pressure diagrams are obtained *in vivo*[9,10] by measuring transairway pressure at the dog's mouth and trans-

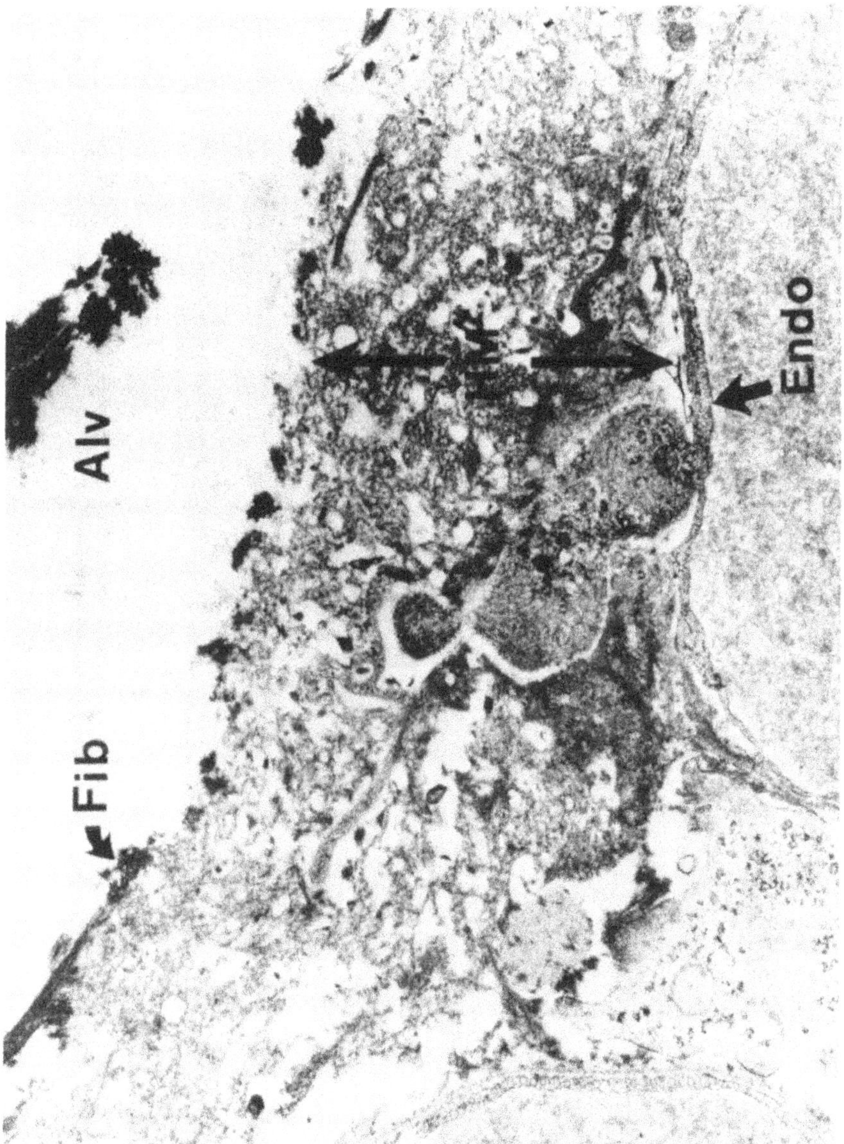

FIGURE 13. Interalveolar septum from the lung of a dog 5 d after injection of nitrosourethane. A hyaline membrane (HM) rests on a denuded basement membrane and is composed of fragments of necrotic epithelial cells admixed with granular material and small amounts of fibrin (Fib). Again, the endothelium (Endo) is intact. (Electron micrograph × 1,450.)

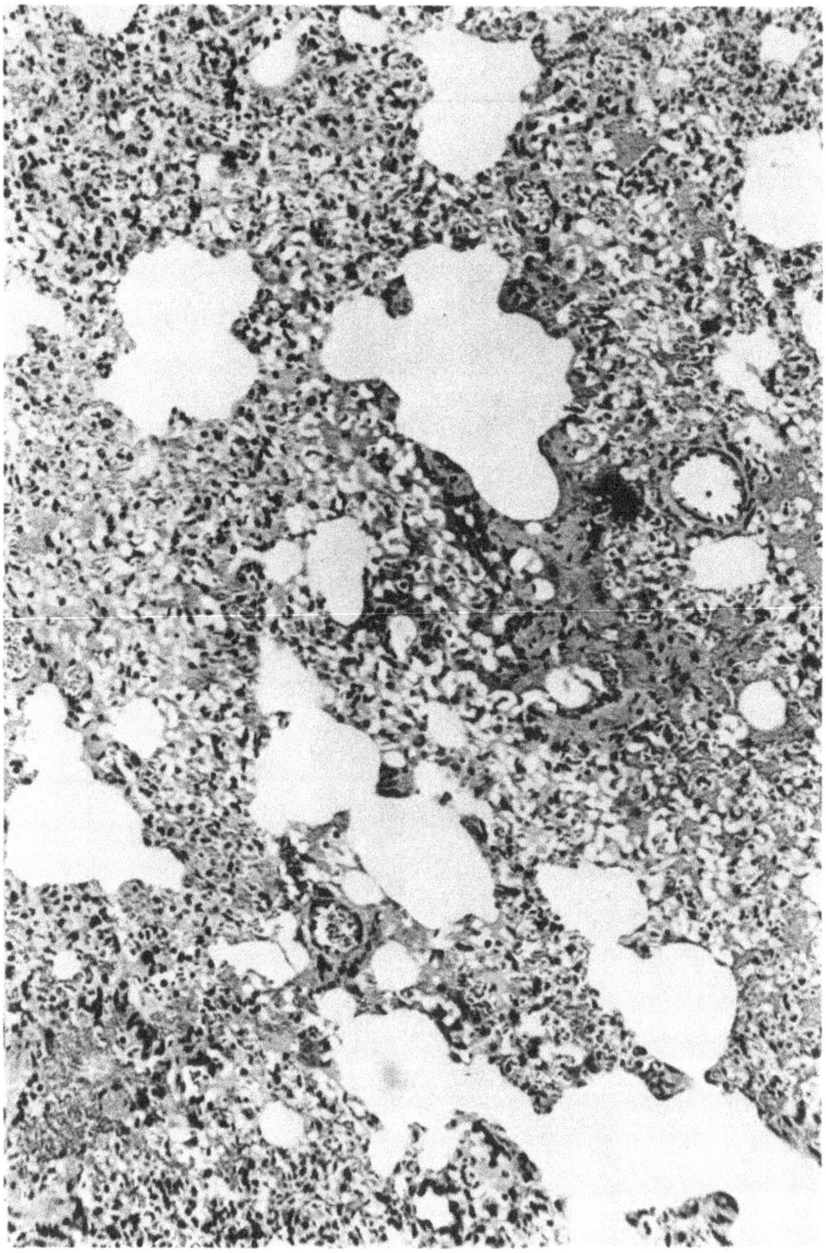

FIGURE 14. Section of lung from a dog 7 d after injection of nitrosourethane. Large patches of alveoli are closed or shrunken and filled with edema fluid, and the airspaces which remain open are unevenly enlarged. (H&E × 120.)

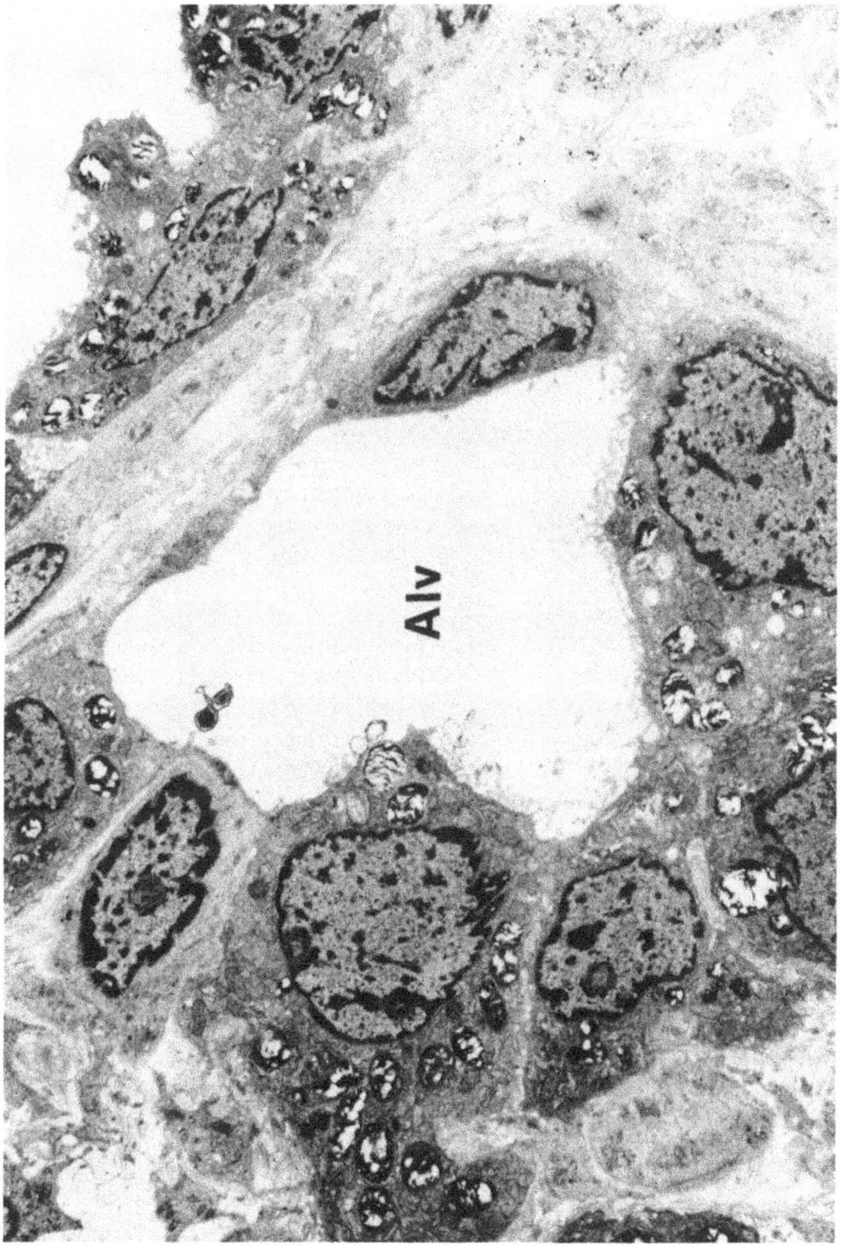

FIGURE 15. An alveolus (Alv) from the lung of a dog 20 d after injection of nitrosourethane. Its wall is thickened and it is almost completely lined by mature-appearing type II cells with many lamellar bodies. (Electron micrograph × 4,800.)

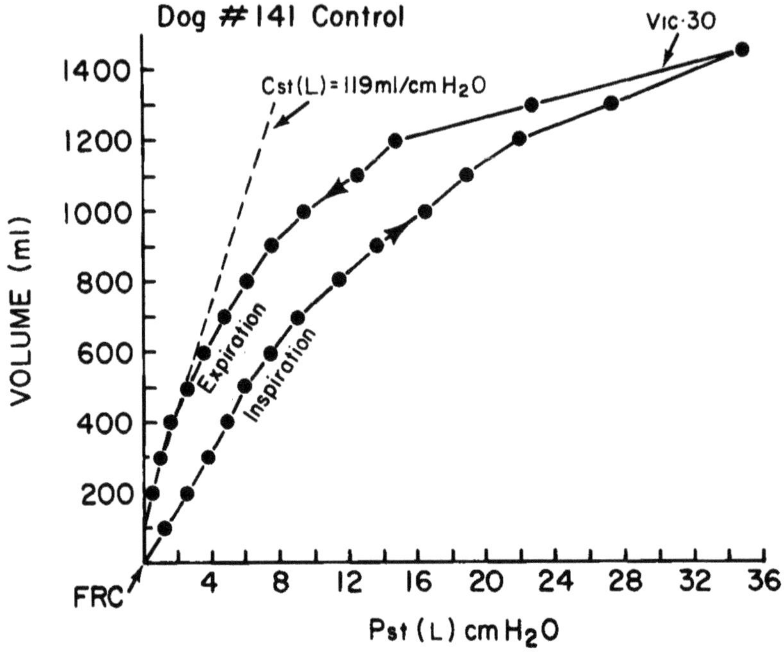

FIGURE 16. Static volume-pressure diagram of a normal anesthetized dog. The slope of the linear portion of the descending limb is shown. Vic-30 refers to the volume at 30 cm H_2O on the descending limb which is added to functional residual capacity to obtain total lung capacity.

pulmonary pressure (PL) using an esophageal ballon positioned in the lower esophagus.[10,11] Lung mechanics are determined during mechanical ventilation by measuring maximal expiratory airway pressure (Paw_{max}) and maximal transpulmonary pressure (PL_{max}) at four increments of tidal volume according to a procedure developed in the laboratory of the College of Physicians and Surgeons.[8,9] Functional residual capacity (FRC) is measured by a closed-circuit helium dilution techniques.[9,12]

Static volume pressure diagrams — Using a 3-l calibrated Hamilton® air syringe, the lungs are inflated progressively by 100-ml volume increments with a 3-s pause at each step until PL exceeds 30 cm of H_2O, after which the lungs are deflated in similar volume decrements. Static lung compliance is obtained from the chord connecting the volume-pressure points along the linear portion of the descending limb of the curve above resting expiratory level.[10] A static volume-pressure diagram from a normal anesthetized dog is shown in Figure 16.

Lung mechanics during mechanical ventilation — During these studies the animals are ventilated at a respiratory frequency of 20/min. Because of the mechanical design of the ventilator used, Paw_{max} and PL_{max} occur at end-inspiration and coincide with the phase of zero flow. Because flow-resistive pressures are virtually excluded. Paw_{max} and PL_{max} reflect static properties of the thorax and lungs. Dynamic compliance of the respiratory system (Crs) and of the lungs (CL) are determined from volume-pressure curves of which ascending and descending limbs are constructed by measuring Paw_{max} and PL_{max} at four different tidal volumes. Paw_{max} and PL_{max} are measured at an initial tidal volume of 100 ml, at increments of 100 ml until tidal volume of 400 ml is reached, and then at 100 ml decrements. Crs represents the slope of the straight line best fitted by the least squares method to plotted points of Paw_{max}. CL is calculated in a similar manner using the points of PL_{max}.

Functional residual capacity — FRC is measured by disconnecting the animal from the respirator and ventilating the lungs at a tidal volume of 200 ml for 15 breaths with a 1500-

ml Hamilton® air syringe containing a mixture of 12 to 14% helium in air. Initial and final helium concentrations are measured by a thermal conductivity meter. The procedure is repeated until two measurements yield FRC values that differ by less than 10%.

Total lung capacity (TLC) — TLC is determined by adding FRC in inspiratory capacity, the volume at a transpulmonary pressure of 30 cm of water on the descending limb of the static volume pressure curve (Figure 16).

Representative protocol — The animal is premedicated with acepromazine, anesthetized with intravenously administered barbiturate, placed supine on the table, intubated with a cuffed endotracheal tube, and ventilated with the respirator. An esophageal balloon is passed into the stomach and the animal is disconnected from the respirator. During spontaneous ventilation, a positive pressure waveform is detected by the esophageal balloon during inspiration, reflecting its intro-abdominal location. The balloon is slowly withdrawn until a negative pressure waveform is detected indicating that the balloon is located in the thoracic cavity. The balloon is withdrawn an additional 10 cm and the catheter position secured at the mouth of the dog. Mechanical ventilation is resumed and succinylcholine administered for muscular paralysis. After inflating the lungs to a transpulmonary pressure of 30 cm of water before each procedure, static volume pressure curves, dynamic lung compliance, and FRC measurements are obtained according to the methods described above.

Results

Clinical features

After the subcutaneous administration of nitrosurethane, the dog shows evidence of respiratory distress within 48 to 72 h. The animal is tachypneic and, when severely affected, may have cyanosis of the tongue, intercostal retractions during inspiration, and grunting during expiration.

Respiratory mechanics

Serial measurements of lung mechanics were performed between 24 h and 42 d after nitrosourethane. Within 48 h after injection, lung compliance decreases and elastic lung recoil increases. These findings coincide with necrosis of the alveolar epithelium, perivascular and interstitial edema, and alveolar collapse. CL and lung recoil abnormalities peak at 8 to 12 d following injection, after which they improve toward control values (Figure 17). Improved lung mechanics coincide with an emerging population of regenerating granular pneumocytes. Late in the recovery phase (after 21 d), elastic lung recoil near the resting expiratory level is not significantly different from that of control values, while that observed at maximum lung inflation remains significantly abnormal (Figure 17). At this stage there is irreversible alveolar closure with dilatation of surrounding small airspaces. Serial changes of lung volume and CL are shown in Figure 18. At peak illness, decreased CL/FRC and FRC indicate that decreased CL is due to abnormal distensibility of patent alveoli with or without concomitant loss of lung volume. Because FRC/TLC is significantly raised, it is likely that at the resting end-expiratory level alveoli are already being inflated at the upper ends of their volume-pressure curves. These abnormalities persist late into the recovery phase, although to a lesser degree (Figure 18).

During the acute phase of nitrosourethane-induced AAI, it was observed that static volume-pressure curves measured *in vivo* shifted downward and to the right with no increase of hysteresis[9] (Figure 19). The effects of increments of positive end-expiratory pressure (PEEP) on lung compliance and FRC after nitrosourethane was studied in 17 dogs.[9] It was found that during increments of PEEP, CL decreased significantly and FRC rose but at a decreasing rate. These data indicate that during increments of PEEP, alveoli are being overdistended and ventilated at the upper ends of their volume-pressure curves.

The characteristics of volume-pressure curves and the response to PEEP observed in the

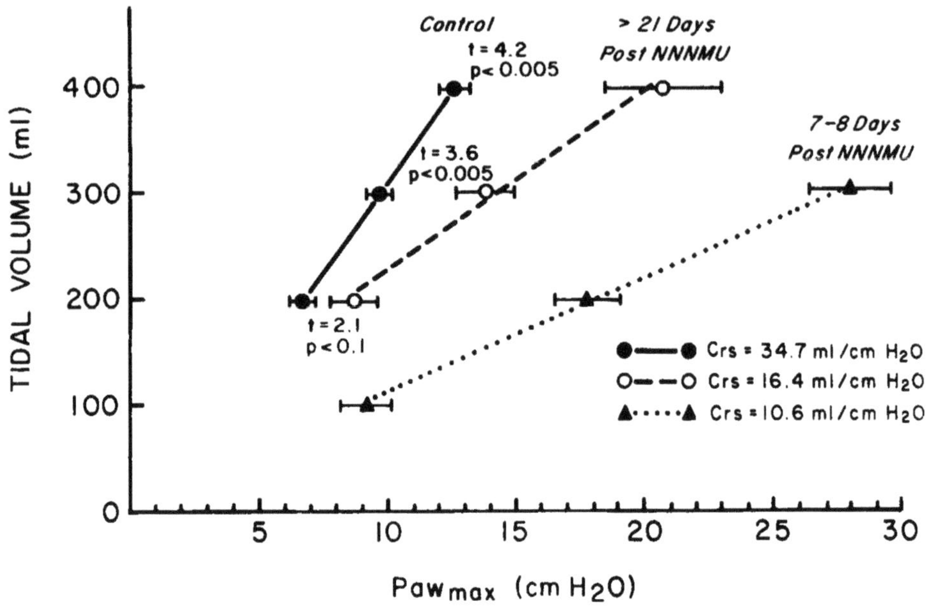

FIGURE 17. Volume-pressure curves constructed during mechanical ventilation. A study of 11 dogs was done 7 to 8 d after nitrosourethane; 5 of these were also studied more than 21 d after injury.

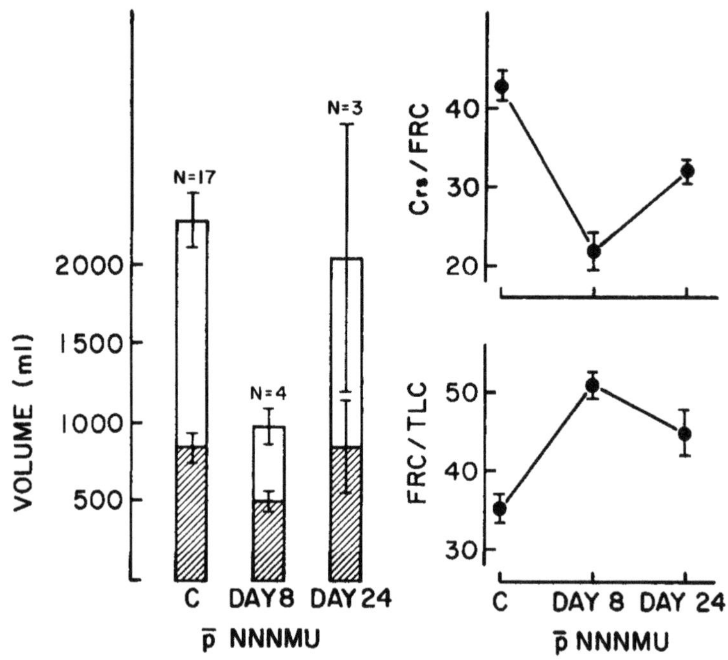

FIGURE 18. Serial changes of FRC, TLC, Crs/FRC, and FRC/TLC in 4 dogs that received nitrosourethane, compared to control data which include 13 additional dogs. The findings after injection are statistically significant except for those of FRC and TLC during recovery. Hatched areas represent FRC and horizontal bars ± SE.

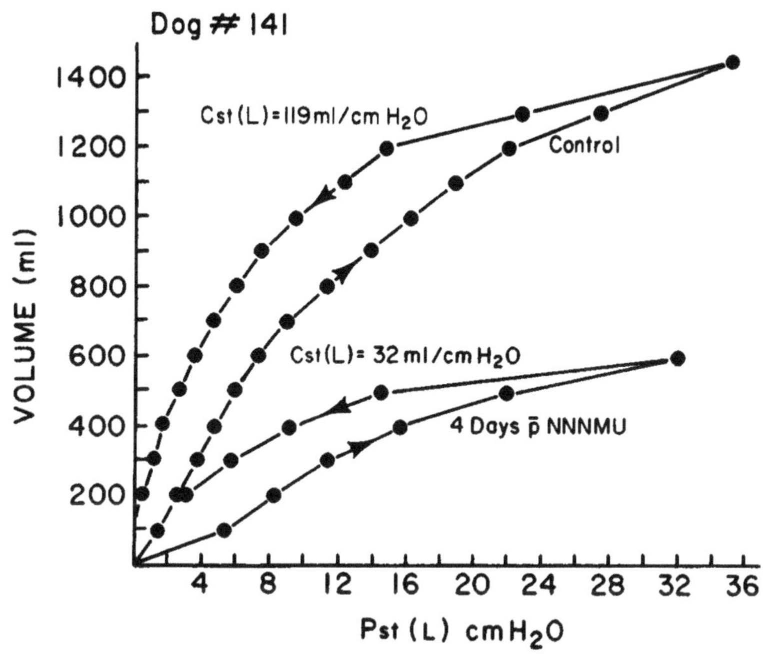

FIGURE 19. Static volume-pressure curves of one dog obtained during the control period and 4 d after injection of nitrosourethane. Values of FRC, TLC, and CL are shown.

nitrosourethane-injured lung differ from those seen in experimental hydrostatic pulmonary edema[13,14] and in lungs injured by oleic acid.[15,16] In these models, increments of PEEP are associated with a rise of lung compliance[13,15] and static volume-pressure curves show increased hysteresis,[14,16] indicating that PEEP causes alveolar recruitment. In both of these experimental models, the physiologic patterns occur when extravascular lung water (Qwl) is markedly increased.[13,17] Studies have shown that during the acute phase of nitrosourethane-induced injury, there is only a 80% rise of Qwl[17a] and abnormal lung mechanics correlates significantly with a quantitative decrease of lung surfactant.[18] Thus, in the surfactant-deficient adult lung, there is no evidence *in vivo* of alveolar recruitment when the lung is inflated above the resting end-expiratory level.

Volume-Pressure and Morphometric Observations

Methods

To further examine the causes of decreased CL during nitrosourethane-induced AAI, volume-pressure diagrams using excised lungs and data derived from them were compared with morphologic observations and with the extent of irreversible alveolar closure.[19]

In these studies, one lung of each dog was used for volume-pressure measurements and the other was prepared for morphologic and morphometric studies as described above except that 10% buffered formalin was used as a fixative and inflation pressure was increased to 40 cm of water. Airspaces which remained closed after liquid inflation fixation at this pressure were defined as irreversibly closed. Airspace sizes in microscopic sections from tissue blocks selected by stratified random sampling were measured by the mean linear intercept (Lm) method. The Lm is the average distance between alveolar surfaces and is directly proportional to the average size of the small airspaces. Only the surfaces of open airspaces were counted as intercepts and thus the Lm increased in direct proportion to the number of irreversibly closed alveoli (Figures 20 and 21).

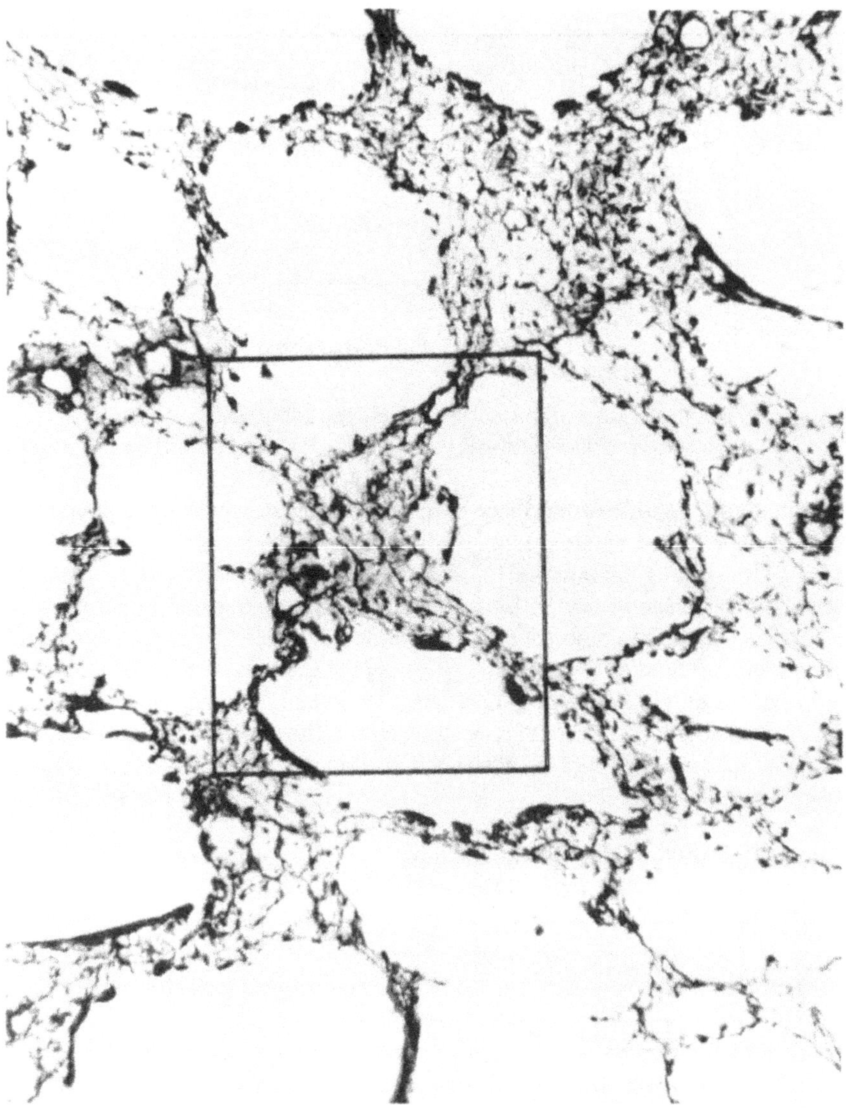

FIGURE 20. Section of lung from a dog 11 d after injection of nitrosourethane. The lung was fixed with intrabronchial formalin at a pressure of 40 cm H_2O. The architecture of the lung is markedly abnormal, with clusters of closed airspaces alternating with enlarged open airspaces. (Silver reticulin stain × 140.)

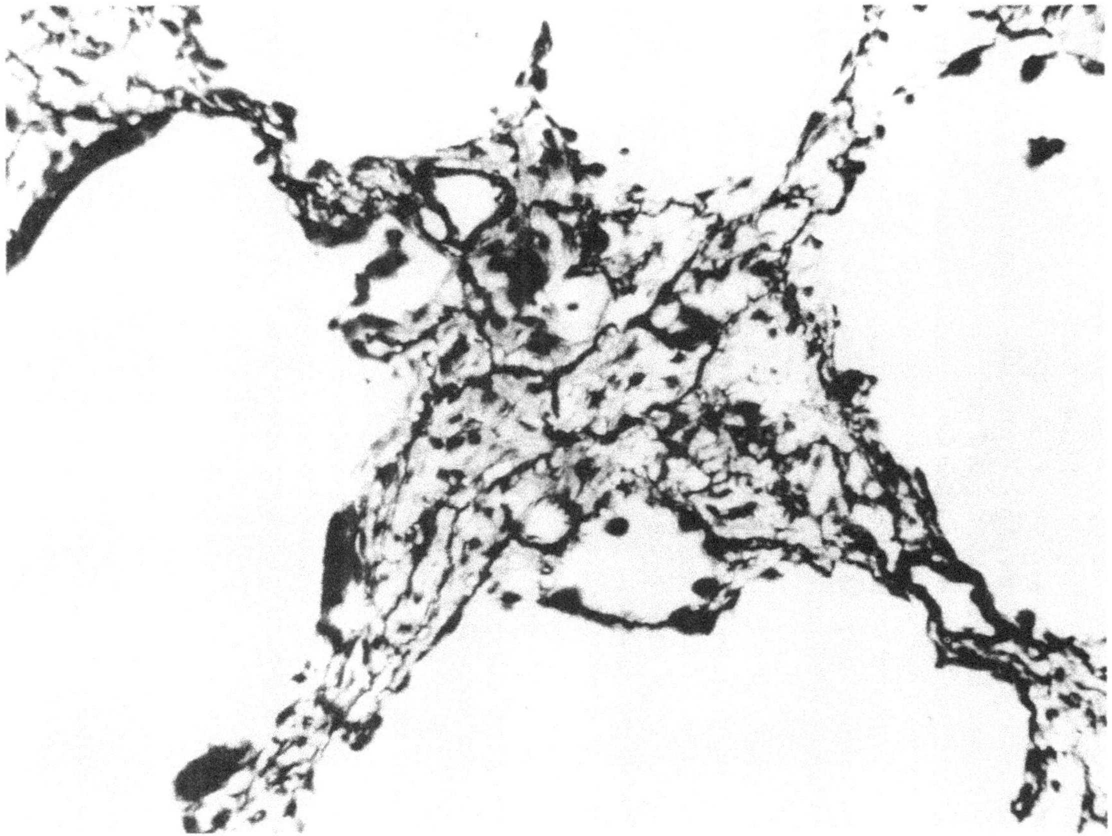

FIGURE 21. Area outlined by the square in Figure 20. A cluster of closed alveoli is surrounded by enlarged open spaces. The connective tissue skeleton of the closed alveoli is clearly recognizable and no fibrosis is present. (Silver reticulin stain × 360.)

Results

Comparisons of the volume-pressure characteristics of the lung during inflation-deflation with air and with saline allow estimation of the contributions of surface forces and of tissue forces to lung recoil because inflation-deflation with saline largely eliminates surface tension in the lung. The normal lung inflated with saline typically reaches total volume at lower pressure and shows less hysteresis than when inflated with air because the contribution of surface tension to recoil is eliminated (Figure 22). Volume-pressure diagrams of dog lungs during deflation after inflation with air at intervals after injection of nitrosourethane are shown in Figure 23. In these curves, volumes are expressed as percent of predicted TLC. There is a progressive downward shift in the entire curve from the control to the 3- to 4-d period and a further significant downward shift at 5 to 7 d. In contrast, the deflation curve after inflation with saline is not significantly different from control during the 3- to 4-d period but shows a marked downward shift at 5 to 7 d.

When the volume-pressure diagrams are corrected for loss of lung volume by expressing lung volume as a percent of observed TLC, only the volume pressure behavior of open or recruitable airspaces is measured and the effect of closed airspaces is excluded. Such curves are shown in Figure 24. The diagrams for air inflation indicate a downward shift at 3 to 4 d with no further reduction, while the diagrams for saline inflation indicate no downward shift from controls.

Lm is slightly increased at 3 to 4 d and steadily increases thereafter (Figure 25). In addition, Lm shows significant inverse linear correlation with total lung volume (the volume of the excised lung measured at a translung pressure of 30 cm of water) and with compliance measured during deflation with saline (Figure 25). These correlations, as well as morphologic examination, eliminate emphysema as a cause of increased Lm and support the conclusion that it is due entirely to irreversible alveolar closure. Microscopic examination rules out obstruction of small airways and decreased tissue distensibility due to fibrosis. The remaining possible causes of loss of lung volume and CL are therefore: (1) increased surface tension, (2) decreased distensibility of recruitable airspaces due to vascular congestion and interstitial edema, and (3) irreversible alveolar closure.

Analyses of the volume pressure and the morphologic and morphometric changes together permit estimation of the relative contributions of each of the three possible causes of increased CL during the evolution of lung injury and repair. During early injury (3 to 4 d), loss of lung volume and distensibility are due almost entirely to increased surface tension because the volume pressure curves on saline deflation are entirely normal while those on air inflation show a marked downward shift. At this time the extent of irreversible alveolar closure is small and interstitial edema (lung weight to body weight ratio) is not yet marked. When the altered volume-pressure behavior during air deflation is corrected for loss of volume by expressing volume as percent of observed TLC, abnormal compliance is still present. Thus, the abnormal compliance on air deflation is due partly to closure of airspaces which cannot be reversed with air inflation to 30 cm H_2O (but can be reversed by saline inflation) and partly to increased surface tension in recruitable alveoli.

Another picture emerges by 5 to 7 d, when loss of lung volume and distensibility can no longer be explained by the effects of increased surface tension alone. At this point the lungs are even less distensible when inflated with air, and they also exhibit decreased distensibility when inflated with saline. Clusters of irreversibly closed alveoli are present in larger numbers and Lm is significantly increased.

At 9 to 14 d, volume-pressure diagrams with air and with saline are not significantly different from those at 5 to 7 d. Morphologically, there are more clusters of irreversibly closed alveoli and Lm is even larger. Inverse correlations between Lm and total lung volume and between Lm and compliance on saline deflation support the concept that irreversible alveolar closure is the principal factor decreasing lung distensibility during the second week

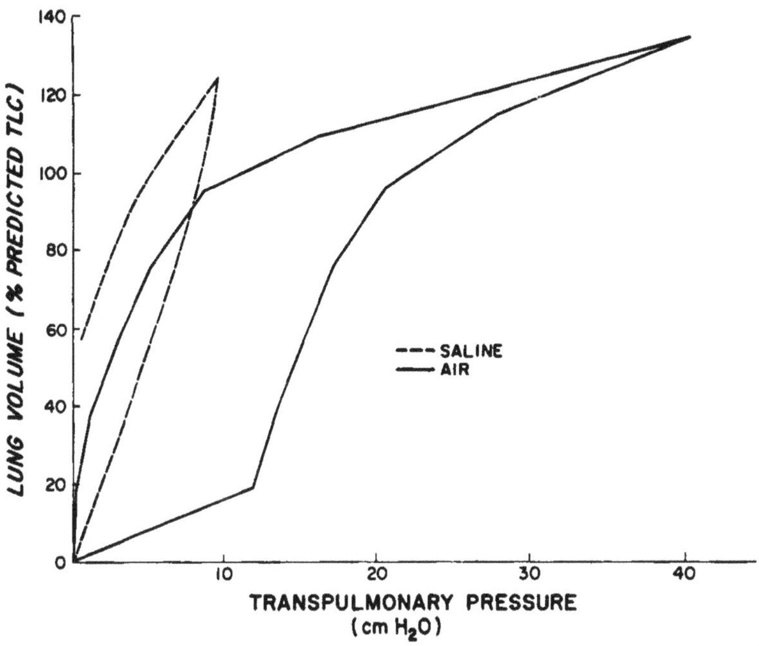

FIGURE 22. Volume-pressure diagrams during inflation and deflation with air and with saline in the excised lung of a normal dog. When inflated with saline the lung reaches TLC at lower pressure and shows less hysteresis than when it is inflated with air.

of nitrosourethane injury (Figure 25). However, a persistent contribution by increased surface tension is demonstrated by the persistent abnormality in the volume-corrected air volume-pressure curve (Figure 24). The normal volume-pressure characteristics of the volume-corrected saline volume-pressure diagram at all phases of injury argue against abnormal tissue distensibility as a significant cause of abnormal compliance even though lung weights steadily increase and interstitial edema is prominent. These latter findings are in agreement with other studies showing that interstitial edema decreases CL only slightly and that the major effect of pulmonary edema is mediated through the increase in surface tension caused by alveolar edema.[13,14,20]

The above observations support the following hypothesis. In the earliest phase of AAI, decreased CL and lung volume are primarily due to alveolar collapse caused by increased surface tension. The contribution of increased surface tension is maximized by 5 to 7 d and remains constant through the following week. Beginning at 5 to 7 d and increasingly thereafter, irreversible closure of alveoli affects lung distensibility. As the number of irreversibly closed alveoli increases, the distensibility of the lung progressively decreases. The number of irreversibly closed alveoli must be important in determining the degree to which CL remains abnormal after recovery.

These observations are reviewed in some detail because it seems highly probable that they pertain to human AAI as well. They support the observations subsequently made in human lungs[1] that irreversible closure of airspaces contributes significantly to volume loss and to architectural revision. They also suggest that increased surface tension is a major cause of mechanical abnormality during early to peak injury, and they call into question the assumption that these abnormalities are due solely to high permeability pulmonary edema.

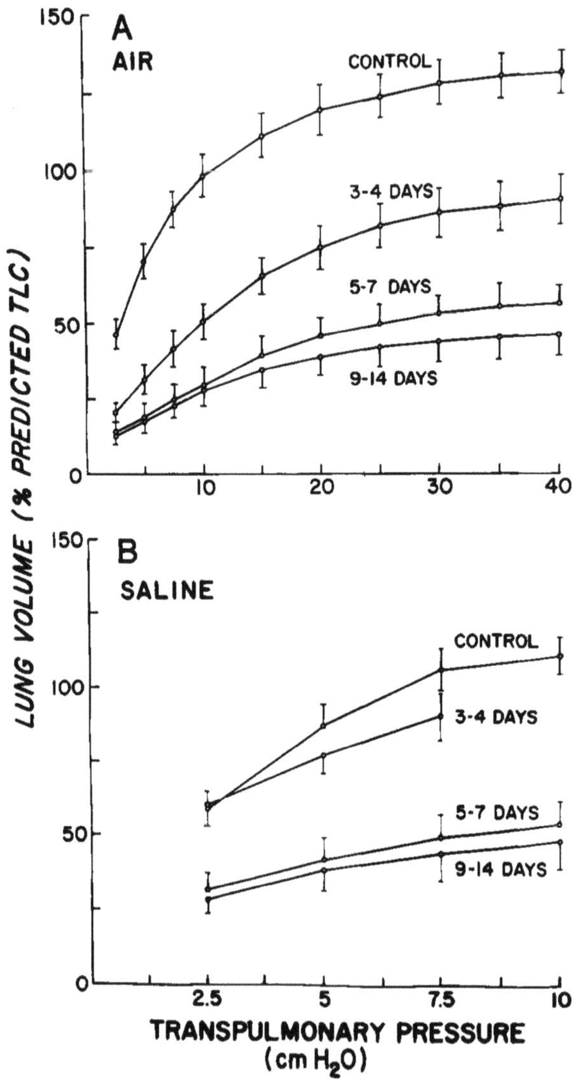

FIGURE 23. Volume-pressure diagrams during deflation after inflation with (A) air or (B) saline in excised lungs of dogs at intervals after injection of nitrosourethane. The points represent mean values. The bars depict ± 1 SE. The volume axis is percent of predicted TLC to normalize for difference in size among the animals.

Biochemical and Biophysical Studies

Methods

Isolation of Surfactant — In these studies of quantitative changes, biochemical abnormalities, and surface behavior, the methods used have been described in detail elsewhere. Briefly, animals were sacrificed at varying intervals following nitrosourethane injection. After sacrifice, the lung was quickly excised, weighed, and degassed in a vacuum. Its bronchus and pulmonary artery were then cannulated and it was lavaged six times via the cannulated bronchus, each time with 10 ml of cold TN buffer (0.01 *M* Tris HCl at pH 7.4, 0.15 *M* NaCl) per gram predicted normal lung weight. The pulmonary artery was then perfused at a pressure of 30 cm of water with 1500 ml of cold TN buffer. The lavages were

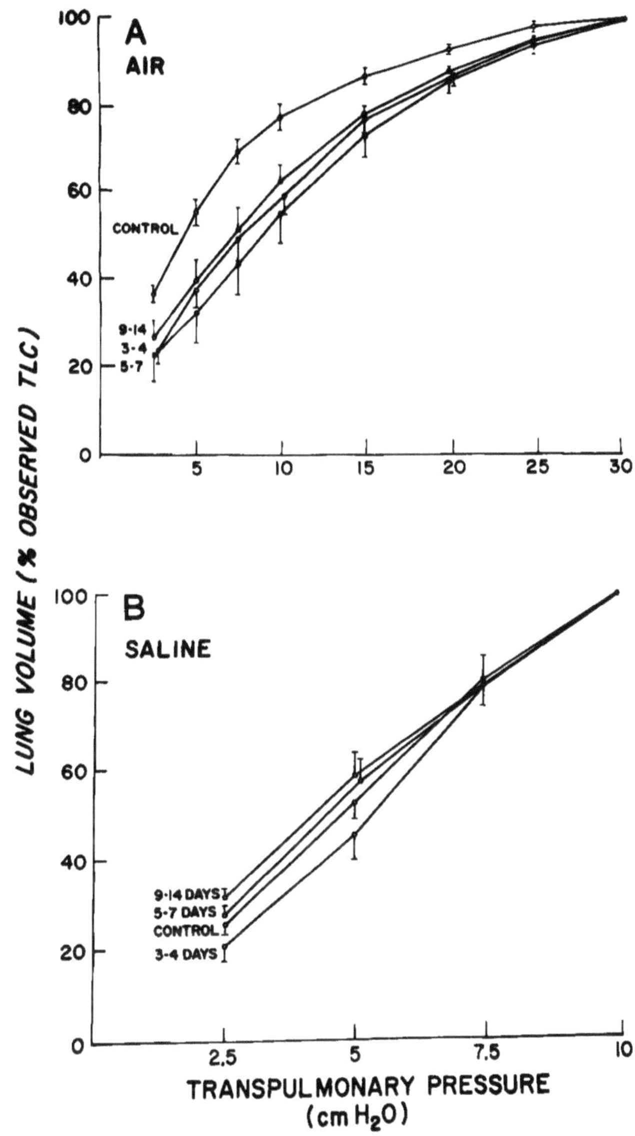

FIGURE 24. Volume-pressure diagrams during deflation after inflation with (A) air or (B) saline. Symbols are same as in Figure 23. Here the volume axis is expressed as percent of observed TLC to correct for loss of lung volume due to alveolar closure.

pooled and centrifuged at 200 × *g* for 10 min at 4°C to remove cells. The cells were washed once with cold TN buffer and the washes were added to the supernatant. The lavage fluid was then either lyophilized for quantitation of lipids or was used for purification of surfactant. After lavage, the lung was homogenized and samples of the homogenized lung tissue were lyophilized. For quantification of phospholipids, the lyophilized lavage and lung tissue were extracted according to the method of Folch and co-workers.[21] The lipid extracts were concentrated on a flash evaporator, transferred into screw-topped tubes, and made to volumes of 20 ml with chloroform:methanol 2:1 (vol/vol). These lipid stock solutions were stored at −20°C and all subsequent lipid analyses were done from them. Phospholipid quantities were then determined after separation by thin layer chromatography.

Purification of surfactant — The alveolar lavage was centrifuged at 27,000 × *g* for

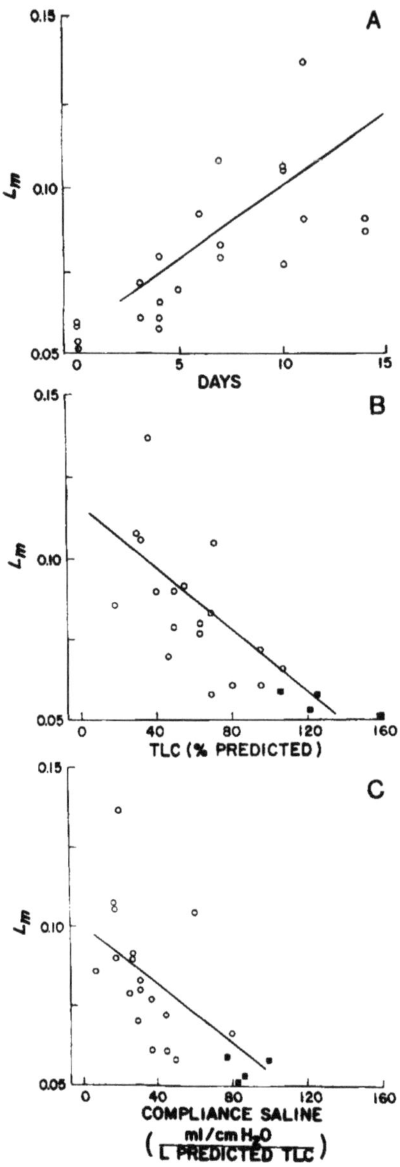

FIGURE 25. Relation between Lm and (A) day of disease, (B) TLC, and (C) compliance with saline. In diagrams B and C, data from control animals are denoted by squares and data from animals with nitrosourethane-induced lung injury are denoted by circles. Correlation coefficients and p values are (A) r = 0.614, p = 0.01; (B) r = 0.611, p = 0.01 (c); (C) r = 0.571, p = 0.02.

2 h at 4°C and the resulting pellet was suspended in TN buffer to yield a suspension containing 5 mg of phospholipids per milliliter.[22-24] This material is referred to as partially purified surfactant. Further purification of the surfactant was carried out on a discontinuous gradient of NaBr using the method of Shelley and coworkers,[25] modified only in that partially purified surfactant rather than lavage was used because of the large volume of lavage. This method was chosen because it has been shown to remove a large proportion of contaminating serum lipids and proteins.

Analysis of surfactant associated proteins — Isolation of surfactant-associated proteins was carried out by the method described by Chan and Knowles.[26] The partially purified and the purified surfactants in TN buffer were extracted with a solution of 4:6 butanol/di-isopropyl ether (vol/vol) at 4°C for 1 h to partition the lipid into the organic phase and the protein into the aqueous phase. After extraction, the mixture was centrifuged at 2,000 rpm for 10 min at 4°C to separate the aqueous phase from the organic phase. The aqueous phase was removed and extracted again with the same volume of extraction solvent. This method has been shown to completely remove the phospholipids and neutrolipids without protein denaturation.[26] The aqueous phase was used for analysis of protein composition by SDS polyacrylamide slab gel electrophoresis. Protein content of partially purified and purified surfactants and of the aqueous phase from the delipidation of the surfactant was determined by the method of Lowry and co-workers[27] with the addition of 1% SDS to the mixture.

Density subfractionation of surfactant — Subfractionation of the surfactant was carried out in a continuous sucrose density gradient.[22] The partially purified surfactant was suspended in TN buffer and layered over a continuous sucrose density gradient (0.1 to 1.0 *M*) in cold TN buffer. The gradient was centrifuged on an SW 27 rotor at 100,000 × *g* for 4 h at 4°C. The visible white band fractions were collected by aspiration, diluted 1:10 with cold TN buffer, and recovered by centrifugation (100,000 × *g* for 1 h at 4°C). The resulting pellet was suspended in TN buffer and the quantities of phospholipid and protein were determined.

Measurement of surface activity — Surface tensions of the surfactant were measured in a modified Wilhelmy® balance at 37°C. A Teflon® trough, 15 × 6 cm (maximal and minimal surface areas of 65 and 13 cm^2) lined with Teflon® tape and a movable Teflon® barrier were used. The trough was covered with a transparent plastic cover. No material was applied until the TN buffer subphase exhibited a clean surface (70 dyn/cm). Surfactant in TN buffer was dried under N_2 and resuspended in 50 μl of isopropanol-water-chloroform 2:1:0.5 (vol/vol/vol). The sample containing 60 μg of total phospholipids was added to the surface with a microsyringe and with the barrier in the fully extended position. After 10 min, the spread films were compressed from 65 to 13 cm^2 at 1 c/100 s. The surface tension vs. surface area was recorded continuously by an X-Y recorder.

Enzyme assays — Details of the methods used for enzyme kinetics are presented in the publications cited.

Results

Quantitative Changes

Quantitative changes in phospholipids in alveolar lavage and postlavage lung tissue and qualitative changes in surfactant isolated from alveolar lavage were studied during all phases of nitrosourethane injury.

The phospholipids in the alveolar lavage were quantified to estimate total synthesis of these lipids by the lung throughout injury and recovery (Table 1). Because the weights of the injured lungs were increased, lipid quantities are expressed in terms of predicted normal lung weight (PLW)[28] to avoid the artifactual decrease that would occur if the observed lung weights were used. The PLWs of the experimental animals from each group were not significantly different from those of control animals. Disaturated phosphatidylcholine (DSPC), expressed as milligrams per gram PLW, began to decline at 2 to 4 d and decreased to a mean of 37% of the control value at 6 to 8 d. It then increased significantly from the peak injury to 80% of control during late recovery. Total phospholipids, phosphatidylcholine (PC), and phosphatidylethanolamine (PE) followed the same pattern, but the changes were slightly smaller. Phosphatidylglycerol (PG) decreased strikingly at early and peak injury but, in contrast to PC and DSPC, this decrease persisted during early and late recovery. The quantity of phosphatidylinositol (PI) at 15 to 20 d was significantly higher than those of control animals and those at 6 to 8 d. The quantities of sphingomyelin (SPH) and

Table 1
QUANTITIES OF PHOSPHOLIPIDS IN ALVEOLAR LAVAGE DURING ACUTE ALVEOLAR INJURY INDUCED BY NITROSOURETHANE

	Post-NNNMU (d)					p		
	0 (8)	2—4 (5)	6—8 (5)	10—12 (5)	15—20 (7)	6—8 vs. 0 (d)	15—20 vs. 0 (d)	15—20 vs. 6—8 (d)
Total phospholipids	2.06 ± 0.41	1.27 ± 0.27	0.97 ± 0.52	1.06 ± 0.71	1.78 ± 0.65	<0.005	NS	<0.05
Phosphatidylcholine	1.56 ± 0.31	0.94 ± 0.20	0.70 ± 0.26	0.80 ± 0.54	1.37 ± 0.63	<0.005	NS	<0.05
Disaturated phosphatidylcholine	0.76 ± 0.20	0.35 ± 0.05	0.27 ± 0.15	0.35 ± 0.27	0.61 ± 0.21	<0.001	NS	<0.01
Phosphatidylglycerol	0.13 ± 0.04	0.04 ± 0.01	0.03 ± 0.02	0.02 ± 0.02	0.02 ± 0.02	<0.001	<0.001	NS
Phosphatidylinositol	0.06 ± 0.02	0.06 ± 0.01	0.06 ± 0.02	0.06 ± 0.02	0.11 ± 0.03	NS	<0.005	<0.005
Phosphatidylethanolamine	0.06 ± 0.02	0.04 ± 0.02	0.03 ± 0.01	0.03 ± 0.02	0.05 ± 0.02	<0.005	NS	NS
Sphingomyelin	0.07 ± 0.02	0.11 ± 0.03	0.08 ± 0.02	0.05 ± 0.02	0.08 ± 0.03	NS	NS	NS
Lysophosphatidylcholine	0.06 ± 0.02	0.05 ± 0.01	0.04 ± 0.02	0.04 ± 0.02	0.07 ± 0.03	NS	NS	NS
Diphosphatidylglycerol	nd	nd	nd	nd	nd			

Note: Values are mean ± SD and are given as milligram per gram predicted normal lung weight. NS = not significantly different and nd = not detectable. Figures in parentheses are the number of dogs in each group.

lysophosphatidylcholine (LPC) did not change significantly. Diphosphatidylglycerol (DPG) was not detectable in lavage, either from experimental or from control animals.

The quantities of PC, DSPC, PG, and DPG in the postlavage tissue were also determined.[29] The quantities of these phospholipids in the tissue represent those in the type II epithelial cells and in other types of cells in the lung. Tissue PC and DSPC, expressed as milligrams per gram PLW, also decreased to low levels at 6 to 8 d and increased steadily to twice those of control lungs during late recovery. Tissue PG decreased sharply coincident with an increase in tissue DPG during early and peak injury. During early and late recovery, tissue PG failed to increase at the same rate as tissue PI and DSPC. In fact, the PG content was only half that of controls, a remarkable difference from the twofold increase found in both tissue PC and DSPC. In contrast, the tissue DPG increased dramatically to threefold control levels during late recovery.

The degree of reduction in lavage DSPC correlated with decrease of C_L, suggesting that deficiency of alveolar surfactant may contribute to the mechanical abnormality.[18] Equivalent reduction of postlavage lung tissue DSPC indicates that impaired lipid synthesis rather than increased degradation is the most likely cause of the deficiency of alveolar surfactant. During recovery from acute alveolar injury, the regenerating epithelial cells had matured. This maturation coincides with the increased quantities of lavage and postlavage lung tissue phospholipids, indicating that surfactant synthesis is renewed. This renewed synthesis with increasing quantities of alveolar phospholipid may be an important mechanism in improving lung function during recovery.

Qualitative Changes in Surfactant

Surface properties and representative surface tension vs. area diagrams of purified surfactant from each of the groups in Table 1 are shown in Table 2 and Figure 26. Increases in minimum surface tension (γ min) and decreases in stability index and adsorption rate (Adπ) were found during early and peak injury, and they recovered toward normal during early recovery. By late recovery, they were almost the same as those of control animals. Surface properties of the partially purified surfactant were similar to those of purified surfactant. Biochemical composition and physical properties of the surfactant from each group were analyzed in an attempt to identify biochemical abnormalities responsible for abnormal surface function.

In the purified surfactant (Table 3), the percentages of PC, DSPC, and PE were similar in the five groups. The percentages of PG were much lower during all phases of injury than in the control animals, and the reverse was true for both PI and LPC. The amount of SPH was higher and thus the PC to SPH ratio (L/S ratio) was lower during early and peak injury than that of control animals; this returned to normal levels during early and late recovery. Although cholesterol ester and triglyceride were different among the groups, total neutral lipids remained unchanged. The composition of lipids of the partially purified surfactant were similar to those of purified surfactant. The low percentage of PG in the purified surfactant (Table 3) during all phases of injury is similar to those isolated from the lungs of prematurely born rabbits or from patients with ARDS.[22,29] PG, the second most abundant phospholipid in the lung surfactant, has been considered to be essential for normal surfactant function.[30-32] However, in this experimental model, the marked reduction of PG at peak injury cannot be responsible for the abnormal surface function of the surfactant because surfactant from the lungs after recovery function normally despite PG levels which were as low as those at peak injury (Table 3).[33] For the same reasons, it is unlikely that increased PI or LPC which persisted during recovery when surface activity had become normal were responsible. Other studies[34,35] in experimental models with PG-deficient surfactant also suggest that PG is not an essential component for normal surface activity of the surfactant system.

Table 2
SURFACE PROPERTIES OF PURIFIED SURFACTANT DURING ACUTE ALVEOLAR INJURY INDUCED BY NITROSOURETHANE

	Post-NNNMU (d)					p		
	0 (5)	**2—4 (5)**	**6—8 (5)**	**10—12 (5)**	**15—20 (5)**	**6—8 vs. 0 (d)**	**15—20 vs. 0 (d)**	**15—20 vs. 6—8 (d)**
γmin (dyn/cm)	3.4 ± 0.8	19.0 ± 2.0	19.6 ± 2.5	10.6 ± 2.2	4.1 ± 1.6	<0.001	NS	<0.001
SC (cm/dyn)	0.075 ± 0.021	nm	nm	0.070 ± 0.002	0.071 ± 0.018		NS	
SI	1.74 ± 0.03	1.01 ± 0.02	1.09 ± 0.04	1.51 ± 0.05	1.73 ± 0.05	<0.001	NS	<0.001
DR (cycle 2:1)	0.89 ± 0.03	0.75 ± 0.09	0.84 ± 0.07	0.86 ± 0.08	0.90 ± 0.07	NS	NS	NS
Adπ (dyn/cm/min)	1.7 ± 0.2	1.4 ± 0.4	1.3 ± 0.3	1.5 ± 0.3	1.6 ± 0.2	<0.05	NS	<0.05

Note: Values are mean ± SD. γmin = minimum surface tension, SC = surface compressibility, SI = stability index, DR = dynamic respreadability, Adπ = surface adsorption rate, and NS = not significantly different. Figures in parentheses are the number of dogs in each group.

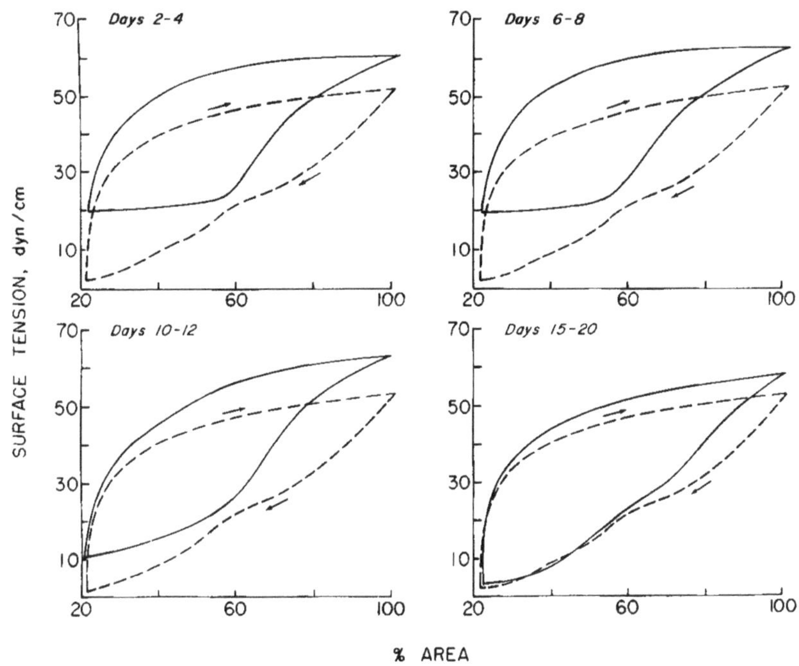

FIGURE 26. Representative surface tension vs. area diagrams of purified surfactants during acute alveolar injury (——) induced by nitrosourethane. The diagram from a control dog (----) is included for comparison in each phase of injury. Each sample, containing 60 μg of phospholipids, was applied to the surface in 50 μl of isopropanol-water-chloroform. Diagrams shown were recorded during the second cycle of compression and expansion of the surface.

Surfactant-associated proteins comprise plasma proteins and surfactant apoproteins.[36-39] Surfactant apoproteins may play a critical role in enhancing surface adsorption[36,40] and in regulating the recycling of surfactant components to type II cells.[41] The quantities and composition of surfactant-associated proteins were analyzed in the partially purified and purified surfactants. The partially purified surfactant contained 20 to 25% protein by weight, whereas the purified surfactant contained 10 to 15 % protein. The partially purified surfactants from normal and injured lungs contained a major apoprotein of 38,000 mol wt, referred to in other studies as apoprotein A,[42,43] and two minor apoproteins with approximately 40,000 and 36,000 mol wt (Figure 27). Apoprotein of low molecular weight (10,000 to 15,000), referred to in other studies as apoprotein B,[42,43] was not found in this preparation. Contamination of plasma proteins was observed (the bands of 94,000, 68,000, 60,000, and 30,000 mol wt) (Figure 27) in all phases of injury and in the normal lung. The quantity of the major apoprotein began to decrease during early injury and reached a nadir at peak injury. It then increased at early recovery and approached normal during late recovery. Similar patterns in the changes of apoprotein were observed with the purified surfactant. Although some contaminating plasma protein remained, it was markedly reduced in the purified surfactant. Thus, abnormal surface function of the surfactant from injured lungs was probably not due to contamination by plasma lipids and proteins. The results suggest that decreased apoprotein may be responsible for or contribute to abnormal surface properties, but this remains to be established.

The density of the surfactant in continuous sucrose density gradients upon ultracentrifugation represents one of its specific physical properties. Two subfractions with isopycnic densities of 1.05 and 1.09 were found in control animals (Figure 28). During early injury, a broad band at d = 1.03 to 1.04 was separated. Only one band with low density (d =

Table 3
COMPOSITION OF PHOSPHOLIPIDS, NEUTRAL LIPIDS, AND PHOSPHOLIPID/PROTEIN RATIO OF PURIFIED SURFACTANT DURING ACUTE ALVEOLAR INJURY INDUCED BY NITROSOURETHANE

	Post-NNNMU (d)					p		
	0 (8)	2—4 (5)	6—8 (5)	10—12 (5)	15—20 (7)	6—8 vs. 0 (d)	15—20 vs. 0 (d)	15—20 vs. 6—8 (d)
Phospholipids								
Phosphatidylcholine	73.5 ± 2.5	72.5 ± 5.9	74.2 ± 2.8	75.3 ± 3.7	77.3 ± 3.3	NS	NS	NS
Disaturated phosphatidylcholine	43.6 ± 2.8	43.7 ± 3.7	44.1 ± 1.6	45.2 ± 2.4	46.1 ± 2.4	NS	NS	NS
Phosphatidylglycerol	12.8 ± 1.3	4.4 ± 0.6	4.2 ± 0.7	4.0 ± 0.9	3.0 ± 0.7	<0.001	<0.001	NS
Phosphatidylinositol	3.3 ± 1.1	5.1 ± 0.8	6.1 ± 1.4	5.7 ± 2.1	6.3 ± 1.6	<0.005	<0.001	NS
Phosphatidylethanolamine	1.9 ± 0.6	2.2 ± 0.6	2.0 ± 0.5	1.5 ± 0.4	1.7 ± 0.6	NS	NS	NS
Phosphatidylserine	nd	nd	nd	nd	nd			
Sphingomyelin	4.6 ± 1.3	10.7 ± 4.8	7.8 ± 1.6	4.4 ± 0.8	4.5 ± 1.1	<0.005	NS	<0.005
Lysophosphatidylcholine	3.9 ± 1.2	4.6 ± 1.5	5.9 ± 2.1	8.7 ± 1.5	7.2 ± 2.0	<0.005	<0.005	NS
Neutral lipids								
Cholesterol	10.4 ± 1.4	11.6 ± 1.9	12.3 ± 2.4	11.6 ± 2.1	10.9 ± 1.8	NS	NS	NS
Cholesteryl ester	1.4 ± 0.3	2.8 ± 1.0	3.1 ± 0.5	1.7 ± 0.4	1.8 ± 0.8	<0.001	NS	<0.001
Triglyceride	8.6 ± 1.2	3.9 ± 0.5	3.6 ± 0.6	4.9 ± 0.5	5.6 ± 0.8	<0.001	<0.001	<0.001
Phospholipid/protein (mg/mg)	7.8 ± 1.2	6.2 ± 1.2	6.2 ± 1.8	6.8 ± 1.4	8.3 ± 1.5	NS	NS	NS

Note: Values are mean ± SD and are given as percent of total phospholipids for each phospholipid and as percent of total lipids (phospholipids + neutral lipids) for each neutral lipid. NS = not significantly different and nd = not detectable. Figures in parentheses are the number of dogs in each group.

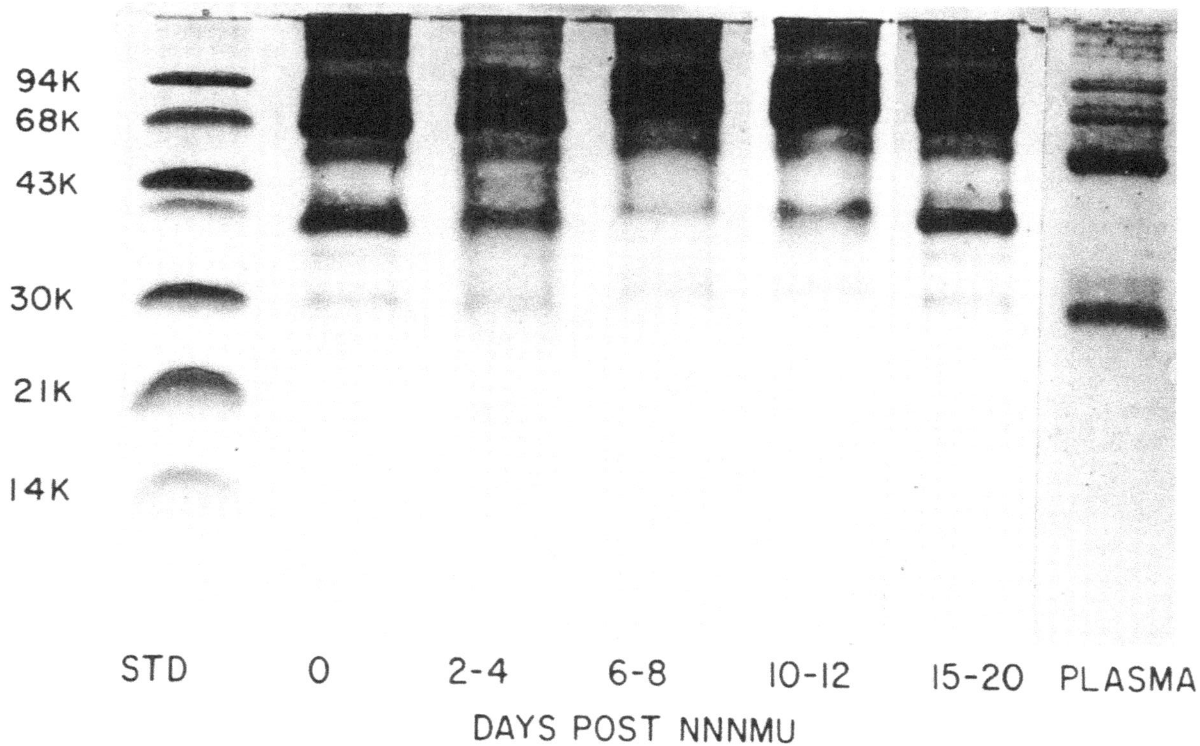

FIGURE 27. Representative SDS-polyacrylamide slab gel electrophoretogram (stained with Coomassie blue) of partially purified surfactants during acute alveolar injury induced by nitrosourethane. Each surfactant sample contained 20 μg total protein. Dog plasma contained 10 μm protein. Standard proteins (STD), 2 μg each. The major apoprotein (38,000) decreased during injury (2 to 4; 6 to 8 d) and increased during recovery (10 to 12; 15 to 20 d).

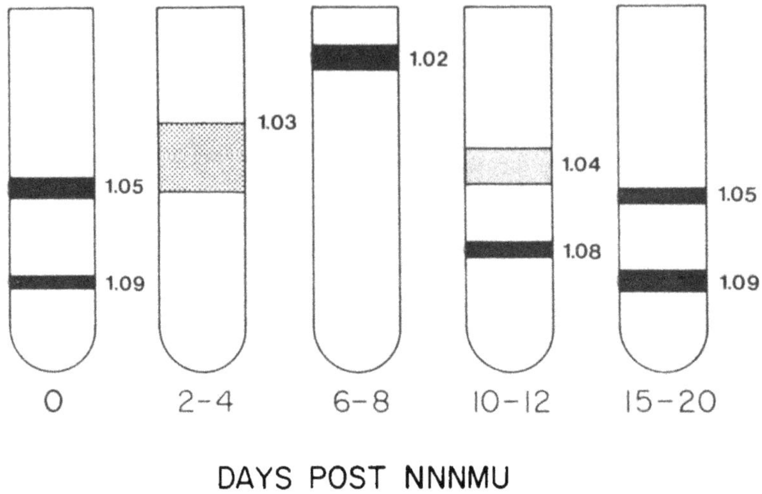

FIGURE 28. Subfractions from continuous sucrose density gradient centrifugation of partially purified surfactants during acute alveolar injury induced by nitrosourethane. The isopycnic densities are indicated at each band. The ratios of phospholipid/protein of each band are given in parentheses: day 0, d = 1.05 (6.6 ± 1.8), d = 1.09 (3.6 ± 1.1); days 2 to 4, d = 1.03 to 1.04 (9.2 ± 2.2); days 6 to 8, d = 1.02 (12.5 ± 3.2); days 10 to 12, d = 1.04 (8.5 ± 2.5), d = 1.08 (3.9 ± 1.6); days 15 to 20, d = 1.05 (6.2 ± 1.4), d = 1.09 (3.5 ± 1.3).

1.02) and high phospholipid to protein ratio was found at peak injury. The two bands found in control animals had reappeared during early and late recovery. These changes in isopycnic density during evolving lung injury may be the result of abnormal lipid to protein ratios in the surfactant material. Changes in surfactant density on continuous sucrose density gradient have also been found in the surfactant from patients with ARDS.[22,23] However, in contrast to the nitrosourethane-injured lung, only part of these surfactants changed their densities to either lower[22] or higher[23] values.

Recent work[38,44] in rabbit lung has suggested that surfactant obtained by alveolar lavage exists in different morphologic forms that have different physical, biochemical, and surface properties. Subfractions of the surfactant obtained by differential centrifugation represent surfactant material at different stages of evolution. The fraction that was easily sedimented contained tubular myelin and was relatively rich in apoproteins of 36,000 and 10,000 mol wt and rapidly adsorbed to an air-water interface, whereas the less sedimentable fraction possessed predominantly vesicular structure, less of these apoproteins, and adsorbed very slowly to an air-water interface. They suggested that the more sedimentable fraction may be a precursor to the less sedimentable fraction, that there is removal of protein during evolution of alveolar surfactant, and that changes in protein content and morphological structure may accompany functional changes in surfactant in the alveoli. The reduced isopycnic density, decreased protein content, and abnormal surface function of the surfactant from nitrosourethane-injured lungs may result from active metabolism of surfactant in the alveoli or from synthesis by regenerating type II cells of abnormal lipids and proteins.

Metabolism of Surfactant

Surfactant DSPC is synthesized in the lung cells through the CDP-choline pathway.[45] This pathway produces primarily unsaturated PC which is remodeled to form DSPC through a deacylation-reacylation reaction. Two of the key enzymes involved in the synthesis of PC and DSPC, cholinephosphotransferase and lysolecithin acyltransferase, involved in the synthesis of PC and DSPC, respectively, were studied in the microsomal fraction of the lungs during all phases of injury.

During early injury, specific activity of cholinephosphotransferase increased significantly to 130% of those of control animals. At peak injury, enzyme-specific activity decreased to about 60% of those of controls. During recovery, enzyme activity increased to the same levels as those during early injury. Similar patterns of enzyme-specific activity were found with lysolecithin acyltransferase. The specific activities of this enzyme correlated with the palmitate proportion in lavage PC.[46] Both the enzyme activity and the proportion of palmitate in alveolar lavage PC increased during the early phase of injury and decreased during peak injury. During early injury, the increase in the specific activity of this enzyme is unexplained. The low enzyme-specific activity and the reduced lavage PC during peak injury were no doubt at least in part a result of reduced numbers of type II cells. However, failure of the existing type II cells to produce normal proportions of palmitate in lavage PC suggests that decreased activity of lysolecithin acyltransferase is responsible for impaired palmitate synthesis. During recovery, specific activity of this enzyme was nearly twice that of control animals, a marked difference from the 30% increase found in cholinephosphotransferase activity. However, the palmitate proportion in the lavage PC was normal. This finding indicates that the enzyme activity per cell was normal and suggests that the increase in specific activity is due to massive regeneration of type II cells and that the enzyme is localized mainly in these cells.

Synthesis of PG was analyzed by the glycerophosphate phosphatidyltransferase activity in the mitochondrial fraction. Changes in specific activity of this enzyme during lung injury and recovery were similar to those of cholinephosphotransferase. The low enzyme-specific activity at peak injury correlated with the low content of PG in both lavage and postlavage lung tissue.[29] During early and late recovery, the enzyme-specific activity increased to levels 30% higher than those of control animals. However, the tissue PG content was only half that of control animals and lavage PG remained at the same level as that of peak injury (15% of that of control animals). The persisted reduced levels of PG in both alveolar lavage and lung tissue were accompanied by elevated levels of PI in alveolar lavage and of DPG in tissue. Studies on the surfactant of the developing fetus revealed that PI initially increases parallel to DSPC, where as PG appears later.[32,47] Competitive synthesis of PG and PI from a common precursor, CDP-diglyceride, was found in the lung and presumably takes place in type II epithelial cells.[48] These findings suggest that inducton of PI synthesis is accompanied by a decrease in PG synthesis and vice versa. The study of biosynthesis of DPG in the nitrosourethane-injured lung suggests that, during recovery, regenerating epithelial cells are the major site of synthesis of PG and DPG. The decreased levels of PG in both alveolar lavage and lung tissue may be due to increased turnover of PG into DPG, thus reducing availability of PG for surfactant in the regenerating type II cells. Presumably, the PG thus synthesized is utilized preferentially to synthesize DPG for membrane construction in regenerating cells. It appears that PG deficiency is characteristic of immature type II cells whether they are fetal or regenerating.

SUMMARY

The injury induced in the lung by nitrosourethane closely resembles acute alveolar injury in the human. The cardinal structural feature of both is necrosis of alveolar epithelial cells followed by regeneration of the cells which at first closely resemble those of the fetal lung and later mature to typical type II cells. Because the injury is dose related and structurally identical in all species so far studied, nitrosourethane provides an ideal model in which to study this injury and its sequellae.

Volume-pressure studies in excised lungs have demonstrated that the decreased compliance typical of this injury is primarily due to increased surface tension during the acute phase, and increasingly to irreversible volume loss during the postacute phase. This volume loss, which has subsequently been documented in the human injury, appears to be due to irreversible closure of totally de-epithelialized alveoli.

Quantitative analysis of key surfactant lipids have shown that the nitrosourethane-injured lung is a surfactant-deficient lung. The degree of reduction in the DSPC in alveolar lavage fluid correlates with the degree of reduction in CL during the acute phase, further supporting the importance of surfactant loss.

Although estimates of DSPC in type II cells suggest that the decreased alveolar surfactant lipid quantity is due to decreased production rather than increased degradation, this reduction is not necessarily due entirely to destruction of type II cells. It is possible that injury only to type I cells may cause decreased surfactant production.

Uninjured type II cells may undergo obligate loss of this synthetic function as they dedifferentiate to divide and reline the alveoli. It is these dedifferentiated cells which resemble the epithelial cells of the fetal lung. The resemblance is structural as well as functional. Subnormal production of PG and the apparent stepwise acquisition of enzymes (e.g., acyltransferase) suggest that the process of biochemical maturation of the regenerating cells also resembles that of the fetal cells.

The role of pulmonary edema as a cause of abnormal lung function in this model has yet to be examined, but preliminary evidence suggests that edema, although doubtless present, may be less important than currently suspected. Not only does positive end-expiratory pressure affect the surfactant-deficient lung and the purely edematous lung differently, but the quantitative increase in extravascular water necessary to cause significant abnormality in hydrostatic edema is not observed in the nitrosourethane-injured lung. In the latter, increases in extravascular water of about 80% are seen in the most severely abnormal lungs, whereas increases of approximately 200% are necessary to cause mechanical abnormalities in hydrostatic edema. However, it is possible that extensive injury to alveolar epithelium in nitrosourethane injury may allow alveolar flooding, the major determinant of mechanical abnormality, at a lower level of total extravascular water than in hydrostatic edema.

This question and others, including compositional cause of the abnormal function of residual surfactant and mechanisms which control cellular regeneration, as well as the mechanism of nitrosourethane injury itself, remain to be studied.

REFERENCES

1. **Ryan, S. F.,** The acute diffuse infiltrative lung diseases and the role of the open lung biopsy in their diagnosis, *Surg. Pathol.*, 5, 209, 1983.
2. **Katzenstein, A. A., Bloor, C. M., and Liebow, A. A.,** Diffuse alveolar damage — the role of oxygen, shock and related factors, *Am. J. Pathol.*, 85(1), 210, 1976.
3. **Pratt, P. C.,** Pathology of adult respiratory distress syndrome, in *The Lung, Structure, Function and Disease*, Thurlbeck, W. and Abell, M. R., Eds., Williams & Wilkins, Baltimore, 1978, 43.
4. **Bachofen, M. and Weibel, E. R.,** Structural atlerations of parenchyma in the adult respiratory distress syndrome, in *Clinics in Chest Medicine*, Vol. 3(1), Bone, R. C., Ed., W. B. Saunders, Philadelphia, 1982, 35.
5. **Herrold, K. M.,** Fibrosing alveolitis and atypical proliferative lesions of the lung: an experimental study in syrian hamsters, *Am. J. Pathol.*, 50, 639, 1967.
6. **Ryan, S. F.,** Experimental fibrosing alveolitis, *Am. Rev. Respir. Dis.*, 105, 776, 1972.
7. **Ryan, S. F., Bell, A. L. L., and Barrett, C. R.,** Experimental injury in the dog — morphologic mechanical correlations, *Am. J. Pathol.*, 82, 353, 1976.
8. **Barrett, C. R., Bell, A. L. L., and Ryan, S. F.,** Alveolar epithelial injury causing respiratory distress in dogs. Physiologic and electron-microscopic correlations, *Chest*, 75, 705, 1979.

9. **Barrett, C. R., Bell, A. L. L., and Ryan, S. F.**, Effects of positive end-expiratory pressure on lung compliance in dogs after acute alveolar injury, *Am. Rev. Respir. Dis.*, 124, 705, 1981.
10. **Robinson, N. E., Gillespie, J. R., Berry, J. D., and Simpson, A.**, Lung compliance, lung volumes, and single-breath diffusing capacity in dogs, *J. Appl. Physiol.*, 33, 878, 1972.
11. **Milic-Emili, J., Mead, J., Turner, J. M., and Glauser, E. M.**, Improved technique for estimating pleural pressure from esophageal balloons, *J. Appl. Physiol.*, 19, 207, 1964.
12. **Avery, W. G. and Sackner, M. A.**, Rapid measurement of functional residual capacity in the paralyzed dog, *J. Appl. Physiol.*, 33, 515, 1972.
13. **Noble, W. H., Kay, J. C., and Oborzalek, J.**, Lung mechanics in hypervolemic pulmonary edema, *J. Appl. Physiol.*, 38, 681, 1975.
14. **Cook, C. D., Mead, J., Schreiner, G. L., Frank, W. R., and Craig, J. M.**, Pulmonary mechanics during induced pulmonary edema in anesthetized dogs, *J. Appl. Physiol.*, 14, 177, 1959.
15. **Cassidy, S. S., Robertson, C. H., Jr., Pierce, A. K., and Johnson, R. L.**, Cardiovascular effects of positive end-expiratory pressure in dogs, *J. Appl. Physiol. Respir. Environ. Exercise Physiol.*, 44, 743, 1978.
16. **Slutsky, A. S., Scharf, S. M., Brown, R., and Ingram, R. H.**, The effect of oleic acid induced pulmonary edema on pulmonary and chest wall mechanics in dogs, *Am. Rev. Respir. Dis.*, 121, 91, 1980.
17. **Dunegan, L. J., Knight, D. C., Harken, A., O'Connor, N., and Morgan, A.**, Lung thermal volume in pulmonary edema, *Ann. Surg.*, 181, 809, 1975.
17a. **Ryan, S. F. and Barrett, C. R.**, unpublished data.
18. **Ryan, S. F., Liau, D. F., Bell, A. L. L., Hashim, S. A., and Barrett, C. R.**, Correlation of lung compliance and quantitation of surfactant phospholipids after acute alveolar injury from n-nitroso-n-methylurethane in the dog, *Am. Rev. Respir. Dis.*, 123, 200, 1981.
19. **Ryan, S. F., Barrett, C. R., Lavietes, M. H., Bell, A. L. L., and Rochester, D. F.**, Volume-pressure and morphometric observations after acute alveolar injury in the dog from n-nitroso-n-methylurethane, *Am. Rev. Respir. Dis.*, 118, 735, 1978.
20. **Johnson, J. W. C., Permutt, S., Sipple, J. H., and Salem, E. S.**, Effect of intra-alveolar fluid on pulmonary surface tension properties, *J. Appl. Physiol.*, 19, 769, 1964.
21. **Folch, J., Lees, M., and Sloane Stanley, G. H.**, A simple method for the isolation and purification of total lipids from animal tissues, *J. Biol. Chem.*, 226, 497, 1957.
22. **Petty, T. L., Reiss, O. K., Paul, G. W., Silvers, G. W., and Elkins, N. D.**, Characteristics of pulmonary surfactant in adult respiratory distress syndrome associated with trauma and shock, *Am. Rev. Respir. Dis.*, 115, 531, 1977.
23. **Hallman, M., Spragg, R., Harrell, J. H., Mose, K. M., and Gluck, L.**, Evidence of lung surfactant abnormality in respiratory failure, *J. Clin. Invest.*, 70, 673, 1982.
24. **Metcalfe, I. L., Enhorning, G., and Possmayer, F.**, Pulmonary surfactant-associated proteins: their role in the expression of surface activity, *J. Appl. Physiol.*, 49, 34, 1980.
25. **Shelley, S. A., Paciga, J. E., and Balis, J. U.**, Purification of surfactant from lung washings and washings contaminated with blood constituents, *Lipids*, 12, 505, 1977.
26. **Chan, B. E. and Knowles, B. R.**, A solvent system for delipidation of plasma or serum without protein precipitation, *J. Lipid Res.*, 17, 176, 1976.
27. **Lowry, D. H., Rosebrough, N. J., Farr, A. L., and Randall, R. J.**, Protein measurement with the Folin phenol reagent, *J. Biol. Chem.*, 193, 265, 1951.
28. **Frank, N. R.**, A comparison of static volume-pressure relations of excised pulmonary lobes of dogs, *J. Appl. Physiol.*, 18, 274, 1963.
29. **Liau, D. F., Barrett, C. R., Bell, A. L. L., Cernansky, G., and Ryan, S. F.**, Diphosphatidylglycerol in experimental acute alveolar injury in the dog, *J. Lipid Res.*, 25, 678, 1984.
30. **Hallman, M. and Gluck, L.**, Phosphatidylglycerol in lung surfactant. I. Synthesis in rat microsomes, *Biochem. Biophys. Res. Commun.*, 60, 1, 1974.
31. **Rooney, S. A., Canavan, P. M., and Motoyama, E. K.**, The identification of phosphatidylglycerol in the rat, rabbit, monkey and human lung, *Biochim. Biophys. Acta*, 360, 56, 1974.
32. **Hallman, M. and Gluck, L.**, Phosphatidylglycerol in lung surfactant. III. Possible modifier of surfactant function, *J. Lipid Res.*, 17, 257, 1976.
33. **Liau, D. F., Barrett, C. R., Bell, A. L. L., and Ryan, S. F.**, Normal surface properties of phosphatidylglycerol deficient surfactant from dogs after acute lung injury, *J. Lipid Res.*, 26, 1338, 1985.
34. **Beppu, O. S., Clements, J. A., and Goerke, J.**, Phosphatidylglycerol-deficient lung surfactant has normal properties, *J. Appl. Physiol.*, 55, 496, 1983.
35. **Hallman, M., Enhorning, G., and Possmayer, F.**, Composition of surface activity of normal and phosphatidylglycerol deficient lung surfactant, *Pediatr. Res.*, 19, 286, 1985.
36. **King, R. J.**, The surfactant system of the lung, *Fed. Proc.*, 33, 2238, 1974.
37. **Katyal, S. L. and Singh, G.**, Analysis of pulmonary surfactant apoproteins by electrophoresis, *Biochim. Biophys. Acta*, 670, 323, 1981.

38. **Wright, J. R., Benson, B. J., Williams, M. C., Goerke, J., and Clements, J. A.,** Protein composition of rabbit alveolar surfactant subfractions, *Biochim. Biophys. Acta,* 791, 320, 1984.
39. **Hawgood, S., Benson, B. J., and Hamilton, R. L.,** Effects of surfactant-associated protein and calcium ions on the structure and surface activity of lung surfactant lipids, *Biochemistry,* 24, 184, 1985.
40. **Suzuki, Y.,** Effect of protein, cholesterol, and phosphatidylglycerol on the surface activity of the lipid-protein complex reconstituted from pig pulmonary surfactant, *J. Lipid Res.,* 23, 62, 1982.
41. **Claypool, W. D., Wang, D. L., Chander, A., and Fischer, A. B.,** An ethanolether soluble apoprotein from rat lung surfactant augments liposome uptake by isolated granular pneumocytes, *J. Clin. Invest.,* 74, 677, 1984.
42. **King, R. J., Martin, H., Mitts, D., and Holstrom, F. M.,** Metabolism of the apoproteins in pulmonary surfactant, *J. Appl. Physiol.,* 42, 483, 1977.
43. **King, R. J. and Macbeth, M. C.,** Physicochemical properties of dipalmitoylphosphatidylcholine after interaction with an apolipoprotein of pulmonary surfactant, *Biochim. Biophys. Acta,* 557, 86, 1979.
44. **Magoon, M. W., Wright, J. R., and Baritussio, A.,** Subfractionation of lung surfactant implications for metabolism and surface activity, *Biochim. Biophys. Acta,* 750, 18, 1983.
45. **Mason, R. J.,** Lipid metabolism, in *The Biochemical Basis of Pulmonary Functions,* Crystal, R. G., Ed., Marcel Dekker, New York, 1976, 127.
46. **Liau, D. F., Barrett, C. R., Bell, A. L. L., Hashim, S. A., and Ryan, S. F.,** Lysolecithin acyltransferase and alveolar phosphatidylcholine palmitate in experimental acute alveolar injury in the dog lung, *Biochim. Biophys. Acta,* 710, 76, 1982.
47. **Hallman, M., Kulovich, M., Kirkpatrick, E., Sugarman, G. R., and Gluck, L.,** Phosphatidylinositol and phosphatidylglycerol in amniotic fluid: indices of lung maturity, *Am. J. Obstet. Gynecol.,* 125, 613, 1976.
48. **Hallman, M. and Epstein, B. M.,** Role of myo-inositoll in the synthesis of phosphatidylglycerol and phosphatidylinositol in the lung, *Biochem. Biophys. Res. Commun.,* 92, 1151, 1980.

IMMUNE COMPLEX INJURY

Jeffrey S. Warren, Peter A. Ward, and Kent J. Johnson

INTRODUCTION

A large number of experimental models have been developed to investigate the putative role of immune complexes in the pathogenesis of tissue injury. Evidence for a pathogenic role for immune complexes in diseases such as rheumatoid arthritis, many glomerulonephritides, and collagen vascular diseases (e.g., systemic lupus erythematosus) is widely accepted. The role of immune complexes in the development of human lung disease has received comparatively less attention. Immune complexes have been implicated in the pathogenesis of diffuse lung injury that is sometimes associated with collagen vascular diseases including lupus erythematosus, scleroderma, rheumatoid arthritis, and dermatomyositis, as well as in the pulmonary vascular lesions observed in periarteritis nodosa and Wegener's granulomatosis. There is also evidence that immune complexes play a pathogenetic role in various types of hypersensitivity pneumonitis (also called extrinsic allergic alveolitis and allergic interstitial pneumonitis) and perhaps in some cases of idiopathic pulmonary fibrosis. Much of the clinical data which suggest that immune complexes are important in these diseases consists of morphologic studies demonstrating immune complexes in alveolar walls (by immunofluorescence and electron microscopy), serologic studies (detection of circulating immune complexes, precipitating antibodies, or complement-fixing antibodies), and positive intermediate (6 to 8 h) skin reactions (Arthus type) to intradermal injections of suspected antigens. It should be emphasized that there are significant gaps in knowledge and that controversy exists regarding the pathogenetic role of immune complexes in lung disease. For instance, circulating immune complexes, complement-fixing antibodies, and precipitating antibodies have been detected in the sera of exposed persons without lung disease. Similarly, there is evidence that type IV (cellular) immune responses may also be important in the evolution of some of these lung diseases. Clearly, many questions must be answered before a detailed understanding of immunologic lung disease is developed.

There is compelling experimental evidence that immune complexes which form in the alveolar septae can trigger an acute inflammatory reaction in the lung. Johnson and Ward have shown that formation of IgG immune complexes in rat alveolar septae results in an acute alveolitis characterized by intra-alveolar hemorrhage and fibrin deposition accompanied by an intense infiltration of neutrophils.[1] This response is initiated by the intratracheal instillation of heterologous antibody and the simultaneous intravascular infusion of antigen. Development of lung injury is both neutrophil- and complement-dependent and resembles the reversed passive Arthus reaction in skin. Depletion of neutrophils by specific antibody or depletion of complement through activation of the alternative pathway with cobra venom factor prevents subsequent development of lung damage. Activation of the classic complement pathway results in the generation of chemotactic peptides, such as C5a, which attract neutrophils from the circulation. Neutrophils, in turn, are activated by immune complexes, resulting in secretion of lysosomal enzymes and generation of superoxide anion ($O_2^{\cdot}$) and other oxygen-derived metabolites such as hydrogen peroxide (H_2O_2), hydroxyl radical ($OH\cdot$), and hypochlorous acid. A large number of *in vivo* and *in vitro* studies have implicated oxygen-derived metabolites and lysosomal enzymes in the pathogenesis of tissue injury.[2] It should be emphasized that the interactions among effector cells, target cells (and extracellular substrates), and various mediator substances (e.g., iron, arachidonate metabolites, and complement components) are complex and the focus of much current research. Suffice it to say that a variety of factors appear to modulate the central immune complex-complement-neutrophil mediated response.

More recently, Johnson et al. described an IgA immune complex lung injury model that has several features which are distinctive from IgG immune complex-mediated injury.[3] Induction of IgA immune complex lung injury in rats is technically nearly identical to induction of IgG immune complex lung injury, but in contrast to IgG lung injury, prior depletion of circulating neutrophils does not abrogate development of lung damage. It appears that in IgA immune complex lung injury, the lung damage is mediated by oxidants produced by lung macrophages rather than by neutrophils.[4] This is an important difference for at least two reasons. First, IgA-induced immunologic lung injury has possible relevance to a different set of diseases than does IgG immune complex lung injury. Second, and perhaps more importantly, IgA immune complex lung injury appears to represent an example of acute, macrophage-mediated tissue injury. Because of the intrinsic functional differences between macrophages and neutrophils, understanding the pathogenesis of macrophage-mediated tissue injury may provide different insights into mechanism of acute tissue injury.

Recognizing the basic differences between these models of immune complex-induced acute lung injury, the following sections of this chapter focus on techniques employed to implement the models, evaluation of the lung injuries, and related experimental findings. This is followed by a brief discussion of research issues relevant to these models of lung disease.

IMPLEMENTATION OF MODELS

The technical aspects of the IgG and IgA immune complex lung injury models are virtually identical.

Animals

Adult male Long Evans rats weighing approximately 350 g are used as the experimental subjects. The rats are maintained in pathogen-free, laminar flow rooms before use in experiments.

Antibody and Antigen Preparations

Rabbit IgG rich in antibody to BSA (Sigma Chemical Company, St. Louis) is prepared from hyperimmune serum by precipitation with 50% saturated ammonium sulfate followed by dialysis against phosphate buffered saline (0.005 *M*, pH 7.4) and passage through a diethylaminoethyl cellulose column. The amount of antibody (expressed as micrograms antibody N per milliliter) is determined by standard quantitative precipitation methods.[1]

In comparative studies between IgA and IgG immune complex lung injury, we have used murine monoclonal IgG and IgA prepared from ascites resulting from the murine plasmacytomas 9E9.1 and MOPC 315, respectively. (The murine plasmacytomas were provided by Litton Bionetics, Kensington, MD, operating under a National Cancer Institute contract NO1-CB-25584). Both antibodies specifically bind to dinitrophenol (DNP)-conjugated BSA. These antibodies are affinity purified and have been characterized by sodium dodecylsulfate-polyacrylamide electrophoresis and enzyme-linked immunosorbent assays as previously described.[3] The amounts of antibody (expressed as microgram antibody) are determined by quantitative precipitation reactions.[1]

To prepare DNP-BSA, 2,4-dinitrobenzenesulfonic acid (Eastman Laboratory and Specialty Chemicals, Eastman Kodak Company, Rochester, NY) is reacted with BSA overnight at room temperature according to the method of Eisen.[5] Unconjugated DNP is removed by exhaustive dialysis and the number of DNP groups per BSA molecule are calculated spectrophotometrically at 360 nm.

Animal Preparation and Antibody:Antigen Administration

Rats are anesthetized with intraperitoneal ketamine prior to initiation of the experimental

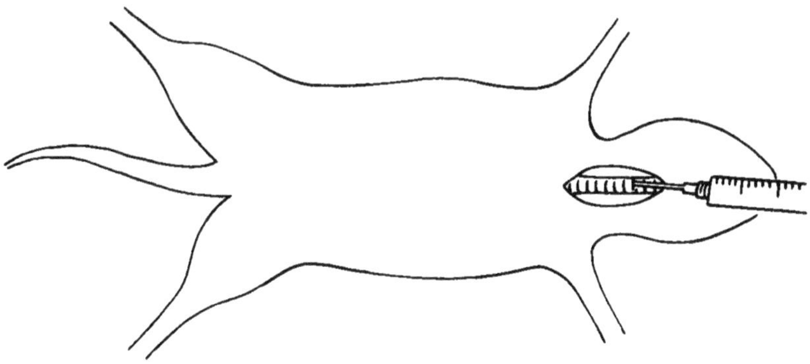

FIGURE 1. A horizontal nick is made between tracheal cartilage rings and a 12-gauge needle is gently inserted (see text).

procedure. After the rat becomes nearly flaccid (usually 5 to 10 min, it is positioned on a 20- × 30-cm board using elastic band restraints for each extremity and a similar restraint around the maxillary incisors in order to position the head and slightly extend the neck (Figure 1). The ventral surface of the neck (mandible to sternal notch) is wetted with 50 to 70% ethanol to prevent hairs from falling into the incision. A 3- to 4-cm midline vertical incision extending from just below the mandible to the sternum is made with a razor blade or scalpel. The ventral neck muscles and fasciae are gently teased to each side to expose the trachea, Care must be taken to avoid horizontal muscle tears because of excessive bleeding and difficult visualization of the trachea. Once exposed, the trachea can be gently held in place with forceps and a horizontal nick made between cartilage rings. The trachea should not be transected and the nick should be just large enough to allow insertion of a 12-gauge needle (Figure 1). In order to optimize the uniform distribution of antibody solution to both lungs, the mounting board should be placed on a 45° head-raised incline. A 27-gauge plastic cannula is threaded into the stationary tracheal needle, gently advanced to the bronchial bifurcation, and then withdrawn a couple of millimeters. The antibody solution is slowly administered (to avoid reflux) via a tuberculin syringe attached to the cannula. It is usually possible to ascertain proper instillation by observing a cough reflex and bubbling sounds. The administration of solution is followed with a 1-ml puff of air to insure intrapulmonary instillation of antibody. The authors have used 0.3 to 0.6 ml of antibody solution containing up to 800 μg monoclonal IgA or IgG and up to 400 μgN rabbit IgG antibody. The precise antibody dose administered will vary somewhat with the inflammatory potency of the particular batch and the experimental objectives. This can be determined through a limited dose-response experiment. Following instillation of antibody, the ventral neck incision is closed with nylon suture in a single layer.

The rat is then released from the mounting board and the antigen plus permeability indicator (^{125}I-labeled protein) is injected intravenously through the penile vein. The authors have generally used 3.3 mg DNP-BSA for IgA lung injury and 10 mg BSA for IgG lung injury. Again, it is important to optimize the lung injury by carrying out dose-response experiments. The antigen solution (1 mg) is spiked with small aliquots of ^{125}I-BSA or ^{125}I-IgG (800,000 cpm) which serve as permeability indicators.

EVALUATION OF MODELS

Quantitation of Lung Injury:Permeability Index

Determination of permeability index (PI) has provided a reproducible, quantitative method for assessing lung injury. After 3 to 4 h, rats are anesthetized with ketamine, precisely 1 ml of caval blood is obtained with a tuberculin syringe, and the inferior vena cava is transected, resulting in exsanguination. Immediately following caval transection, the thoracic cavity is opened and the still-beating heart and lungs are removed *en bloc*. The pulmonary vasculature is flushed free of blood with saline (10 mg) injected via the right ventricle or pulmonary artery. Care must be taken to avoid pulmonary artery rupture and to assure that the lungs blanch, thereby indicating a vascular washout. It should be noted that injured lungs may have a hemorrhagic mottled appearance but that gross correlation with degree of injury is crude. The lungs are then removed from the heart. The PI is derived from the ratio of total lung ^{125}I cpm divided by ^{125}I cpm present in 1 ml of blood. Transection of the vena cava is important because it minimizes the pooling of blood that otherwise occurs during *en bloc* removal of the heart and lungs. It is important to exercise care during removal of the heart and lungs to avoid accidental laceration of the pulmonary vasculature, since this makes it difficult or impossible to flush the vascular space free of ^{125}I-containing blood. Collection of 1 ml of blood from each animal serves to normalize PI values from rat to rat since slight technical variation in ^{125}I indicator injection may occur. The data derived from this method compare favorably with lung weight determinations.

Morphologic Assessment:Transmission Electron Microscopy

Ultrastructural examinations of injured rat lungs have provided useful information in delineating the pathogenesis of IgA and IgG immune complex-induced lung injury. Transmission electron micrographs (TEM) are prepared from rat lungs which have been injured as described above (except that ^{125}I is omitted). At the time of sacrifice, the *en bloc* heart-lung preparation is gently inflated with 4% glutaraldehyde in cacodylate buffer (0.1 *M*, pH 7.3) via the trachea which is then ligated. The preparation is then submerged in a beaker of glutaraldehyde and allowed to fix for 2 to 3 h. After adequate fixation, 1-mm^3 sections of lung parenchyma from representative areas of the lungs are processed for TEM.

Interventional Studies

A variety of interventional studies have provided important insights into the pathogenesis of these immune complex lung injury models. Neutrophil-depletion studies have been carried out using rabbit anti-rat PMN serum to reduce the number of circulating neutrophils to less than 400 per cubic millimeter.[6] In these studies, anti-PMN serum (0.5 to 1.0 ml, depending on antiserum potency) is injected intravenously 1 d prior to use of the rat in the experiment. Peripheral blood is examined before the rats are used in order to assure profound neutropenia and that there are no changes in the other blood cell compartments.

Complement-depletion studies have been carried out using cobra venom factor derived from *Naja naja* cobra venom as described by Cochrane et al.[7]

The potential roles of oxygen-derived metabolites have been examined using specific antioxidant enzymes, free radical scavengers, and metal chelators.[4,8] Proteolytic enzyme antagonists have been utilized to help elucidate potential contributions to the development of immune complex lung injury by lysosomal enzymes. Superoxide dismutase (SOD) (Data Diagnostics, Inc., Mountainview, CA) and catalase (Sigma Chemical Company, St. Louis) specifically degrade $O_2^{\cdot}$ and H_2O_2, respectively. In studies where inactivated SOD or catalase are used as controls, the enzymes are inactivated by reduction and alkylation as previously described.[8] In order to assess the role of OH·, several reagents have been used, including the iron chelator desferoxamine (Ciba-Geigy Corporation, Summit, NJ), iron-binding apoproteins such as apolactoferrin, and OH· scavengers such as dimethylsulfoxide (DMSO) and dimethylthiourea.[9]

Studies of Cells Obtained by Bronchoalveolar Lavage

In vitro examination of pulmonary inflammatory cell functions (e.g., oxidant generation, enzyme secretion, phagocytosis, mediator production, and biochemical studies) can be carried out by harvesting cells via bronchoalveolar lavage carried out either before or during the ongoing development of lung injury.[5,10]

EXPERIMENTAL FINDINGS

This section briefly summarizes the chief morphological and pathogenetic differences between these models of IgG and IgA immune complex-induced lung injury.

As shown in Figure 2, there is a marked contrast in the predominant type of inflammatory cell present in IgA immune complex-induced lung injury vs. IgG immune complex-induced lung injury. Despite the potential difficulties which exist in deriving quantitative conclusions from bronchoalveolar lavage differential cell counts, such studies strongly corroborate the morphologic differences observed in Figure 2. Neutrophils normally constitute less than 5% of the total (less than 10^7) cells which can be obtained by bronchoalveolar lavage of a 350 g rat which has been maintained in a pathogen-free environment.[10] In contrast, $3^1/_2$ h after initiation of IgG-BSA immune complex-induced lung injury, more than 3×10^7 cells, largely neutrophils, can be obtained by lavage. Similar studies carried out in IgA immune complex lung injury have shown increased yields of macrophages from 6.9×10^6 to 9.6×10^6 per rat.[4] In addition, there is a nearly fourfold increase in nonstimulated $O_2^{\cdot}$ production by alveolar macrophages obtained from rat lungs after development of IgA immune complex lung injury vs. macrophages obtained from rats given only intratracheal IgA. Less than 3% of the cells obtained by bronchoalveolar lavage of rats after IgA-induced injury are neutrophils.

The results of neutrophil- and complement-depletion experiments and studies employing antioxidant compounds, oxygen radical scavenger, and iron chelators are summarized in Table 1. Neutrophil-depletion studies in rats injured with IgG immune complexes suggest that the predominant effector cell in this model is the neutrophil since there is a 60 to 80% reduction in the degree of lung injury that develops in these animals.[4] In contrast, there is virtually no protection afforded to rats with IgA immune complex lung injury following neutrophil depletion. *In vitro* studies indicate that preformed IgA immune complexes selectively activate alveolar macrophages but not neutrophils (unpublished data).

Complement-depletion studies indicate that while IgG and IgA immune complex lung injuries appear to be primarily mediated by different effector cells, both models require an intact complement system for full development of lung injury to occur. Complement depletion prior to induction of IgG immune complex lung injury results in a greater than 80% reduction in lung injury.[1,4] Similar complement-depletion procedures in the IgA immune complex lung injury model result in a 50% reduction in subsequent lung injury.[4]

As shown in Table 1, interventional studies using specific antioxidant substances and transition metal chelators have provided data which suggest that both IgG and IgA immune complex lung injuries are mediated by oxygen-derived metabolites.[4,8,11] Treatment of rats with parenteral SOD before induction of IgG immune complex induces lung injury results in suppression of lung injury at 2 h.[11] By 4 h, however, there is full development of injury suggesting that $O_2^{\cdot}$ has a transient effect in the development of IgG immune complex lung injury. Concommitant morphologic studies indicate that there is little neutrophil influx into the pulmonary parenchyma at 2 h, but that by 4 h the lungs cannot be distinguished from lungs of rats not treated with SOD. These observations are consistent with the *in vitro* studies which have demonstrated the generation of arachidonate-derived lipids which are chemotactic for neutrophils.[12,13] The chemotactic lipid results from incubation of plasma with an $O_2^{\cdot}$ generating system (xanthine/xanthine oxidase). Intratracheal catalase largely suppresses the development of IgG immune complex-induced lung injury in the rat.[8] SOD and catalase provide substantial, dose-dependent protection against development of acute lung injury in

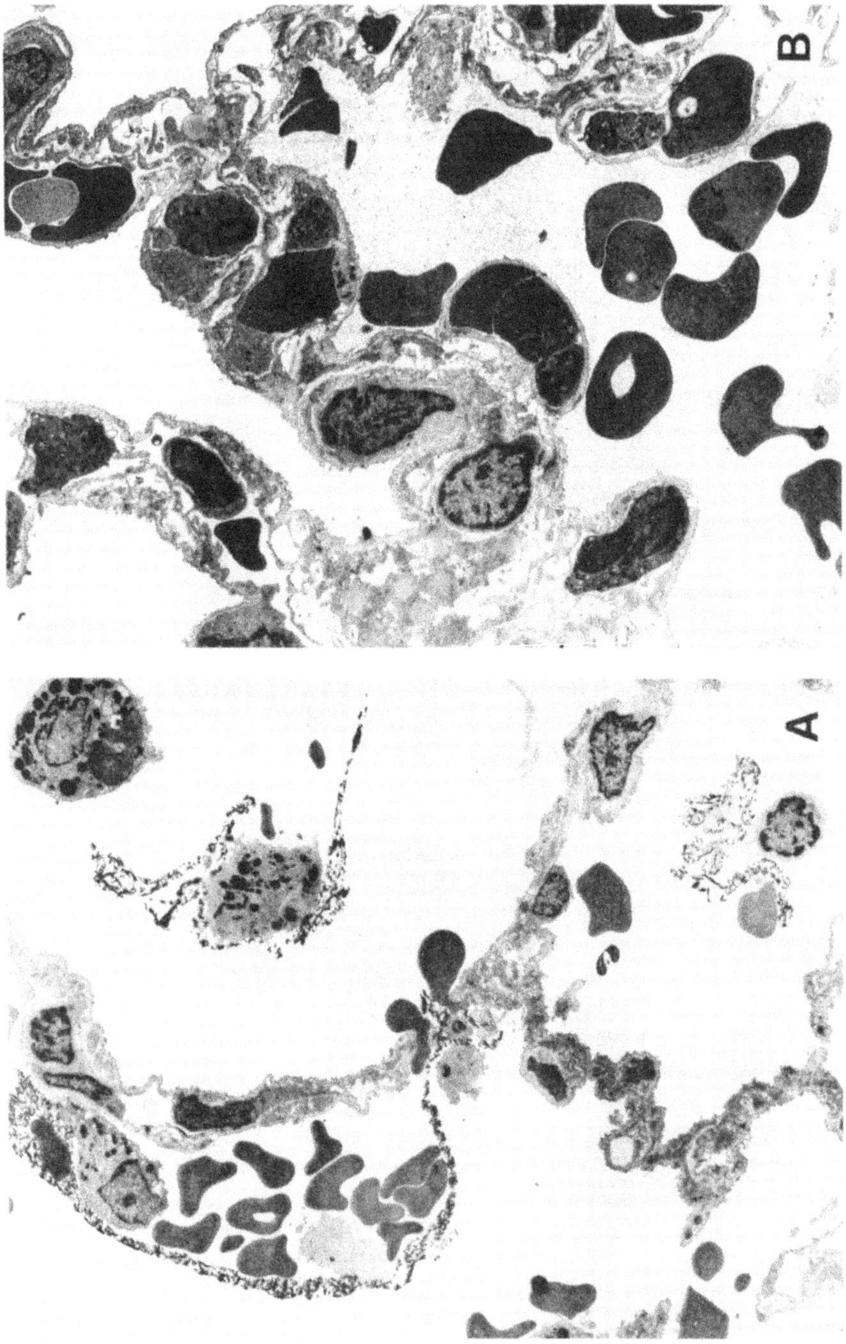

FIGURE 2. (A) In IgA immune complex-induced lung injury there are increased numbers of macrophages accompanied by intra-alveolar hemorrhage and fibrin deposition. (Magnification × 1750.) (B) In IgG immune complex-induced lung injury there is a large influx of neutrophils accompanied by intra-alveolar hemorrhage. (Magnification × 2850.) (From Johnson, K. J. et al., *J. Clin. Invest.*, 74, 358, 1984. With permission.)

Table 1
COMPARISON OF IgG AND IgA IMMUNE COMPLEX LUNG INJURY

	IgG immune complex	IgA immune complex
Predominant effector cell	Neutrophil*[a]	Macrophage
Complement dependence	>80%	~50%
Suppression with		
SOD	Early**[b]	+
Catalase	+	+
Desferoxamine	ND	+
DMSO	ND	+
Soybean trypsin inhibitor	—	ND
Alpha-1-antitrypsin inhibitor	—	ND

Note: ND = not done.

[a] Neutrophil-depletion of IgG immune complex-injured rats significantly, but not completely, abolishes increases in PI, suggesting that other effector (macrophages?) mechanisms may also contribute. *In vitro* studies[10] indicate that preformed IgG immune complexes can directly activate alveolar macrophages.

[b] The protective effect of SOD in IgG immune complex-mediated lung injury diminishes with delay in administration of SOD and with prolongation of the injury development period.

the IgA immune complex lung model.[4] In the IgA model, these studies have been extended by data which show protection against lung injury in rats treated with the iron chelator desferoxamine and the OH· scavenger DMSO.

DISCUSSION

There is hope that the foregoing models of immune complex-induced acute lung injury will help provide insights into the pathogenesis of acute tissue injury. Not only may these distinct models of acute immune complex lung injury have relevance to different groups of human disease, they will hopefully contribute to understanding of basic differences that undoubtedly exist between neutrophil-mediated and macrophage-mediated inflammatory processes.

Clearly, many questions regarding the pathogenesis of immune complex-mediated lung disease remain to be answered. For instance, what information derived from these experimental models will prove to be applicable to human disease? Which human lung diseases most closely mirror these IgG and IgA immune complex-induced acute lung injury models? What are the roles of nonoxidant inflammatory mediators and how might these substances modulate oxidant mechanisms of tissue injury? What processes or mechanisms are unique to the lungs and which can be generalized to other organ systems? It seems certain that the development and understanding of relevant animal models of lung disease will continue to provide insights into the mechanisms of human lung disease.

REFERENCES

1. **Johnson, K. J. and Ward, P. A.,** Acute immunologic pulmonary alveolitis, *J. Clin. Invest.*, 54, 349, 1974.
2. **Fantone, J. C. and Ward, P. A.,** Role of oxygen-derived free radicals and metabolites in leukocyte-dependent inflammatory reactions, *Am. J. Pathol.*, 107, 397, 1982.
3. **Johnson, K. J., Wilson, B. S., Till, G. O., and Ward, P. A.,** Acute lung injury in rat caused by immunoglobulin A immune complexes, *J. Clin. Invest.*, 74, 358, 1984.
4. **Johnson, K. J., Ward, P. A., Kunkel, R. G., and Wilson, B. S.,** Mediation of IgA induced lung injury in the rat. Role of macrophages and reactive oxygen products, *Lab. Invest.*, 54, 499, 1986.
5. **Eisen, H. N.,** Preparation of purified anti-2,4-dinitrophenol antibodies, *Methods Med. Res.*, 10, 94, 1964.
6. **Ward, P. A. and Cochrane, C. G.,** Bound complement and immunologic injury of blood vessels, *J. Exp. Med.*, 121, 215, 1965.
7. **Cochrane, C. G., Muller-Eberhard, H. J., and Aidin, B. S.,** Depletion of plasma complement in vivo by protein of cobra venom: its effect on various immunologic reactions, *J. Immunol.*, 105, 55, 1970.
8. **Johnson, K. J. and Ward, P. A.,** Role of oxygen metabolites in immune complex injury of lung, *J. Immunol.*, 126, 2365, 1981.
9. **Ward, P. A., Till, G. O., Kunkel, R. G., and Beauchamp, C.,** Evidence for role of hydroxyl radical in complement and neutrophil-dependent tissue injury, *J. Clin. Invest.*, 72, 780, 1983.
10. **Ward, P. A., Duque, R. E., Sulavik, M. C., and Johnson, K. J.,** *In vitro* and *in vivo* stimulation of rat neutrophils and alveolar macrophages by immune complexes: production of $O_2^{\overline{\cdot}}$ and H_2O_2, *Am. J. Pathol.*, 110, 297, 1983.
11. **McCormick, J. R., Harkin, M. M., Johnson, K. J., and Ward, P. A.,** Suppression by superoxide dismutase of immune complex-induced pulmonary alveolitis and dermal inflammation, *Am. J. Pathol.*, 102, 55, 1981.
12. **Petrone, W. F., English, D. K., Wong, K., and McCord, J. M.,** Free radicals and inflammation: the superoxide dependent activation of a neutrophil chemotactic factor in plasmas, *Proc. Natl. Acad. Sci. U.S.A.*, 77, 1159, 1980.
13. **Perez, H. D. and Goldstein, I. M.,** Generation of a chemotactic lipid from arachidonic acid by exposure to a superoxide generating system, *Fed. Proc.*, 37, 1170, 1980.

Interstitial Fibrosis

BLEOMYCIN-INDUCED PULMONARY FIBROSIS

Jerome O. Cantor

INTRODUCTION

Pulmonary fibrosis is characterized by a progressive increase in the connective tissue content of lung parenchyma.[1] The development of animal models to study the disease has resulted in new insight into its pathogenesis. While these experimental models do not completely mimic human pulmonary fibrosis, they nevertheless provide a means of studying the morphological and biochemical events of the disease under controlled, reproducible conditions.

The experimental model of pulmonary fibrosis described in this chapter is induced by a single intratracheal injection of bleomycin into the lungs of hamsters. The drug is a mixture of glycopeptides derived from a strain of *Streptomyces verticillus* and is used in the treatment of a variety of neoplasms. The major cytotoxic action of bleomycin is inhibition of DNA synthesis.[2] However, it may also cause cellular damage by inducing the formation of oxygen radicals.[3]

Pneumonitis, progressing in some instances to fibrosis, occurs in approximately 10% of patients receiving the drug.[4] The lung is especially susceptible to the cytotoxic effects of bleomycin because of the relatively long half-life of the drug in that organ. In particular, bleomycin is retained by alveolar epithelial cells.[5]

Based on the clinical observation of pulmonary toxicity, a number of investigators have utilized bleomycin to produce pulmonary fibrosis in a variety of animal species.[5-7] Multiple intravenous or intraperitoneal injections of the drug induce the formation of lung lesions which morphologically resemble human interstitial pumonary fibrosis. More recently, it was shown that a single intratracheal injection of bleomycin can cause a similar lung reaction.[8]

There are several advantages in using the intratracheal model. Compared to the intravenous and intraperitoneal modes of delivery, intratracheal instillation of bleomycin produces lung lesions more rapidly and requires a much lower dose of the agent. It also permits better characterization of the sequence of cellular events leading to pulmonary fibrosis since the disease arises from a single insult. These features have resulted in widespread use of this experimental model in lung research.

IMPLEMENTING THE MODEL

Animal Species

Syrian Golden hamsters have been used exclusively in studies at the College of Physicians and Surgeons, although a number of other species may be substituted, including rats, mice, and dogs. Smaller animals require less bleomycin and are therefore more economical. However, they provide the investigator with a smaller volume of tissue for study. For biochemical analysis of less abundant lung components, the availability of a large amount of starting material may outweigh cost considerations.

Hamsters are particularly well-suited for these types of lung studies because they rarely develop spontaneous pneumonia and recover rapidly from the surgical procedures involved in inducing pulmonary fibrosis. In general, animals (male or female) weighing approximately 120 g are used. Hamsters in this weight range may be obtained from animal suppliers such as Charles River (North Wilmington, MA).

Handling of Animals

Although usually docile, hamsters may on occasion become sufficiently aroused to inflict a rather deep and painful bite. To prevent this, the animal should always be grasped from

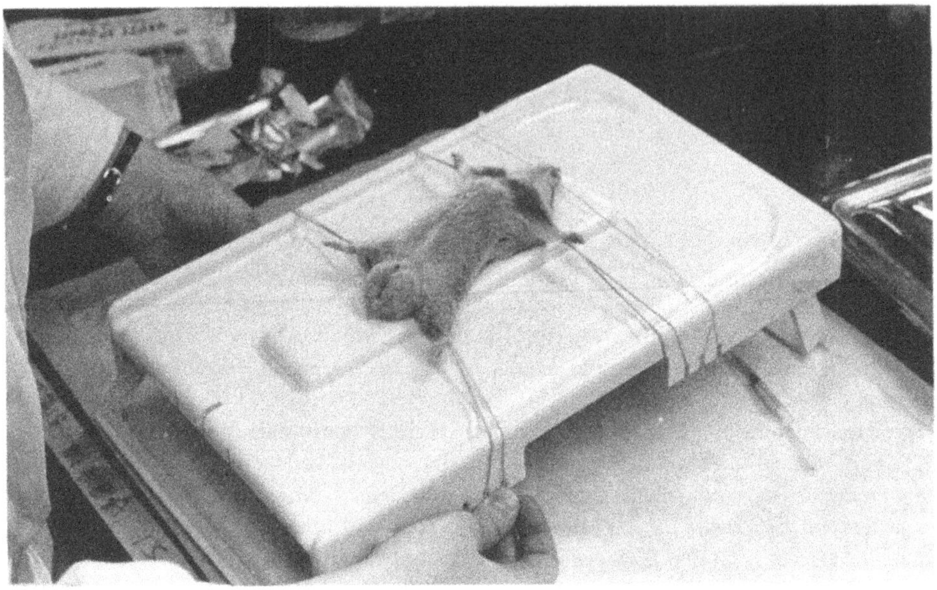

FIGURE 1. The hamster is placed on the surgical table in a supine position. The limbs and head are restrained with elastic bands.[28]

behind, using the thumb and forefinger to pull the neck skin back firmly, thereby maintaining the head in the forward position. The remaining fingers can then hold the loose fur of the back to establish control of the animal. A firm grip will discourage the animal from struggling. It is advisable to practice this technique with gloves before handling the animals with bare hands.

Surgical Procedures

Anesthesia

It is preferable to induce only mild anesthesia in the animal so that strong respiratory drive is maintained. This insures that the bleomycin is deeply inhaled and reduces the chance of cardiopulmonary arrest during the operative procedure. Suitable anesthesia may be induced by intramuscular injection of 0.05 to 0.1 ml of ketamine hydrochloride (Ketaset; Bristol Laboratories, Syracuse, NY) into the hind limb. The animal should be sufficiently anesthetized for surgery after 5 min.

Tracheostomy

Following induction of anesthesia, the animal is placed in a supine position on a small plastic surgical table (9 x 15 in.; Harvard Bioscience, South Natick, MA). Elastic bands are used to restrain the limbs and an additional band is hooked around the upper teeth to prevent head movement (Figure 1).

The fur overlying the trachea is then swabbed with alcohol and shaved with a razor or scalpel. The shaved area is stretched taut using the thumb and forefinger and a 2-cm vertical incision is made with the scalpel, extending through the skin and subcutaneous fat (Figure 2). The underlying muscle is then lifted with forceps and cut crosswise with scissors to expose the trachea (Figure 3), The scissors should be held parallel to the body of the animal when making the incision through the muscle to avoid cutting the trachea.

Instillation of Bleomycin

Bleomycin (Blenoxane; Bristol Laboratories, Syracuse, NY) is supplied by the manufacturer in ampules containing 15 mg (15 U) of lyophilized material. After dissolving the

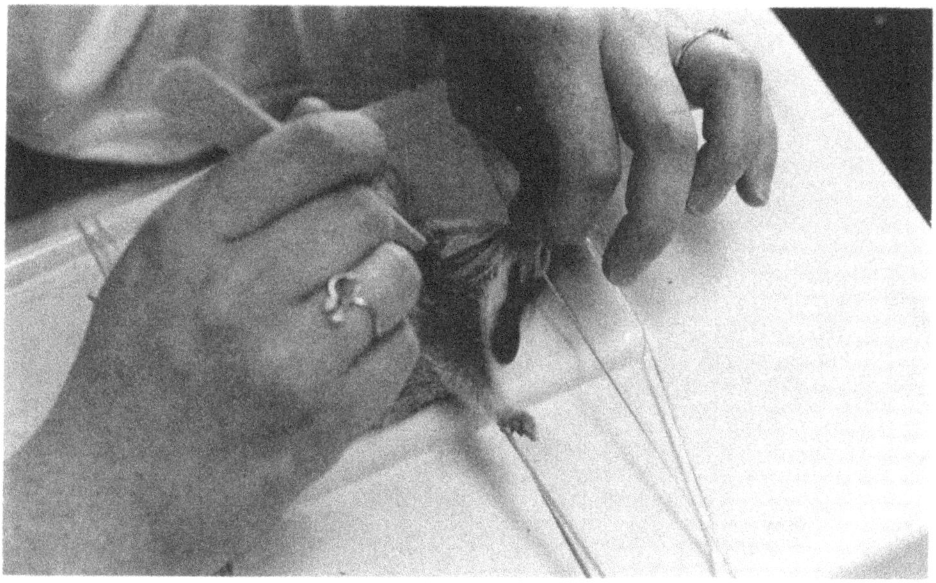

FIGURE 2. The neck skin is stretched taut with the thumb and forefinger and an incision is made with the scalpel to expose the underlying muscle.[28]

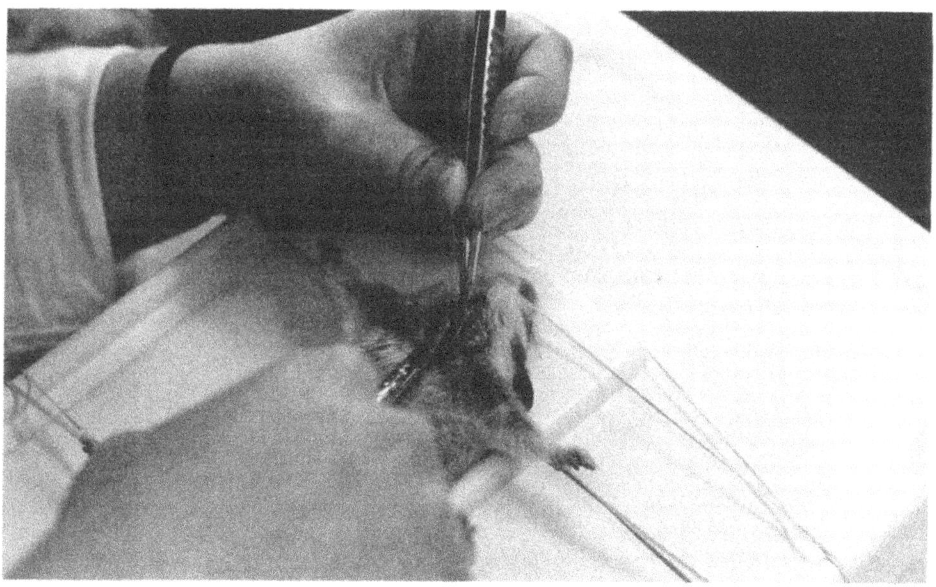

FIGURE 3. The muscle overlying the trachea is lifted with forceps and cut crosswise, holding the scissors parallel to the body of the animal to avoid severing the airway.[28]

bleomycin by addition of 3 ml of sterile saline, a portion is transferred to a 1-ml syringe capped with a 26- or 27-gauge needle. To facilitate insertion into the trachea, the needle is bent at a right angle to the syringe. It is placed inside the trachea between the rings of cartilage and gently lowered into the lumen to avoid snagging the mucosa (Figure 4). A total of 0.2 ml (1 mg) of bleomycin is delivered into the trachea in small increments, permitting the inspirations of the animal to draw the fluid into the lungs. If the animal begins

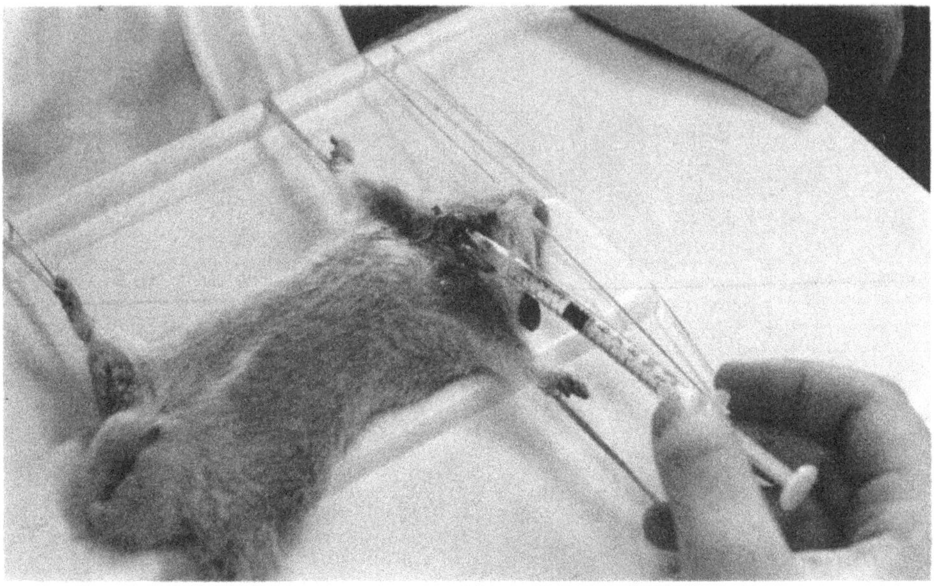

FIGURE 4. Bleomycin is delivered to the lungs via a 26- to 27-gauge needle attached to a 1-ml syringe. The needle is bent at a right angle to the syringe to facilitate insertion into the trachea. A total of 0.2 ml of saline, containing 1 mg of bleomycin, is delivered in small increments, permitting the inspirations of the animal to draw the fluid into the lungs.[28]

to hold its breath, instillation of the bleomycin should be temporarily halted until respiration is reinitiated.

When delivery of the bleomycin is completed, the needle is withdrawn from the trachea and the skin incision is closed with several sutures or staples. The animal should recover from surgery within 1 h. Only a very low level of mortality is associated with this procedure.

Controls

As with the bleomycin-treated animals, tracheostomies are performed and 0.2 ml of sterile saline is instilled into the lungs.

EVALUATING THE MODEL

Preparation of Lungs for Histological Examination

Following sacrifice of the animal by intraperitoneal injection of pentabarbital, the lungs are fixed *in situ* by inserting a catheter into the trachea and allowing 10% neutral-buffered formalin to inflate the tissue at a pressure of 20 cm H_2O. The apparatus used for fixation consists of a 50-ml syringe attached to a length of small-bore tubing, the distal end of which is connected to a catheter that fits snugly into the trachea (Figure 5). Formalin is added to the syringe to a height of 20 cm above the animal and allowed to drain through the catheter before being placed in the trachea. After 1 to 2 h, the catheter is removed and the trachea is tied off. It is advisable to have a knotted string around the trachea as the catheter is withdrawn to minimize deflation of the lungs. Both the lungs and heart are then removed from the thorax as a single block and fixed for several days in 10% neutral-buffered formalin.

After fixation, the lungs are dissected from the heart and extraparenchymal structures such as the trachea, mainstem bronchi, and major vasculature are removed. The lobes of the lungs are separated and one or more sagittal cuts are made through each. These pieces are then dehydrated in alcohol, embedded in paraffin, and 5-μm sections are cut on a

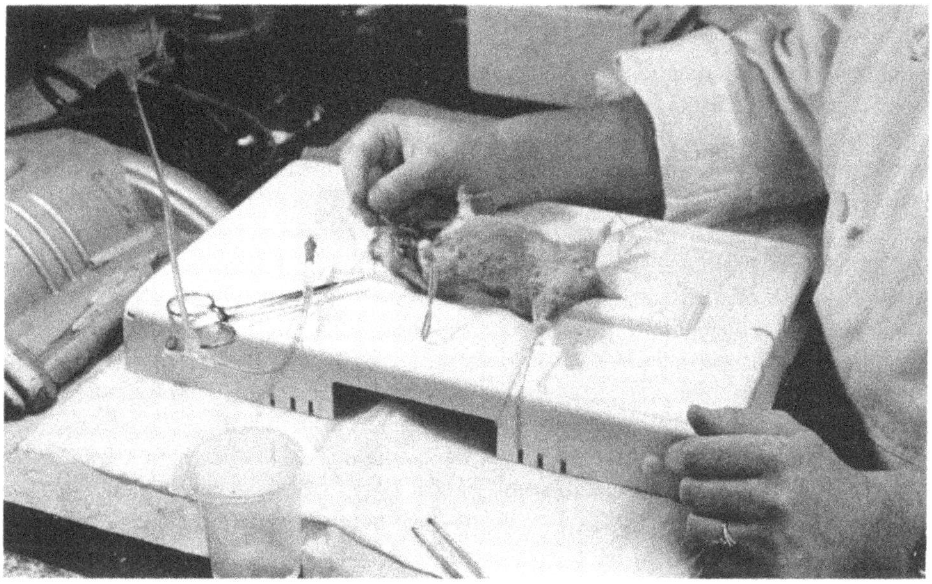

FIGURE 5. Fixation of the lungs is performed *in situ*. Ten percent neutral-buffered formalin is instilled via the trachea at a pressure of 20 cm H_2O.[28]

standard rotary microtome and mounted on glass slides. In those cases where thinner, more-detailed sections are required, tissues should be embedded in methacrylate rather than paraffin.

With regard to histological stains, hematoxylin-eosin or hematoxylin-phloxine-safran are useful for general examination of cellular relationships. Visualization of connective tissue components may be obtained with the trichrome (collagen), Verhoeff (elastin), and alcian blue (glycosaminoglycan) stains. Stepwise procedures for applying these and other stains are found in the *Manual of Histological Staining Techniques of the Armed Forces Institute of Pathology*.[8a]

Estimation of Lung Disease

Semiquantitative measurement of lung lesions may be performed with a light microscope at magnification × 40, using an eyepiece grid composed of 36 equal-sized squares. Slide sections are scanned under the grid and the number of squares containing disease (see morphological description in "Experimental Findings" below) is divided by the total number studied to yield a percentage termed the "disease index."[8] In order to obtain a reasonable estimate of lung injury, at least 1000 squares should be examined.

Bronchoalveolar Lavage

The inflammatory response induced by bleomycin may be monitored by determining the types of cells present in lung lavage fluid. Following sacrifice, tracheostomies are performed as described in "Implementing the Model — Surgical Procedures" above. A total of 25 ml of saline is instilled into the lungs in 5-ml aliquots via a tracheal catheter, using a pair of syringes connected to a three-way stopcock to alternately add and gently withdraw fluid from the lungs.

Total leukocyte counts are obtained with a standard hemocytometer. Differential counts are performed on cytocentrifuge smears treated with Wright-Giemsa stain. For each specimen, 500 cells are counted.

Biochemical Studies

Collagen Synthesis

Hamsters are given injections of ^{3}H-proline (New England Nuclear, Boston) for 4 consecutive days to obtain adequate labeling of newly synthesized collagen (1 mCi/kg/d). Animals are sacrificed 24 h after the final injection of ^{3}H-proline. The lungs are resected from the thorax and freed of extraparenchymal structures, including the trachea, mainstem bronchi, and major vasculature. Tissues are then cut into small pieces, washed twice in saline, and homogenized. The homogenates are dialyzed against water to remove remaining unincorporated ^{3}H-proline. Following dialysis, the samples are lyophilized.

Aliquots of the dried tissue, 5 to 10 mg in size, are digested twice with 50 U/mg of purified bacterial collagenase (Form III, Advance Biofactures, Lynbrook, NY) overnight at 37°C in 0.01 *M* Tris buffer, pH 7.4, supplemented with 10 m*M* $CaCl_2$, 10 m*M* KCl, and 1.5 m*M* $MgCl_2$. To inhibit the activity of any contaminating proteases, 0.001 *M* N-ethylmaleimide is added to the buffer solution. The supernatants from each digestion are combined and precipitated with 10% trichloroacetic acid overnight at 4°C. Following centrifugation, aliquots of the supernatants are measured for radioactivity in a liquid scintillation spectrometer. Collagen synthesis is expressed as counts per minute per milligram dry lung or as counts per minute per lung.

Measurement of the pool-size of ^{3}H-proline in the lung is required to insure that differences in labeling of experimental and control groups are not artifactual.[9] In addition, hydroxyproline analysis of the undigested lung tissue is useful in determining the efficacy of collagenase treatment.[10]

Synthesis of Other Connective Tissue Components

Procedures for measuring the synthesis of elastin and glycosaminoglycans have been applied to the bleomycin model and are described in detail in the literature.[11,12]

Explant Studies

Explant cultures provide an alternative to *in vivo* studies when tighter control of experimental conditions is needed. All procedures involved in establishing the cultures are performed under sterile conditions. Following sacrifice of the animal, the chest fur is swabbed with alcohol and a thoracotomy is performed. The lungs are dissected out and washed in phosphate-buffered saline to remove blood. Portions of the tissue are transferred to fresh solution and gently cut into 1- to 2-mm^2 pieces with a scalpel. These fragments are placed on 1-cm squares of filter paper (three to four per square) which are mounted on wire rafts and placed in culture dishes. Medium (e.g, F-12 supplemented with 10% fetal calf serum and antibiotics) is added to the dishes to the level of the filter paper. The cultures are incubated at 37°C in an atmosphere of 5% CO_2 and 95% air. Tissue viability can be maintained under these conditions for at least 72 h.

EXPERIMENTAL FINDINGS

Morphological

Instillation of bleomycin results in a rapid sequence of morphological changes in the lung. An acute alveolitis is seen by 24 h. Lavage fluid obtained at this time reveals a marked increase in the percentage of neutrophils (42% of bleomycin-treated lungs vs. 5% for controls).[13] Over the next several days, the interstitium becomes thickened due to the combination of fibrinous edema, mononuclear cell influx, and alveolar epithelial hyperplasia (Figure 6). Labeling of cell nuclei with ^{3}H-thymidine indicates that proliferation of epithelial cells reaches a maximum at 8 d following administration of bleomycin and quickly declines thereafter (Figure 7).[14] Termination of the acute tissue reaction is further demonstrated by a return to normal in the neutrophil content of lavage fluid after 2 weeks.[13]

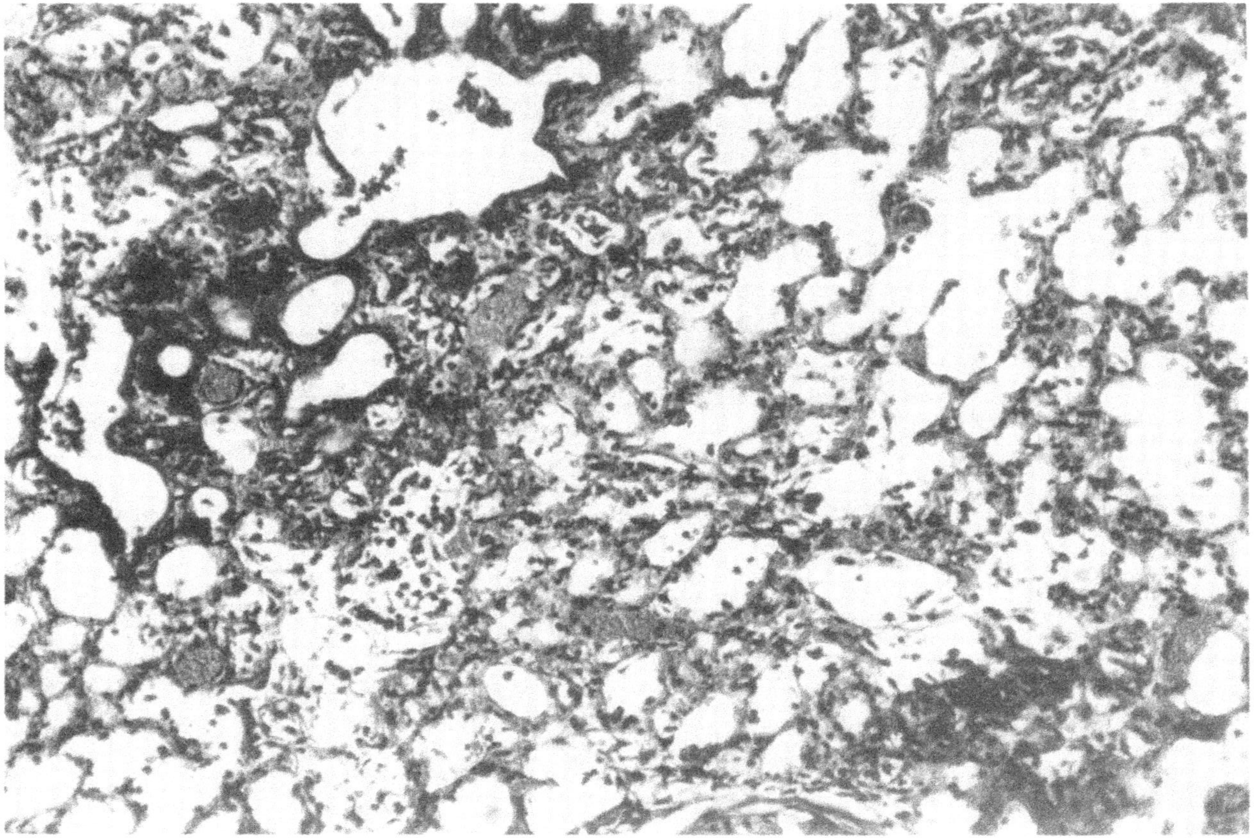

FIGURE 6. Photomicrograph of a hamster lung, 5 d after intratracheal instillation of bleomycin. Interstitial thickening is due to fibrinous edema, inflammatory cell influx, and epithelial hyperplasia. Cellular exudates are present in the alveolar spaces. (Hematoxylin and eosin; magnification × 270.)

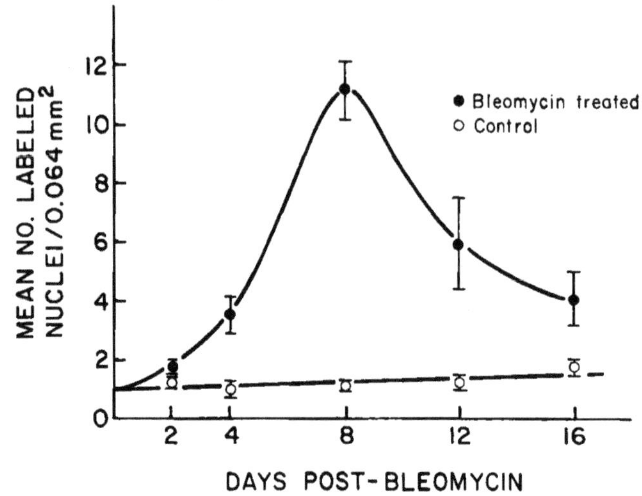

FIGURE 7. Graph depicting alveolar epithelial hyperplasia, as measured by incorporation of ^{3}H-thymidine. Proliferation reaches a peak at 8 d and declines rapidly during the following week.

Fibrosis is readily discernible in the interstitium by 14 d and increases in amount over the next several weeks (Figure 8). Those portions of the lung most damaged by bleomycin show markedly dilated terminal bronchioles and alveolar ducts, both of which may become lined by cuboidal epithelium. In general, the disease process is patchy with normal parenchyma separating areas of interstitial injury.

Biochemical

Pulmonary fibrosis is characterized by profound changes in the connective tissue matrix of the lung. There are significant increases in a number of different matrix constituents, including collagen,[9] elastin,[11] and glycosaminoglycans,[12] Most of these alterations occur during the first month following intratracheal instillation of bleomycin (Figure 9).

Glycosaminoglycans are one of the earliest matrix components to respond to the bleomycin insult, undergoing marked proliferation during the first week (Figure 10).[12] Collagen and elastin synthesis, in contrast, is maximal between 1 and 3 weeks following induction of injury (Figure 11).[9,11] Coinciding with the increase in collagen and elastin production is a proportionate rise in synthesis of the dermatan sulfate glycosaminoglycan (Figure 12).[12] Since similar increases in dermatan sulfate have been observed in other experimental models of fibrosis, it has been suggested that this glycosaminoglycan influences the organization of collagen and/or elastin in the extracellular matrix.[15]

To determine the importance of connective tissue deposition in the pathogenesis of pulmonary fibrosis, chemical agents have been used to block the synthesis of one or more matrix constituents. In one study, a proline analogue, L-3-4-dehydroproline, reduced pulmonary collagen content in rats when given daily between 15 and 28 d after intratracheal instillation of bleomycin.[16] Attenuation of the fibrotic response resulted in an increase in both lung compliance and vital capacity.

The cross-link inhibitor penicillamine has similarly been used to limit bleomycin-induced pulmonary fibrosis.[17] Deposition of both collagen and elastin was decreased following daily administration of the drug. Although penicillamine does not directly affect glycosaminoglycans, a reduction in this component was also observed, suggesting that its synthesis may be linked to that of collagen and/or elastin.

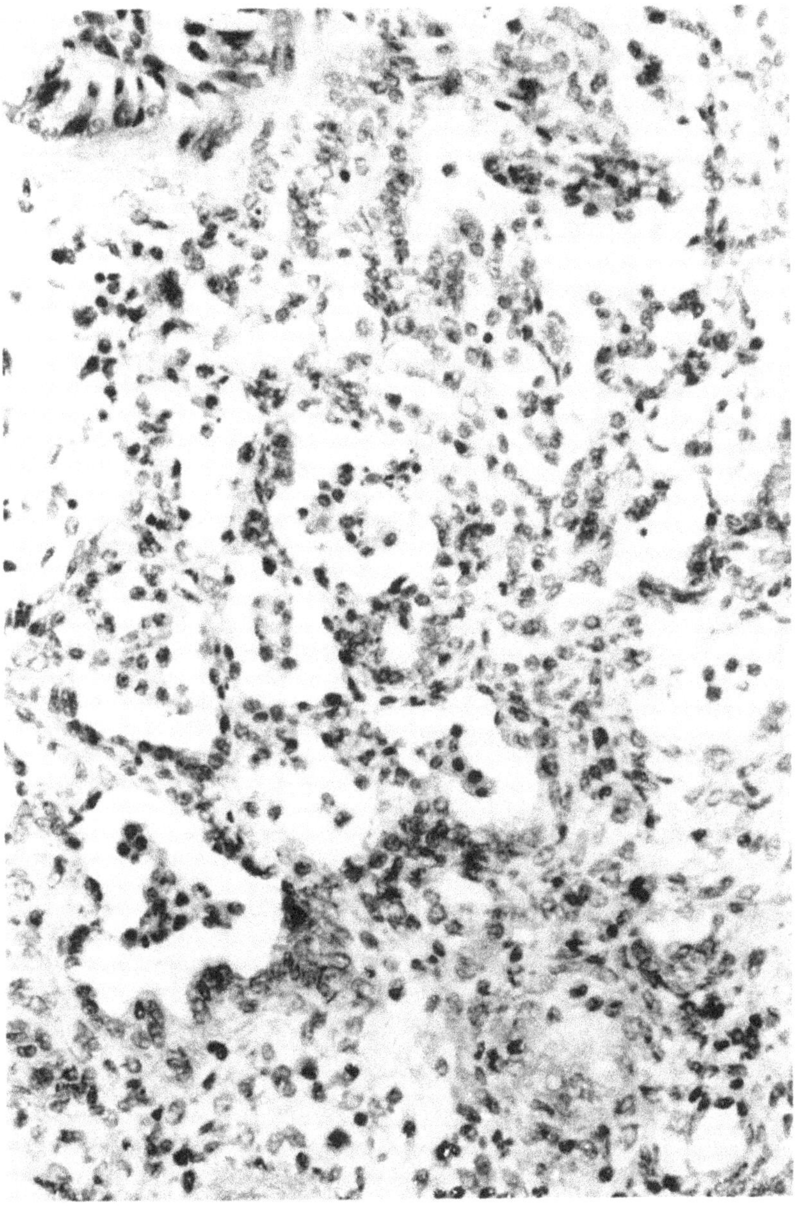

FIGURE 8. Photomicrograph of a hamster lung, 15 d after intratracheal instillation of bleomycin. The markedly thickened interstitium contains a mixture of fibrous tissue, hyperplastic epithelium, and mononuclear inflammatory cells. (Hematoxylin and eosin; magnification × 292.)

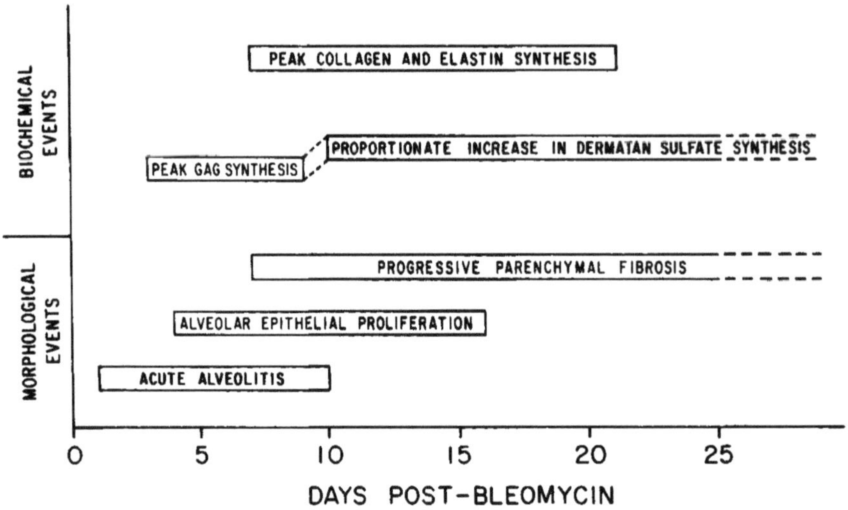

FIGURE 9. Graph indicating the sequence of morphological and biochemical events following intratracheal instillation of bleomycin.[28]

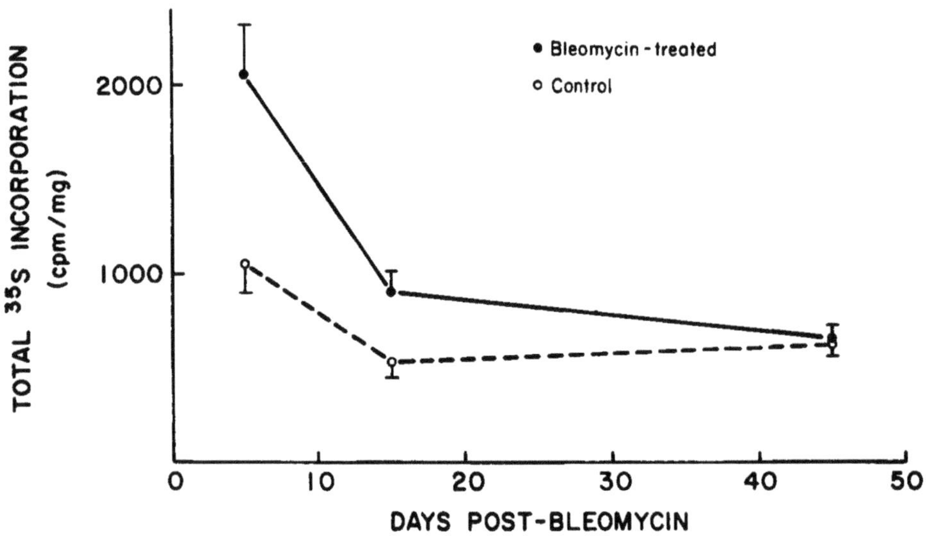

FIGURE 10. Graph depicting the increased synthesis of glycosaminoglycans shortly after administration of bleomycin (as measured by ^{35}S-sulfate incorporation). (From Cantor, J. O., Cerreta, J. M., Osman, M., Mott, S. H., Mandl, I., and Turino, G. M., *Proc. Soc. Exp. Biol. Med.*,, 174, 172, 1983. With permission.)

Immunological

A number of studies have shown that inactivation of specific humoral components of the immune response can reduce the magnitude of bleomycin-induced fibrosis. Depletion of serum complement with cobra venom factor, for example, decreases collagen deposition following instillation of bleomycin.[18] Inhibition of prostaglandin synthesis with indomethacin has a similar effect.[19]

With regard to inflammatory cells, experimental evidence suggests that neither neutrophils nor T-cells are essential components of the response of the lung to bleomycin.[20,21] T-cells may, however, play a role in determining the severity of the resultant fibrosis.[22] Such regulation might depend on interaction with macrophages, which secrete factors that regulate

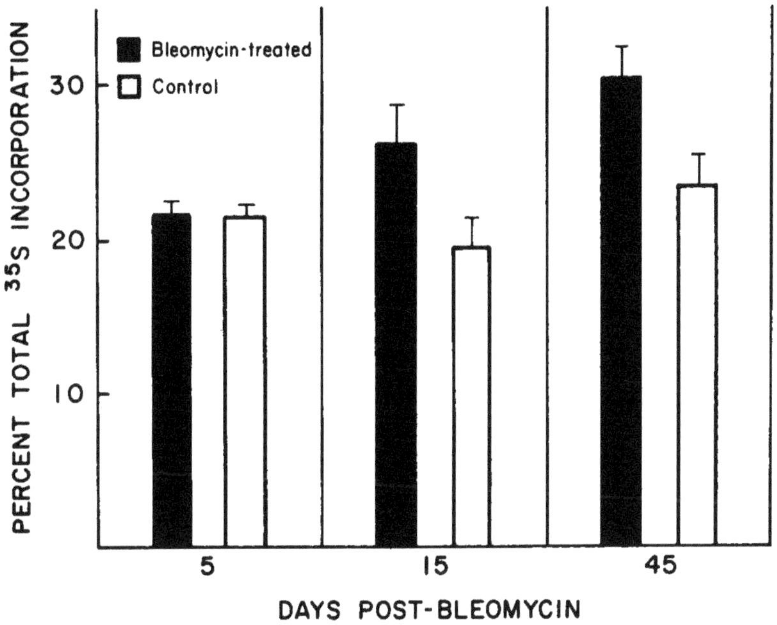

FIGURE 11. Graph depicting the increased synthesis of elastin following administration of bleomycin (as measured by ^{14}C-lysine incorporation into the cross-linking amino acids desmosine and isodesmosine). (From Cantor, J. O., Osman, M., Keller, S., Cerreta, J. M., Mandl, I., and Turino, G. M., *J. Lab. Clin. Med.*, 103, 384, 1984. With permission.)

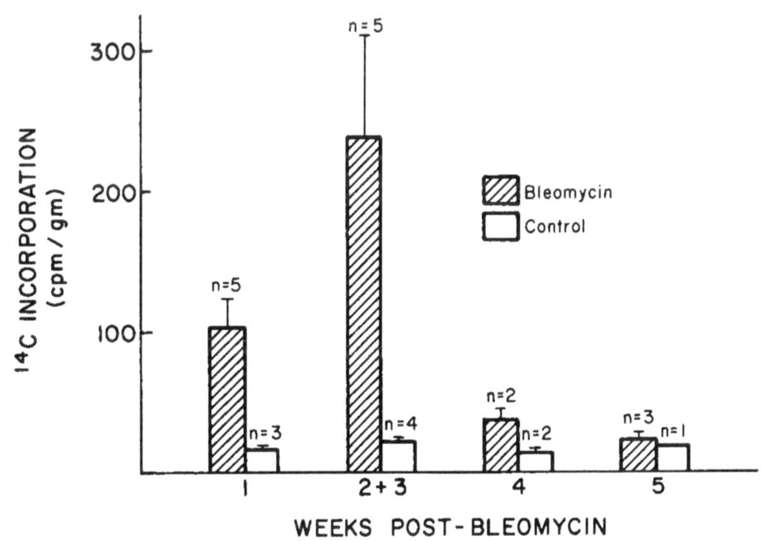

FIGURE 12. Graph indicating the proportionate increase in dermatan sulfate synthesis following instillation of bleomycin. (From Cantor, J. O., Cerreta, J. M., Osman, M., Mott, S. H., Mandl, I., and Turino, G. M., *Proc. Soc. Exp. Biol. Med.*, 174, 172, 1983. With permission.)

collagen synthesis by fibroblasts.[23] Inflammatory cells may further influence fibrosis by releasing enzymes that degrade collagen or other matrix components.[20]

DISCUSSION

The lung lesions induced by bleomycin fulfill the anatomic criteria proposed by Carrington for experimental models of pulmonary fibrosis.[24] These are (1) a mixed cellular exudate in the interstitium, (2) protein exudate in air spaces with or without leukocytes, (3) proliferation of lining epithelium, (4) gradual progression to fibrosis and honeycombing, (5) continuing evidence of cellular reaction, even when the lung is partly fibrotic, and (6) diffuse distribution with the possibility of skip zones.

While the morphological changes in this experimental model have been well characterized, the mechanisms responsible for the damage to lung tissue remain poorly understood. It is possible that injury may be initiated by the action of bleomycin-induced oxygen radicals on alveolar lining epithelium.[25] Resultant cell necrosis and denudation of the underlying basement membranes may then act as a stimulus for inflammation, proliferation of type II cells, and enhanced deposition of extracellular matrix.[26]

Failure of the lung to return to normal following bleomycin injury may be at least partly the result of changes occurring in the connective tissue matrix. The influence exerted by the extracellular environment has been best demonstrated in embryonic tissues. Alteration of glycosaminoglycan synthesis with β-d-xylopyranoside, for example, markedly inhibits salivary gland morphogenesis.[27] Similarly, bleomycin-induced matrix disturbances may interfere with the cellular processes needed for reconstitution of normal tissue architecture following injury.

With regard to treating pulmonary fibrosis, insight into the interplay between cells and their surrounding matrices will be particularly important. Most cases of the disease become clinically apparent well after the initial inflammatory events have subsided, when the lung may already be "programmed" to undergo fibrosis by virture of basic changes in matrix relationships. Attempts at therapeutic intervention may therefore have to focus on modifying the responses of lung cells to the extracellular environment or on altering the connective tissue matrix itself. Developing these types of treatment will require the continued use of animal models of pulmonary fibrosis.

REFERENCES

1. **Fulmer, J. D. and Crystal, R. G.,** The biochemical basis of pulmonary function, in *The Biochemical Basis of Pulmonary Function*, Crystal, R. G., Ed., Marcel Dekker, New York, 1976, 419.
2. **Haidle, C. W.,** Fragmentation of deoxyribonucleic acid by bleomycin, *Mol. Pharmacol.*, 7, 645, 1971.
3. **Johnson, K. J., Fantone, J. C., III, Kaplan, J., and Ward, P. A.,** *In vivo* damage of rat lungs by oxygen metabolites, *J. Clin. Invest.*, 67, 983, 1981.
4. **Blum, R. H., Carter, S. K., and Agre, K.,** A clinical review of bleomycin — a new antineoplastic agent, *Cancer*, 31, 903, 1973.
5. **Adamson, I. Y. R. and Bowden, D. H.,** Bleomycin-induced injury and metaplasia of alveolar type 2 cells, *Am. J. Pathol.*, 96, 531, 1979.
6. **Fleischman, R. W., Baker, J. R., Thompson, G. R., Schaeppi, U., Illievski, V. R., Cooney, D. A., and Davis, R. D.,** Bleomycin-induced interstitial pneumonia in dogs, *Thorax*, 26, 675, 1971.
7. **Bedrossian, C. W. M., Greenberg, S. D., Yawn, D. H., and O'Neal, R. M.,** Experimentally induced bleomycin sulfate toxicity, *Arch. Pathol. Lab. Med.*, 101, 248, 1977.
8. **Snider, G. L., Hayes, J. A., and Korthy, A. L.,** Chronic interstitial pulmonary fibrosis produced by intratracheal bleomycin: pathology and stereology, *Am. Rev. Respir. Dis.*, 117, 1099, 1978.

8a. **Luna, L. G., Ed.,** *Manual of Histological Staining Techniques of the Armed Forces Institute of Pathology*, McGraw-Hill, New York, 1968.

9. **Clark, J. G., Overton, J. E., Marino, B. A., Uitto, J., and Starcher, B. C.,** Collagen biosynthesis in bleomycin-induced pulmonary fibrosis, *J. Lab. Clin. Med.*, 96, 943, 1980.
10. **Woessner, J. F., Jr.,** the determination of hydroxyproline in tissue and protein samples containing small proportions of this amino acid, *Arch. Biochem. Biophys.*, 93, 440, 1961.
11. **Cantor, J. O., Osman, M., Keller, S., Cerreta, J. M., Mandl, I., and Turino, G. M.,** Measurement of crosslinked elastin synthesis in bleomycin-induced pulmonary fibrosis using a highly sensitive assay for desmosine and isodesmosine, *J. Lab. Clin. Med.*, 103, 384, 1984.
12. **Cantor, J. O., Cerreta, J. M., Osman, M., Mott, S. H., Mandl, I., and Turino, G. M.,** Glycosaminoglycan synthesis in bleomycin-induced pulmonary fibrosis: biochemistry and autoradiography, *Proc. Soc. Exp. Biol. Med.*, 174, 172, 1983.
13. **Zupnick, H., Cantor, J. O., Osman, M., and Turino, G. M.,** Alveolar macrophages from bleomycin treated hamsters secrete increased amounts of interleukin 1, *Am. Rev. Respir. Dis.*, 131 (Abstr.), A31, 1985.
14. **Cantor, J. O., Cerreta, J. M., Osman, M., Mott, S., Mandl, I., and Turino, G. M.,** Alveolar epithelial cell proliferation in bleomycin-induced pulmonary fibrosis, *Am. Rev. Respir. Dis.*, 125 (Abstr.), 238, 1982.
15. **Ivaska, K.,** Effect of extracellular glycosaminoglycans on the synthesis of collagen and proteoglycans by granulation tissue cells, *Acta Physiol. Scand. Suppl.*, 494, 1, 1981.
16. **Kelley, J., Newman, R. A., and Evans, J. N.,** Bleomycin-induced pulmonary fibrosis in the rat: prevention with an inhibitor of collagen synthesis, *J. Lab. Clin. Med.*, 96, 954, 1980.
17. **Fedullo, A. J., Karlinsky, J. B., Snider, G. L., and Goldstein, R. H.,** Lung statics and connective tissues after penicillamine in bleomycin treated hamsters, *J. Appl. Physiol.*, 49, 1083, 1980.
18. **Phan, S. H. and Thrall, R. S.,** Inhibition of bleomycin-induced pulmonary fibrosis by cobra venom factor, *Am. J. Pathol.*, 107, 25, 1982.
19. **Thrall, R. S., McCormick, J. R., Jack, R. M., McReynolds, R. A., and Ward, P. A.,** Bleomycin-induced pulmonary fibrosis in the rat: inhibition with indomethacin, *Am. J. Pathol.*, 95, 117, 1979.
20. **Thrall, R. S., Phan, S. H., McCormick, J. R., and Ward, P. A.,** The development of bleomycin-induced pulmonary fibrosis in neutrophil-depleted and complement-depleted rats, *Am. J. Pathol.*, 105, 76, 1981.
21. **Szapiel, S. V., Elson, N. A., Fulmer, J. D., Hunninghake, G. W., and Crystal, R. G.,** Bleomycin-induced interstitial pulmonary disease in the nude athymic mouse, *Am. Rev. Respir. Dis.*, 120, 893, 1979.
22. **Schrier, D. J. and Phan, S. H.,** Modulation of bleomycin-induced pulmonary fibrosis in the BALB/c mouse by cyclophosphamide-sensitive T cells, *Am. J. Pathol.*, 116, 270, 1984.
23. **Clark, J. G., Kuhn, C., III, McDonald, J. A., and Mecham, R. P.,** Lung connective tissue, *Int. Rev. Connect. Tissue Res.*, 10, 249, 1983.
24. **Carrington, C. B.,** Organizing interstitial pneumonia: definition of the lesion and attempts to devise an experimental model, *Yale J. Biol. Med.*, 40, 352, 1968.
25. **Freeman, B. A. and Tanswell, A. K.,** Biochemical and cellular aspects of pulmonary oxygen toxicity, *Adv. Free Radical Biol. Med.*, 1, 133, 1985.
26. **Bowden, D. H.,** Unraveling pulmonary fibrosis: the bleomycin model, *Lab. Invest.*, 50, 487, 1984.
27. **Thompson, H. A. and Spooner, B. S.,** Inhibition of branching morphogenesis and alteration of glycosaminoglycan biosynthesis in salivary glands treated with b-d-xyloside, *Dev. Biol.*, 89, 417, 1982.
28. **Cantor, J. O.,** Experimental pulmonary fibrosis, in *Handbook of Animal Models for the Rheumatic Diseases*, Greenwald, R. A. and Diamond, H., Eds., CRC Press, Boca Raton, FL, 1988, 205.

AMIODARONE-INDUCED PULMONARY FIBROSIS

Joseph M. Cerreta

INTRODUCTION

The heterogenous group of pulmonary disorders known as fibrotic lung diseases represent 20% of the noninfectious diseases of the lung.[1] The etiology is often unknown, although it is now understood that a variety of insults may set in motion a series of pathogenetic events resulting in end-stage fibrosis. Agents with apparent fibrogenic ability include inorganic dust,[2] bacteria,[3,4] radiation,[5] drugs,[6-8] toxic chemicals,[9,10] and, on an experimental basis, immune complexes.[11] Pulmonary fibrosis is characterized by having changes in lung connective tissue distribution and content.[12-15] The pathogenesis of fibrosis involves a sequence of events that includes an inflammatory response, proliferation of alveolar epithelium and fibroblast, and elevated synthesis of connective tissue fibers which leads to alterations of normal lung architecture. This last change is often referred to as honeycombing.

The functions of connective tissue of lung have been well documented.[16-18] First, it acts as a supporting network for airways, blood vessels, and alveoli.[16] Second, it provides some of the elastic recoil needed to deflate the lung during expiration. The connective tissue consists of a cellular component, predominantly fibroblast, and an extracellular matrix which is principally protein. The connective tissue proteins of lung parenchyma (in order of decreasing amounts in the tissue) are collagen, elastin, acidic structural glycoproteins (ASG), and proteoglycans. Collagen and elastin are arranged as fibers, while the other proteins are usually found as an amorphous matrix in which the rest of the connective tissue components sit. Collagen fibers course through the lung interstitium in circular and longitudinal arrays.[17] These fibers assume a helical structure as they encircle the lumens of the respiratory bronchioles and alveolar ducts. The collagen fibers splay as individual fibrils into the alveolar wall. Elastic fibers are closely associated with collagen as the latter courses through the lung, especially at the level of the respiratory bronchioles and alveolar ducts.[18] The two components of connective tissue provide a flexible network of support for respiratory structures. Alterations in collagen and elastin and in their relationship to one another can have profound physiological effects, creating uneven stresses on the lung parenchyma. This may lead to the closing off of smaller airways and blood vessels, resulting in uneven ventillation, perfusion, and diffusion.

Recent investigations have studied the types of connective tissue abnormalities found both in patients with fibrotic lung disorders and in experimental models of the disease. Morphological studies indicate that collagen is altered in both human lung fibrosis and animal models of this disease.[13-15,19-24] In humans, histological sections show an increase in collagen, but this increase has been difficult to demonstrate by biochemical analysis.[13] Animal models, however, have shown ample evidence that such increases do occur.[19-24] The discrepancy may be attributable to the limitations of the human material in terms of sample size, stage of the disease studied, and possibility of treatment of the patient with steriods.

The elastin component has been found to be elevated in animal models of interstitial lung disease.[19] In the laboratory of St. John's University, synthesis of the elastin-specific cross-linking amino acid desmosine has been found to increase from normal in the bleomycin model of pulmonary fibrosis.[24] New and sensitive methods for measuring both elastin synthesis[25-27] and elastin breakdown[28-33] have facilitated analysis of this connective tissue component in various lung disease processes despite its insolubility in the mature form and its relatively slow rate of turnover.

Only a few studies have measured the changes of the glycosaminoglycans (GAGs) in interstitial lung disease.[34-36] Although GAGs constitute only a small fraction (less than 1%) of the connective tissue mass of the lung, these studies indicate that they are elevated in the disease process.

A new model of pulmonary interstitial fibrosis has recently been developed.[7] Recognizing from clinical reports[37-42] the potential of the antiarrhythmic drug amiodarone to cause lung injury, a model of fibrosis in hamsters was developed. Amiodarone, a benzofuran, is capable of causing pulmonary fibrosis in hamsters after a single intratracheal treatment. This model fulfills the four basic anatomic criteria that should be met by an experimental model of that disease process. These are (1) an early acute inflammatory response which is exudative in nature, (2) proliferation of alveolar epithelial lining, (3) alteration in the connective tissue leading to increases in fiberous components, and (4) alteration in the structure of the lung which leads to consolidated areas interspersed around zones of normal lung parenchyma.

IMPLEMENTING THE MODEL

Cell Proliferation Studies: Animal Treatment

Female, 3-month-old Syrian hamsters (110 to 120 g each) were divided into experimental and control groups. Experimental animals were given a single dose of 0.1 ml of amiodarone (Sanofi Pharmaceuticals, Paris, France) at a concentration of 12.5 mg/ml by intratracheal insufflation.[7] Control animals received 0.1 ml of a solution of normal saline containing 1% benzyl alcohol and 2% Tween® 80. The latter two substances are the carriers for amiodarone. The insufflation was carried out under ketamine hydrochloride (Bristol Laboratories, Syracuse, NY) anesthesia using sterile instruments and solutions. To expose the trachea, a small incission was made in the ventral midline of the neck after which a small-bore hypodermic needle was inserted into the airway and the amiodarone or control vehicle slowly injected. Following injection, the wound was closed using metal skin clips. Sets of five treated animals and five control animals were killed at 2, 4, 6, 8, 10, 12, 14, 21, and 28 d postinsufflation. Each of these animals received 0.02 μCi/g body weight of ^{3}H-thymidine by intraperitoneal injection 1 h prior to sacrifice. The lungs from these animals were used for the morphologic examinations described below.

Analysis of Collagen: Animal Treatment

This portion of the study used 40 female Syrian hamsters weighing 130 to 150 g each. Animals were treated with either amiodarone or a control vehicle as described above. Sets of five treated and five control animals were killed at 7, 14, 21, and 28 d postinsufflation. Both treated and control animals were labeled with ^{14}C-proline (2 μCi/g/d) for 3 consecutive days prior to sacrifice.

Analysis of Elastin: Animal Treatment

Syrian hamsters weighting 130 to 150 g each were divided into two groups, treated and control. The treated animals received 0.2 ml of a solution which contained amiodarone at a concentration of 12.5 mg/ml, while the controls received an equal volume of normal saline. Animals were sacrificed in groups at 8, 15, 22, and 29 d postinsufflation. Each animal was treated with radiolabled lysine 5 d prior to sacrifice. The ^{14}C-lysine was administered by intraperitoneal injection at a dose of 0.02 μCi/g for 3 consecutive days and the animals were killed 2 d following the last treatment.

EVALUATING THE MODEL

Morphologic Methods

Fixation of the Lung

Lungs were fixed *in situ* after sacrifice. The trachea of the animal was exposed by dissection and incised to allow the insertion of a cannula. The lungs were collapsed by cutting the diaphram. The fixative was instilled via the tracheal cannula at 20 cm of water pressure using 10% neutral buffered formalin for light microscopy and 2.5% glutaraldehyde prepared in 0.1 *M* sodium cacodylate for electron microscopy.

Preparation of Tissue for Light Microscopy

Following fixation in neutral buffered formalin, the lungs were excised and entire lobes cut in half in the sagittal plane. One half of each lobe was dehydrated in alcohol and embedded in paraffin prior to sectioning. Sections were cut on a standard rotary microtome and stained using hematoxylin and eosin. Additional histochemical staining techniques such as Verhoff's or Musto's[43] stain for elastic fibers and Mallory's trichrome for collagen were used to visualize specific connective tissue structures.

Where increased resolution and thinner sections were required (i.e., for autoradiography), the remaining half of each lobe was cut into blocks, dehydrated in alcohol, and embedded in glycolmethacrylate. These blocks were cut on a DuPont® JB-4 microtome, and the sections placed on glass slides and stained using Gill's hematoxylin and phloxine. The advantage of glycolmethacrylate sections is that, unlike paraffin sections, they can be cut at thicknesses as small as 0.5 μm. This allows for greater resolution and easier identification of structures within the lung.

Preparation of Tissue for Transmission Electron Microscopy

Inflated lungs fixed in glutaraldehyde were cut into blocks approximately 1 mm.[3] These blocks were postfixed in 1% osmium tetroxide prepared in 0.1 *M* sodium cacodylate buffer, stained as a whole with 0.5% uranyl acetate, dehydrated in acetone, and embedded in Epon-Araldite mixture. Following polymerization, thin sections were cut from these blocks, stained with uranyl acetate and lead citrate, examined, and photographed with an electron microscope.

Autoradiographic Procedures for the Light Microscope

Glycolmethacrylate sections were cut using a JB-4 microtome, sections were placed on glass slides, and the sections dried for 2 h at 60°C. These slides with their tissue sections were dipped in Kodak® NTB-2 autoradiographic emulsion and, after drying, were placed in lighttight boxes. These boxes were stored in the cold for varying lengths of time for exposure of the emulsion to the tissue radiation source. Following exposure, the slides were developed in Kodak® Dektol and fixed. Sections were stained with Gill's hematoxylin and phloxine, processed to xylene, and then coverslipped before being examined and photographed for further evaluation.

Analysis of Autoradiographs

Autoradiographs were examined at magnification × 400 under a light microscope equipped with a counting grid. The labeled cells from two compartments, epithelial and interstitial, were counted and differences between treated and control groups determined.

Analysis of Collagen

Collagen Analysis

The labeled lungs were removed from the thorax, freed of extrapulmonary structures, cut into small pieces, washed, and then homogenized. The homogenates were dialyzed and lyophilized. Weighed portions were suspended in Tris buffer, pH 7.4 with 10 m*M* $CaCl_2$ and digested with bacterial collagenase which was free of extraneous proteinase activity.[35] Digestion was carried out at 37°C and repeated two additional times to insure removal of nearly all (greater than 90%) collagen.[35] The digests were pooled and assayed for radioactivity. The specific radioactivity of the free proline pool in the lung was determined.

Estimation of Proline Pool Size

Experimental and control animals were given intraperitoneal injections of ^{14}C-proline (0.02 μCi/g) 7 d postinsufflation of amiodarone or control vehicle and sacrificed 1 h later. Lungs were removed, homogenized, and lyophilized. Weighed samples were resuspended in 2 ml of 10% trichloroacetic acid (TCA) and precipitated overnight at 4°C. Following precipitation, the sample was centrifuged, the pellet was washed twice with 5% TCA, and the three supernatants pooled.[8] A 200-μl sample was then counted in a liquid scintalation spectrophotometer to determine the radioactivity of the free proline pool. A portion of the supernatant was then assayed for total proline content. The radioactivity of the proline pool was then determined by dividing the radioactivity by the total proline content of the sample.

Analysis of Elastin Synthesis

Elastin synthesis in the amiodarone model has been examined by Cantor and colleagues.[44] The following description is based on their study.

Elastin Analysis

The lungs were excised, freed of nonparenchymal structures, cut into pieces, homogenized, and dialyzed. After lyophilizing the homogenate, weighted samples were hydrolyzed in 6 *N* HCl at 110° overnight. To remove the HCl, the hydrolysates were evaporated over a steam bath. These dried samples were reconstituted with distilled water, filtered, and lyophilized again. These lyophilized samples were reconstituted once again in 1 ml of distilled water. Aliquots of each sample were spotted on Whatman® 3-MM filter paper and chromatographed in butanol, glacial acetic acid, and water (4:1:1). Samples were repeatedly chromatographed (magnification × 4) to increase the separation of material from the origin. The origins of each chromatograph were cut out and eluted in distilled water. These eluates were filtered and lyophilized to concentrate the sample. The lyophilized samples were reconstituted with distilled water to a standard volume. This sample was spotted on a thin layer cellulose plate (MN300, Brinkmann Instruments, Westbury, NY). the samples were then subjected to electrophoresis in pyridine, glacial acetic acid, and water (1:10:189) at 600 V for 90 min.[25] After locating desmosine/isodesmosine through the use of comigrating standards, the material was scraped from the plate and the radioactivity measured in a liquid scintillation spectrophotometer.

The resultant values were normalized against the dry weight of the starting sample. Total incorporation of ^{14}C-lysine into the original hydrolysates was measured as well as the specific activity of the free lysine pool in the lungs of both experimental and control animals.

Estimation of Lysine Pool-Size

Lysine pool size was determined in a fashion similar to that used to estimate the pool size of proline, with the substitution of ^{14}C-lysine for proline, and measuring the total amount of lysine in the sample using the method outlined above.

EXPERIMENTAL FINDINGS

Cell Proliferation Studies

Lungs from both experimental and control animals were examined for alteration in inflammatory events, cell proliferation, and ultimately for changes in lung architecture. Animals killed 2 d post-amiodarone treatment had an abundance of pulmonary macrophages (Figure 1) in alveolar airspaces accompanied by a large number of neutrophils. Electron microscopy clearly revealed eosinophils associated with alveolar septa (Figure 2). Alveolar epithilial cell morphology was largely undisturbed, while increased width of the septa was largely attributable to edema fluid. In a few instances, early fibrotic events were evident in

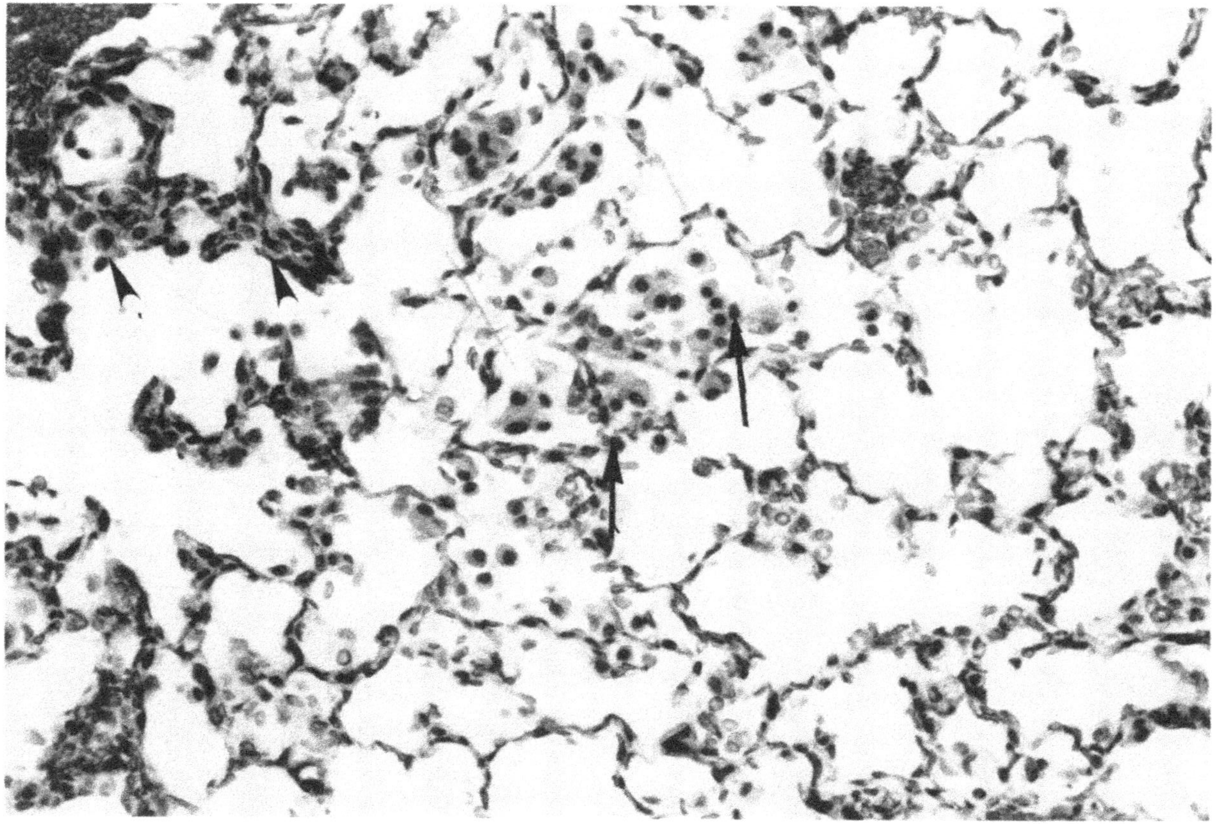

FIGURE 1. Photomicrograph of a lung section from a hamster treated with amiodarone. Note the large numbers of alveolar macrophages in alveolar spaces (arrows). Arrowheads indicate an area of early fibrosis. (Magnification × 285.)

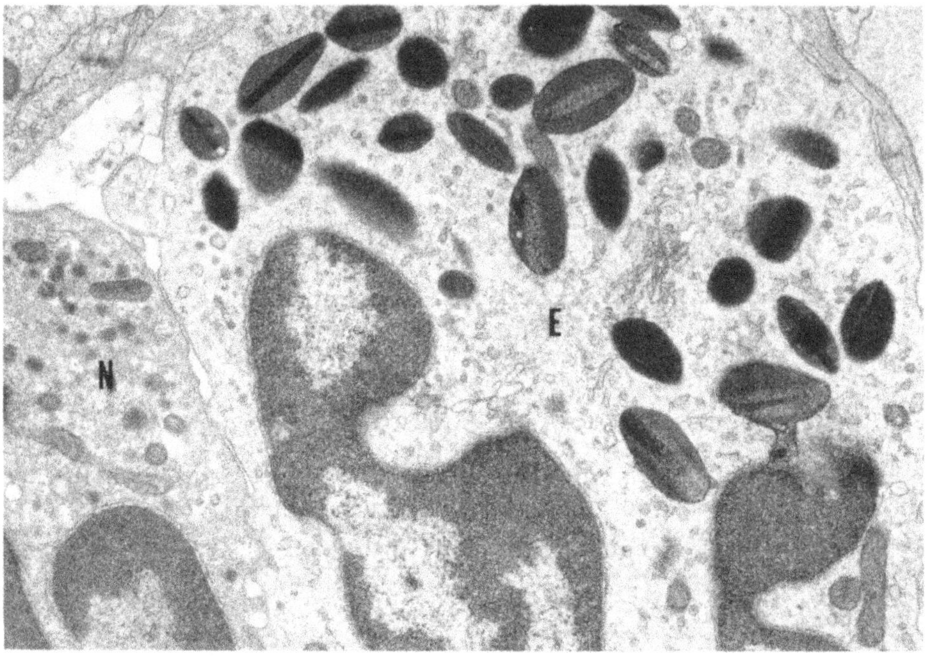

FIGURE 2. Electron micrograph of an interstitial area from a hamster lung 2 d posttreatment with amiodarone. The center of the micrograph contains an eosinophil (E). The cell marked N is a neutrophil, which were also numerous in the septa of treated animals. (Magnification × 19,400.)

areas surrounding large airways (Figure 1). Counts measuring the number of cells incorporating ^{3}H-thymidine into their nuclei show marked statistical significant increases in treated animals when compared to controls. This was true for both epithelial cells and interstitial compartments. Controls from this time period had increased numbers of pulmonary macrophages and slight septal edema, but lacked increased numbers of eosinophils (Figure 3). Control animals lacked regions of early fibrosis in their lungs.

Treated animals killed 4 d postinsufflation with amiodarone still contained large numbers of alveolar macrophages in airspaces (Figure 4). In addition to these cells, lymphocytes were clearly visible in thickened areas of the parenchyma. These areas of thickened parenchyma with widened alveolar septa were visible dispersed between relatively normal regions of lung tissue (Figure 5). The wall-width increase appeared to be the result of increased number of cells rather than from edema fluid. Cell counts from this time period clearly support this observation. Cells with nuclei which had incorporated the tritiated thymidine were increased in number in both epithelial and interstitial compartments as compared to controls. These increases in both compartments were statistically significant. However, the number for radioactive cells in both compartments had by this time begun to decline from the levels observed at 2 d postinsufflation. Electron microscopy revealed large numbers of eosinophils, with some in the process of disintergration (Figure 6). Lungs from control animals at this time period show a persistance of inflammatory cells, although their numbers appear to have declined. There were no areas of increased septal thickening as a result of cellular proliferation, with average cell counts for both epithelial and interstitial compartments unchanged from earlier controls.

At 6 d postinsuffluation, the treated animals showed unmistakable signs of pulmonary fibrosis (Figure 7). The structure of the alveoli in some areas of the lungs was compeltely obliterated, with the parenchyma taking on the typical honeycombed appearance seen in human lung fibrosis (Figure 7). When these fibrotic regions were examined closely, the

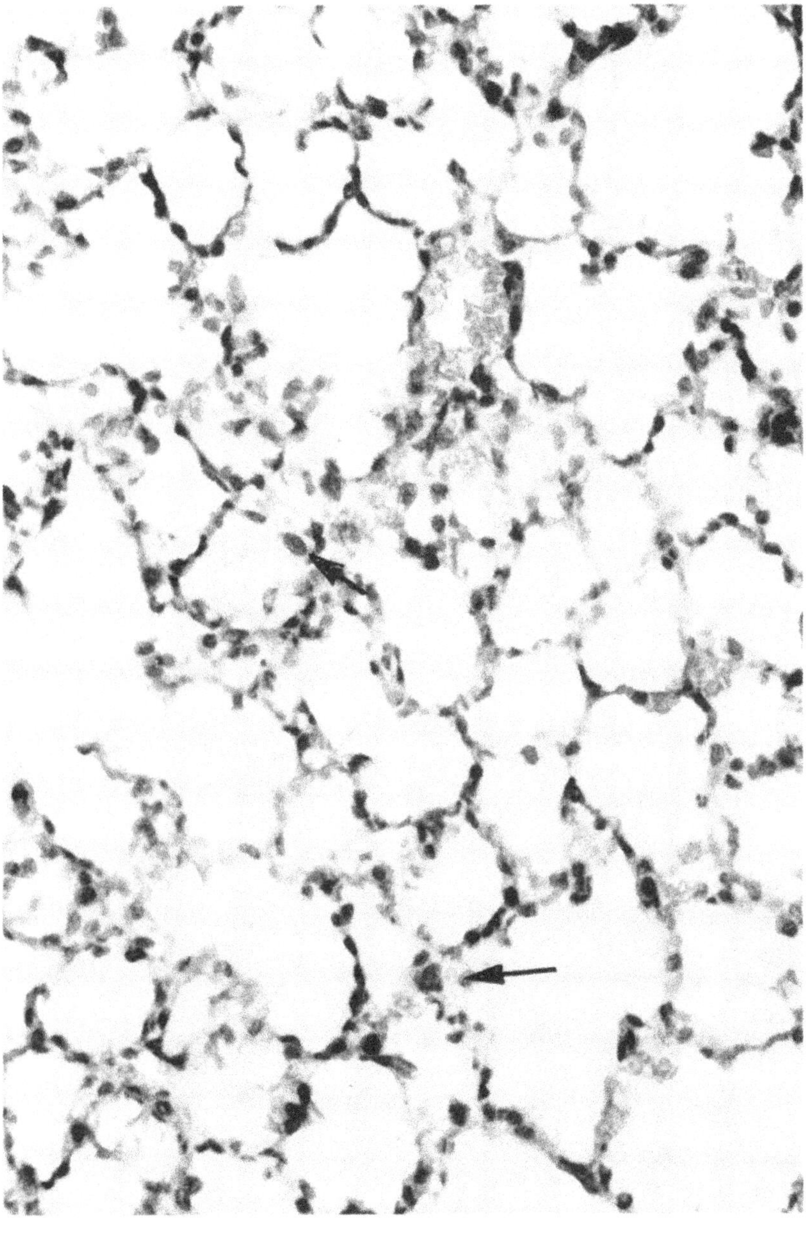

FIGURE 3. Photomicrograph of a lung section from a 2-d control animal without areas of pulmonary fibrosis. Note that only a few alveolar macrophages are present (arrows). (Magnification × 285.)

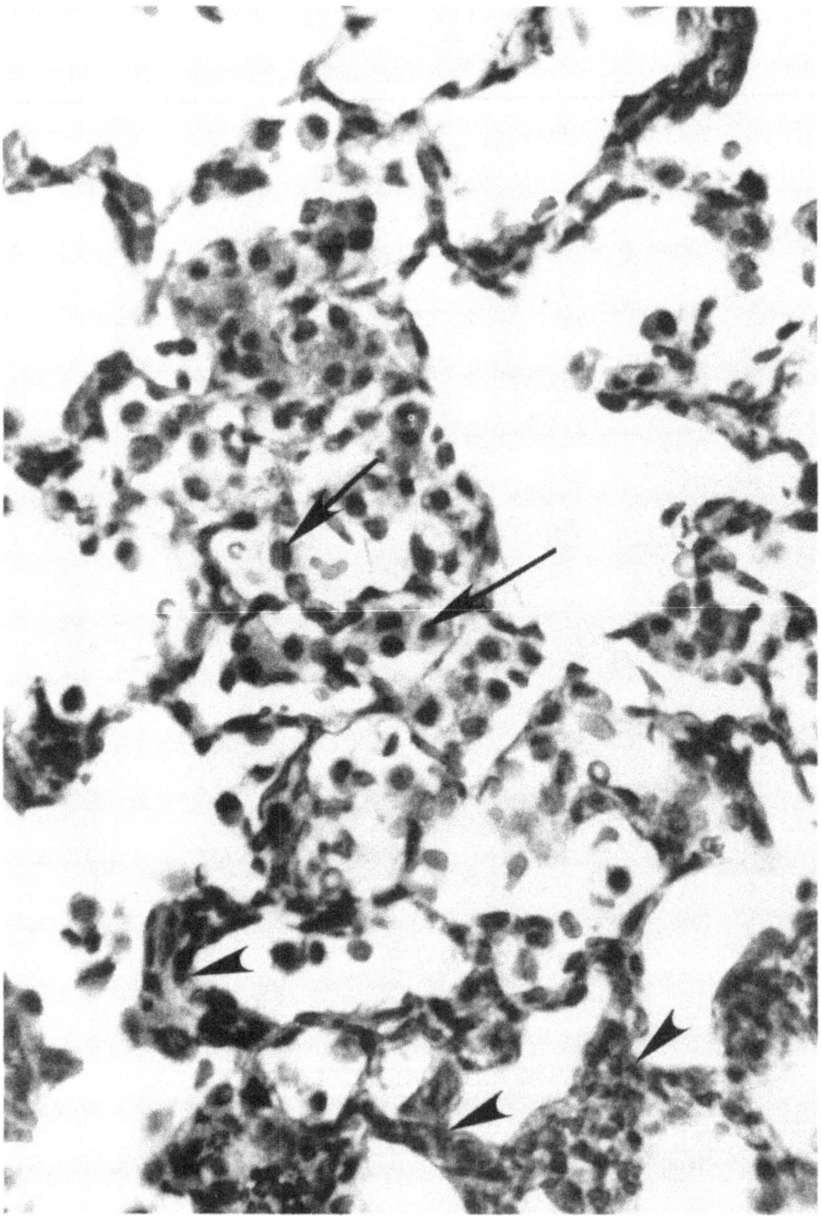

FIGURE 4. Photomicrograph of a lung section from a hamster 4 d posttreatment. Many of the alveolar spaces are almost completely filled with alveolar macrophages (arrows). Areas marked with arrowheads are in the early stages of fibrosis. (Magnification × 570.)

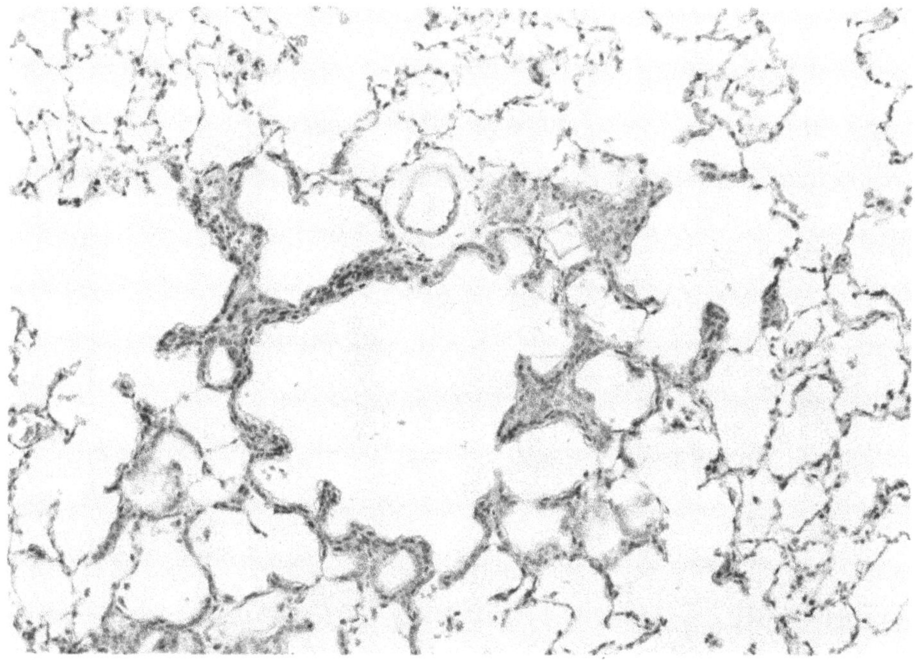

FIGURE 5. Photomicrograph of a lung section from a hamster 4 d postamiodarone treatment. The alveolar septa are thicker around an alveolar duct, with almost complete loss of surrounding alveoli. The thickened septa are clearly due to increased cellularity. (Magnification × 90.)

honeycombed areas were lined with a cuboidal epthelial (Figure 8) which replaced the normal squamous epthelial cells. This cuboidal epithelium lacked lamellar bodies typical of type II pneumocytes probably being derived from mitotic cells noted within the epithelium (Figure 9). Sections stained to show collagen or elastic fibers revealed an increase of these proteins in the thick fibrotic regions. Fibrotic regions also contained large numbers of inflammatory cells. These fibrotic areas were interspersed between large regions of normal pulmonary tissue (Figure 7). Examination of the lungs for cells which had incorporated the tritiated thymidine revealed statistically significant elevated counts in both epithelial and interstitial compartments when compared to controls. These lungs demonstrated a continuation of the changes noted earlier. Eosinophils were still present in fibrotic areas, although their numbers declined slightly. The number of these cells which had undergone lytic-like changes increased. In addition to these changes, alterations were observed in vascular endothelium and in interstitial compartments. The most common change in the vascular endothelium were swirls in the membrane (Figure 10). Interstitial compartments had what appeared to be an increase in septal width (Figure 11). These septal alterations appear to be the result of cell proliferation, infiltration of inflammatory cells, and increases in fibers and ground substance of the connective tissue matrix (Figure 11). Lungs from control animals of this period revealed that inflammatory events observed in 2 and 4 d had subsided and the lungs appeared to be indistinguishable from those of untreated animals.

At 8 d postinsufflation, the lungs of hamsters treated with amiodarone revealed a continuation of the processes initiated at the onset of treatment. Cell proliferation continued, but at a slower rate with slightly reduced counts of incorporated thymidine in both compartments counted. Alterations in lung architecture, however, continued to increase with larger areas of the lung becoming involved in the fibrotic process. These areas revealed regions of honeycombing along with the continued presence of inflammatory cells, especially pulmonary macrophages.

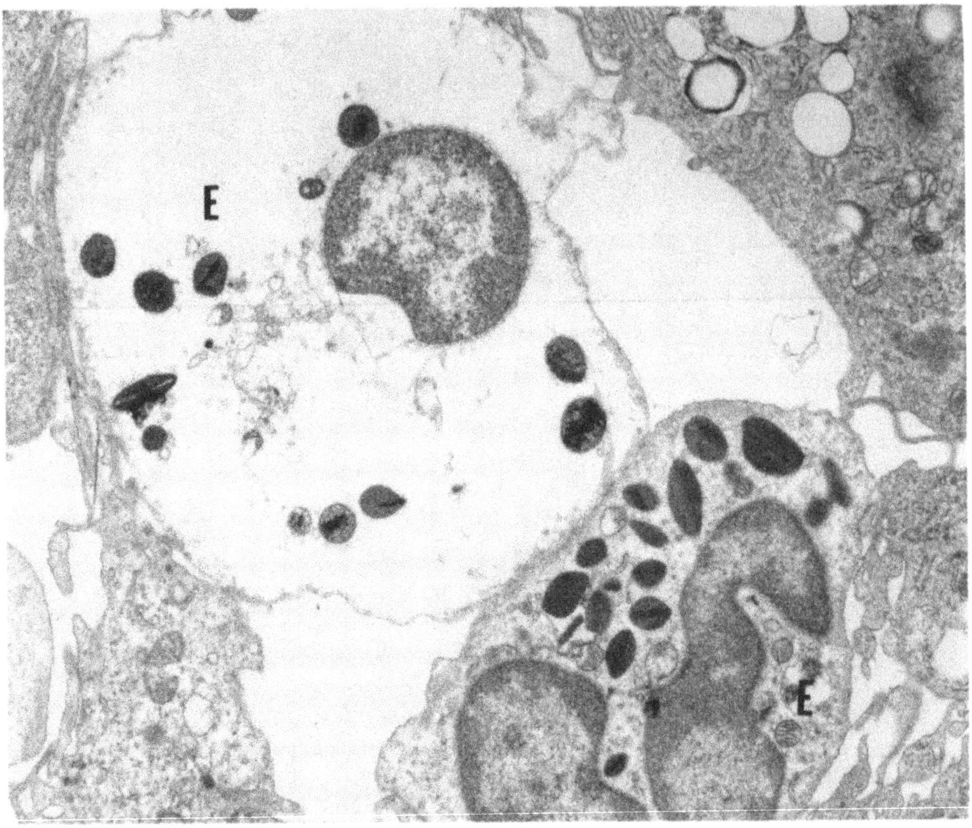

FIGURE 6. Electron micrograph of a lung section from a hamster 4 d postamiodarone treatment. Interstitial areas at this time period contain large numbers of eosinophils, some of which are in various stages of disintergration as seen in this micrograph. E marks eosinophils. (Magnification × 11,650.)

Ultrastructural examination demonstrated a continuation of changes noted at earlier stages. There were increased alterations in epithelial cells with some type II pneumocyte degeneration (Figure 12). Eosinophils were present in interstitial areas (Figure 13), with some in various states of disintegration. In some areas, the interstitial areas of alveoli from the lungs of treated animals contained large numbers of cuboidal-shaped cells (Figure 13). Changes noted in endothelial cells declined in number and severity.

The remaining test periods showed a decline in the numbers of cells which incorporated the ^{3}H-thymidine in both epithelial and interstitial compartments. At the later time periods examined (21 and 28 d), these counts were indistinguishable from those of the control animals. Despite the decline in radioactive thymidine incorporation, the areas of fibrosis within the lung continued to expand, involving greater areas of the lung (Figure 14).

Ultrastructural changes observed during these remaining study periods showed little alteration from earlier stages, with the exception that there appeared to be an increase in the thickness of alveolar septal regions. This increase appeared to be associated with expansion of the connective tissue matrix. These thickened interstitial areas were overlayed by a thick layer of epithelium which in some instances was ciliated (Figure 15).

Collagen Analysis

Biochemical analysis of tissue samples for collagen content was based on incorporation of ^{14}C-proline in the collagen molecules. Scintillation counts of collagenase digested tissue samples revealed that at 7 d postinsufflation of amiodarone, collagen synthesis was increased

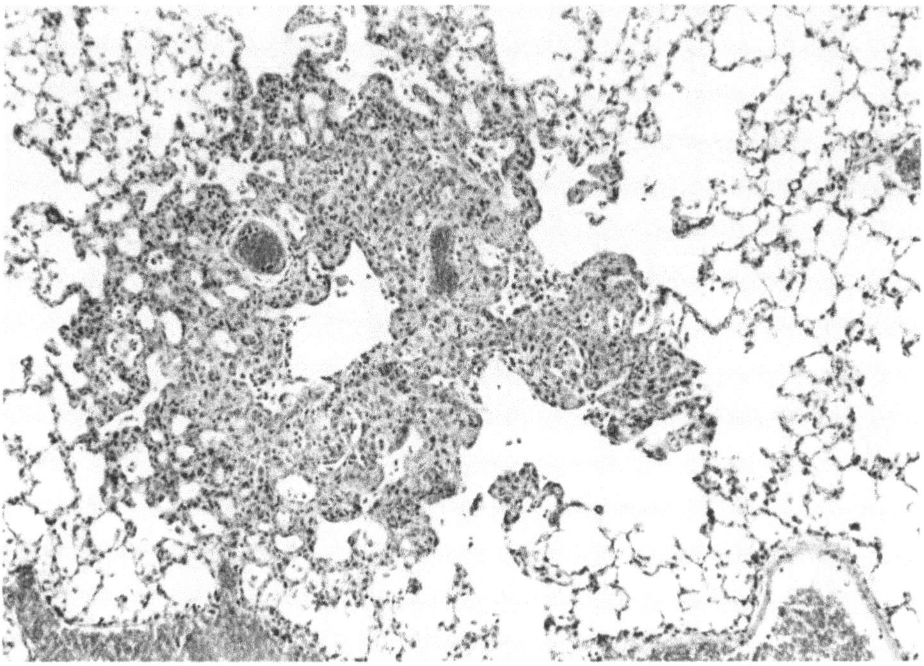

FIGURE 7. Photomicrograph of a lung section from a hamster 6 d postamiodarone treatment. Areas of fibrosis are much more extensive than at earlier periods. A large area of this lobe is occupied by fibrotic tissue which has a typical honeycomb appearance. Adjacent areas are clear of fibrosis, although pulmonary macrophages can be seen in many of these apparently normal areas. (Magnification × 90.)

approximately 60% above controls. This difference which was statistical significant had a p value greater than 0.001. When subsequent time periods were examined (i.e., 14, 21, and 28 d), the increased synthesis observed at 7 d continued. At 14 d the increase was 56%, at 21 d 62%, and at 28 d it was 74% when compared to controls. All these increases were statistically significant.

Elastin Analysis

The data for elastin synthesis is extracted from the work of Cantor and co-workers.[44] They reported that elastin synthesis as measured by the level of specific cross-links, desmosine/isodesmosine, was increased at 1 week postamiodarone treatment. The level of these cross-links remained elevated throughout the entire test period of 28 d. Cantor et al.[44] reported that the relationship of labeled lysine to total lysine in the amino acid pool of the lung was not significantly different between experimental control groups. They also reported that the total elastin content of the lungs in treated animals measured at 14 d postamiodarone insufflation was 32% greater than that of controls. This alteration was statistically significant.

DISCUSSION

Examination of the amiodarone model of pulmonary fibrosis reveals that there are a number of events which lead to an end-stage lung. These events fulfill the four basic criteria which any experimental model of lung fibrosis must meet: inflammation, cellular proliferation, increased matrix synthesis, and lung remodeling. The earliest stage of the injury process involves an acute inflammatory reaction. This reaction is characterized by large accumulations of pulmonary macrophages and infiltration of neutrophils which may be responsible for the damage of lung parenchyma. An interesting part of this inflammatory reaction is the

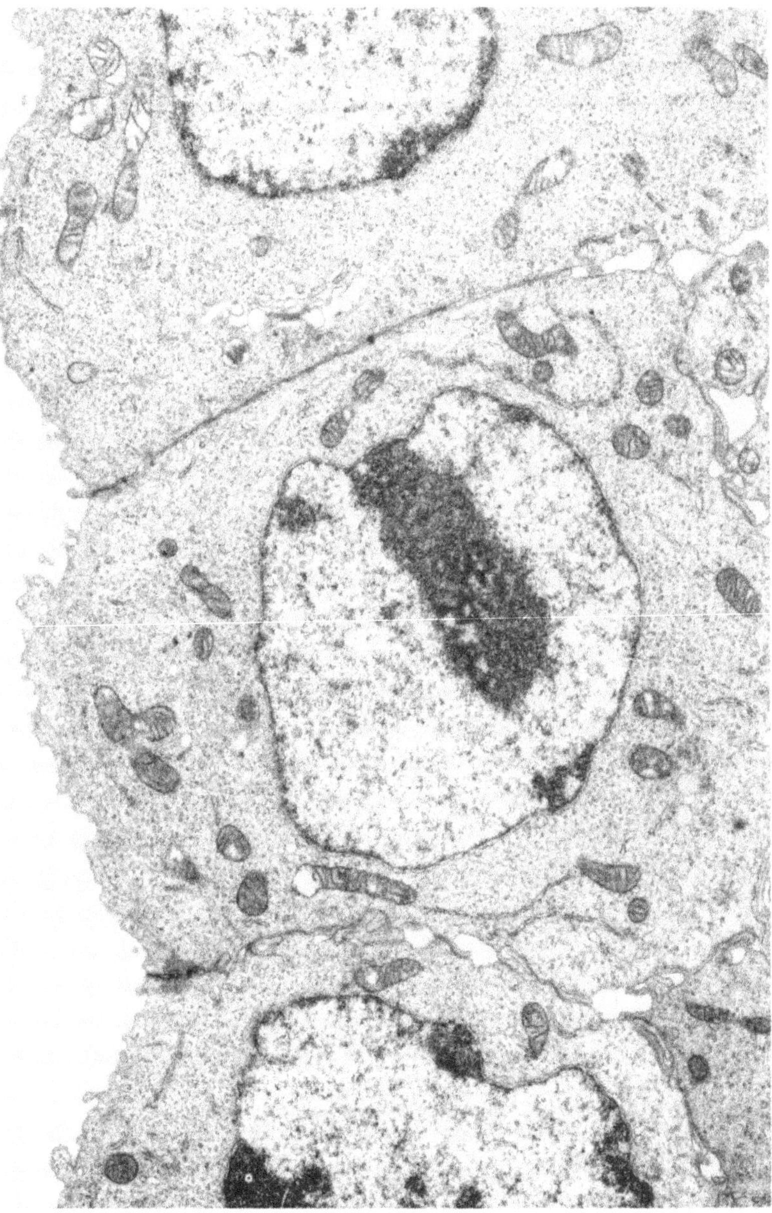

FIGURE 8. Electron micrograph of a lung section from a hamster 6 d postamiodarone treatment. This alveolar epithelium is composed of cuboidal cells. This is typical of the linings of the honeycomb regions of fibrotic lesions. (Magnification × 11,650.)

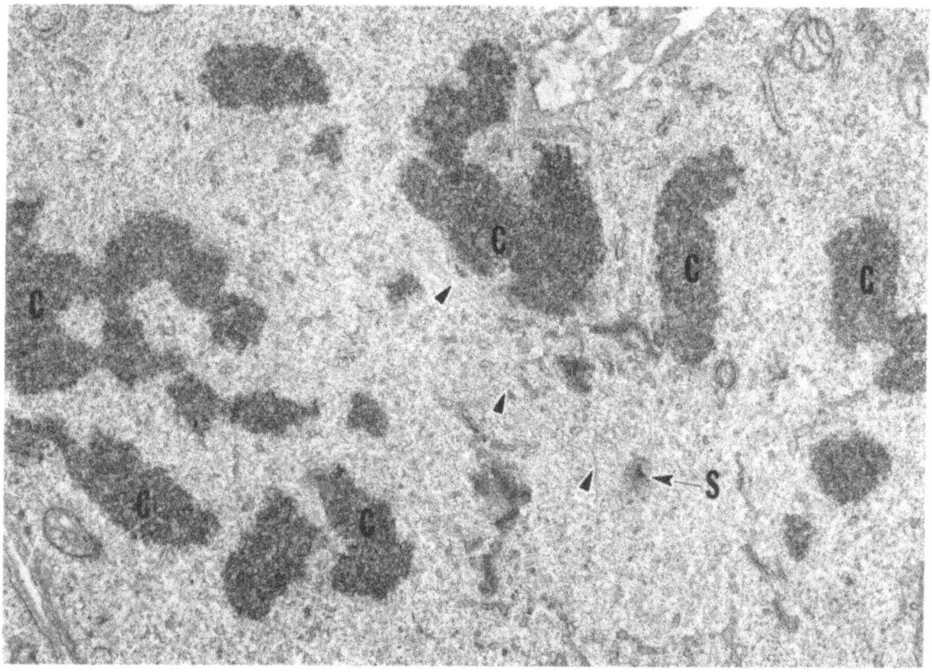

FIGURE 9. Electron micrograph of a lung section from a hamster 6 d postamiodarone treatment. The micrograph is of the alveolar epithelium and reveals a cell in late anaphase. Note that chromosomes (C) are converging on what appears to be a spindle pole (S) with numerous spindle fibers (arrowheads) leading to the pole. (Magnification × 13,800.)

increase in numbers of eosinophils which are associated with this model. Eosinophils are associated with allergic reactions and their presence may in part be related to an exogenous irritant, amiodarone, within the lung.[45] The disintegration of these cells in fibrotic areas may be specifically associated with the remodeling of the lung. Eosinophils are known to contain collagenase[45-49] which may be released during these lytic events. This enzyme could be responsible for breakdown of "old" collagen, allowing for alterations in lung architecture associated with pulmonary fibrosis.[45,48,49] The initial inflammatory stage resolves within a few days of treatment. The eosinophils noted during this early stage, however, continue to increase in numbers and are elevated throughout all the time periods studied.

Cell proliferation begins during the inflammatory period. This event is characterized by increased numbers of mitotic figures seen in both light and electron microscopy. Further, counts of cells which incorporate the ^{3}H-thymidine label indicate that both epithelial and interstitial compartments are undergoing cell increases. The highest numbers of labeled cells are seen in the early stages of the injury. Cell proliferation occurs in the epithelial compartment which is associated with cell injury and replacement. Newly produced epithelial cells are evident throughout areas of injury as new epithelial linings are produced. These cells are typically cuboidal in shape and lack any specific inclusions. They are found surrounding areas where the alveoli have been disturbed, and regeneration of these cellular components leads to the honeycombed architecture which is typical of pulmonary fibrosis. These newly proliferated cells are thus associated with lung remodeling.

Cells of the connective tissue compartment also demonstrate an increased incorporation of ^{3}H-thymidine when compared to controls. This increased incorporation parallels that seen in the epithelial compartment. The cells of this compartment were not separated into individual types (i.e., fibroblasts, smooth muscle cells, etc.) due to difficulties in cell identification which occur because of disturbances in cellular and tissue architecture in areas of

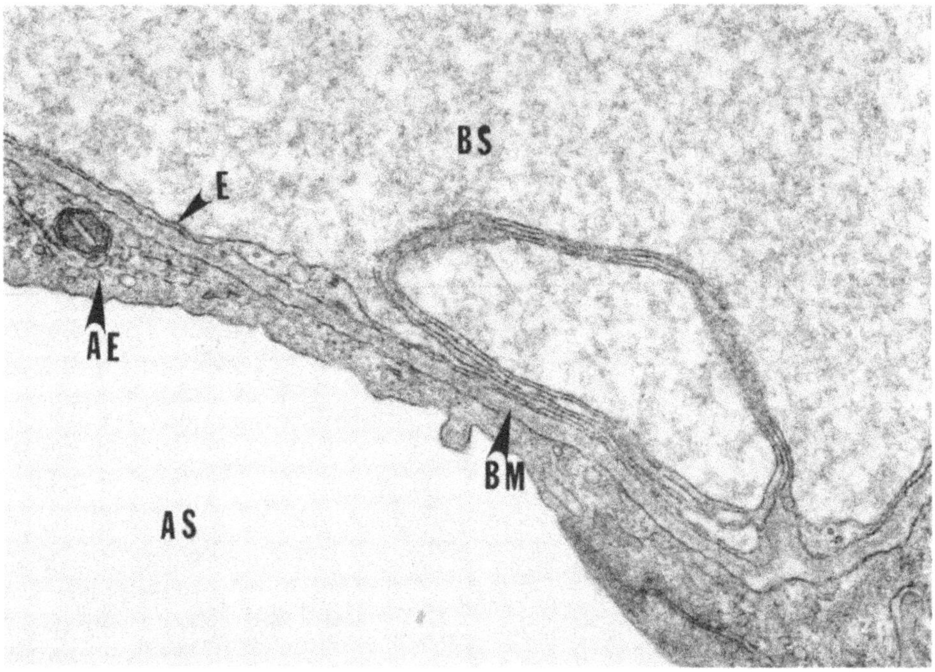

FIGURE 10. Electron micrograph of a lung section from a hamster 8 d postamiodarone treatment. Note that the endothelial lining of this blood vessel is thrown into a series of concentric rings. AS = airspace, BS = blood space, E = blood vessel endothelium, BM = basement membrane, and AE = alveolar epithelium. (Magnification × 36,800.)

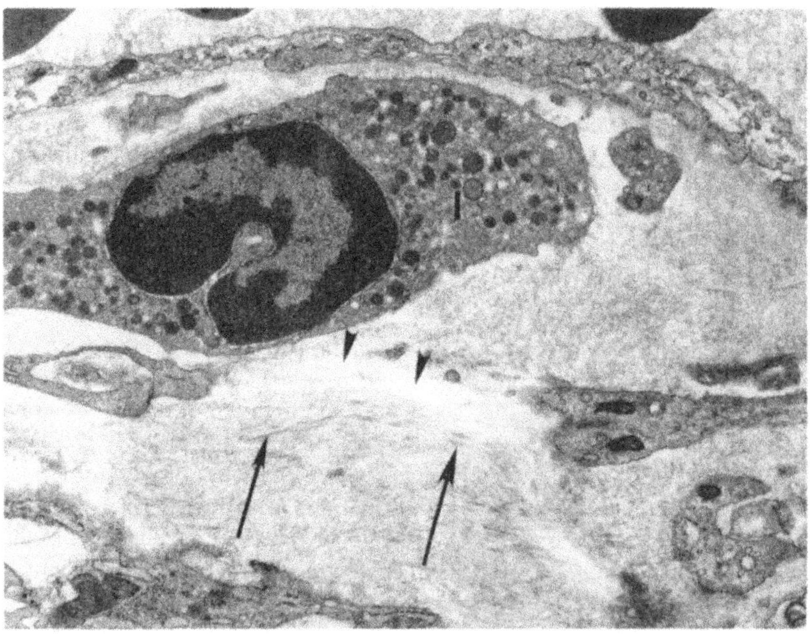

FIGURE 11. Electron micrograph of a lung section from a hamster 8 d postamiodarone treatment. This septum contains a broad area of connective tissue matrix with a large inflammatory cell (I). The matrix contains collagen fibers (arrows) and elastic fibers (arrowheads). (Magnification × 19,400.)

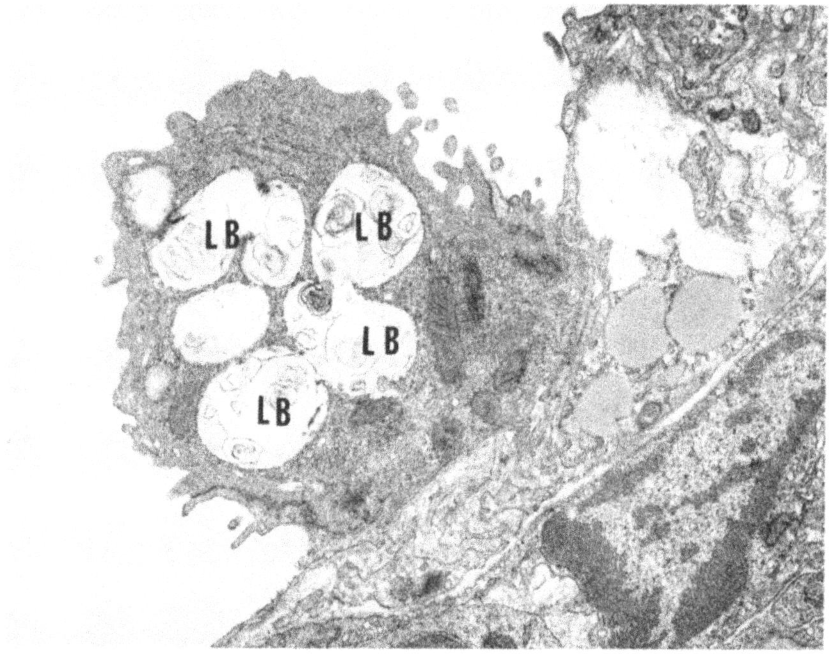

FIGURE 12. Electron micrograph of a lung section from a hamster 8 d postamiodarone treatment. This is a section through a portion of a type II pneumocyte. Note that the lamellae bodies (LB) have lost their typical lamellae and appear to be coalescing to form one large structure. (Magnification × 19,400.)

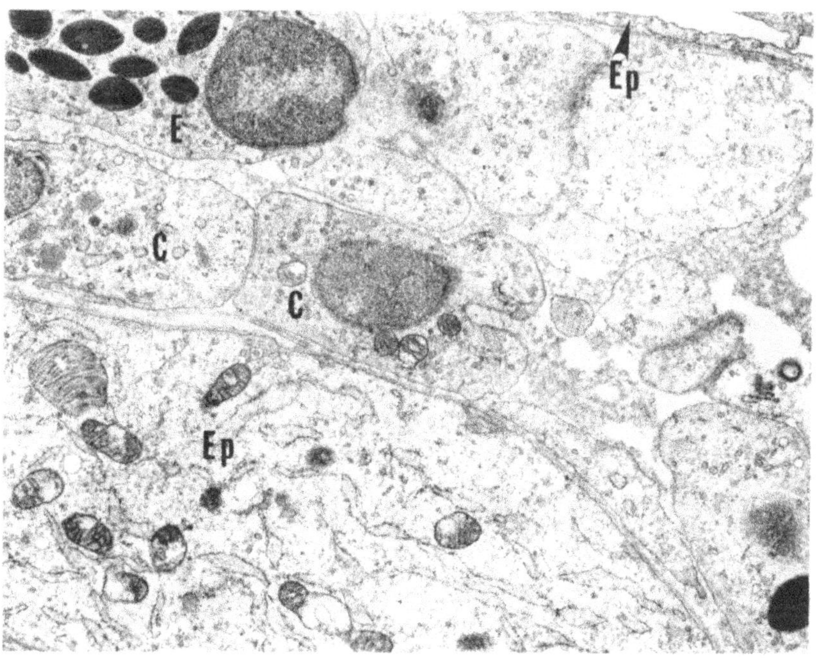

FIGURE 13. Electron micrograph of a lung section from a hamster 8 d postamiodarone treatment. This micrograph shows an interstitial area between two epithelial cells (Ep) which is occupied by cuboidal cells (C). Eosinophil = E. (Magnification × 13,600.)

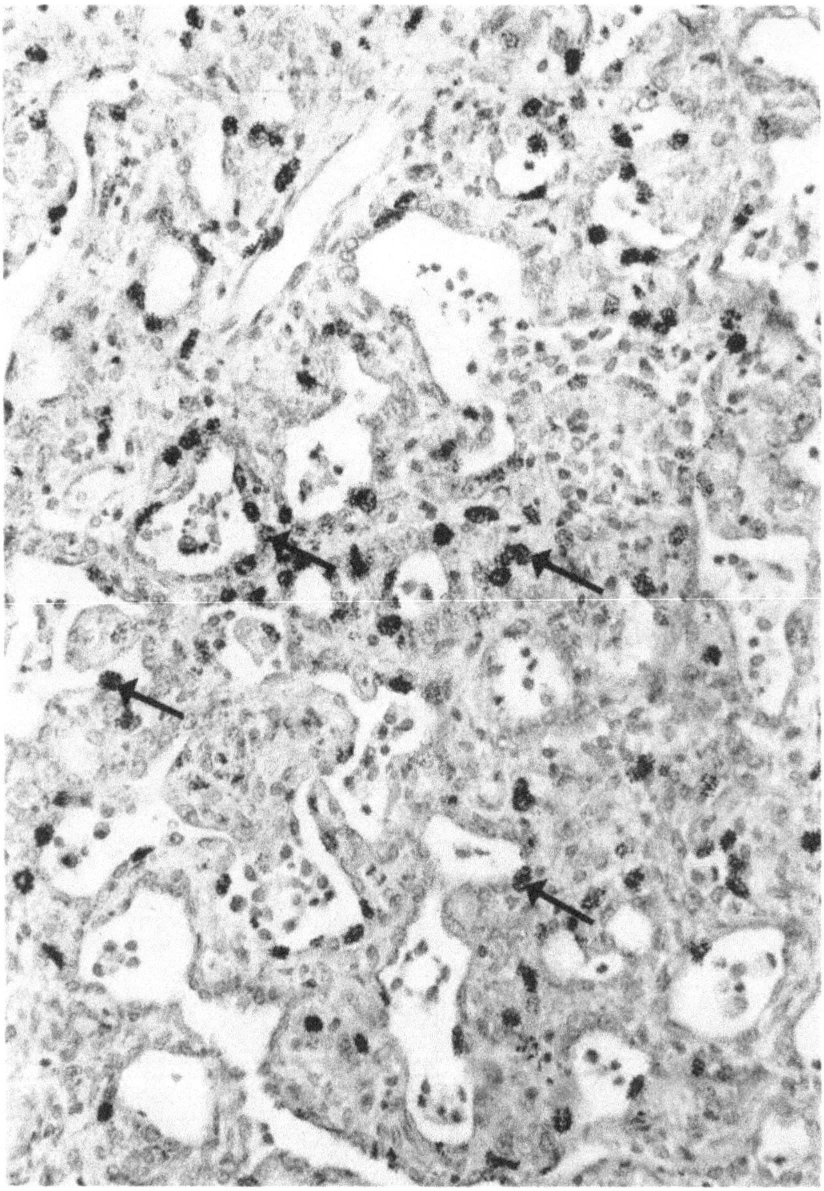

FIGURE 14. Photomicrograph of a lung section from a hamster 14 d postamiodarone treatment. This section was used for an autoradiographic study and contains numerous cells which have incorporated the ^{3}H-thymidine label in their nuclei (arrows). Note also the almost complete loss of normal architecture along with the honeycomb appearance of the lung. (Magnification × 285.)

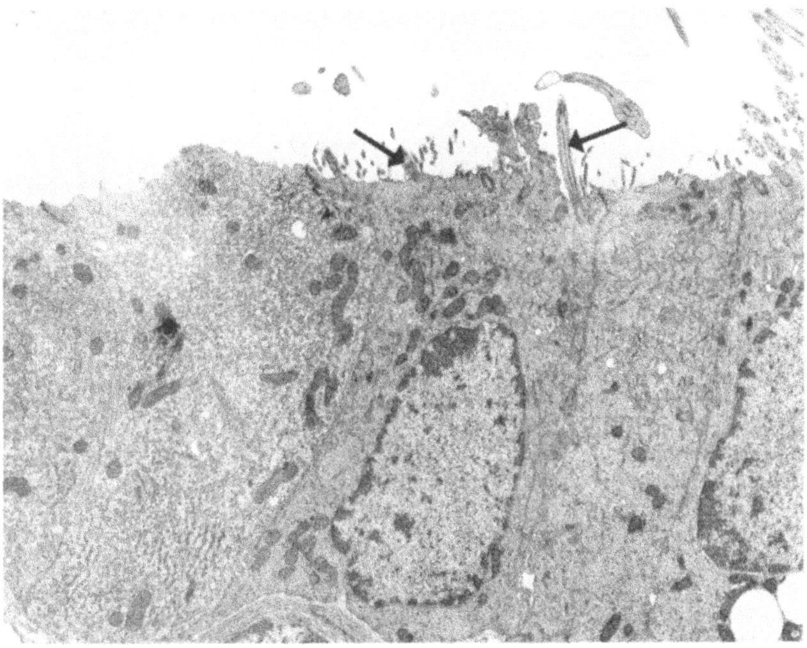

FIGURE 15. Electron micrograph of a lung section from a hamster 21 d postamiodarone treatment. This is an epithelial cell from a honeycomb region. Note that this epithelium is composed of columnar cells, some of which are ciliated (arrows). (Magnification × 6000.)

fibrosis. Increases in overall numbers of cells within this compartment are seen and are a significant part of the disease process. The increases in interstitial cells may in part be responsible for elevated levels of collagen and elastin synthesis observed. While this early cell proliferation declines with time, these events probably set the stage for subsequent continued synthesis of connective tissue matrix observed throughout this experiment.

The final event of the disease process is the increase in connective tissue matrix material. Although the exact time of the onset of increased matrix synthesis was not determined, it is clear that this production begins at an early stage of the fibrotic process. Both newly formed collagen and elastin are increased in amiodarone-treated animals.[44,50] The levels of these components are significantly higher than in control animals. This increase is most probably a reflection of synthesis by greater numbers of intersitial cells produced during the proliferative stage of the disease process. One of the more striking features of this model is the rapid onset of the fibrotic process. Fibrosis is produced at an earlier time period than occurs in the bleomycin model. This fibrotic process continues to spread throughout all time periods monitored by these experiments. The events described above, acute inflammation, cell proliferation, and matrix synthesis, ultimately lead to lung remodeling. This remodeling continues for an extended period of time. Treated animals kept for 2 to 4 months have lungs with substantially more fibrosis than those from animals at 1 month. This increased fibrotic involvement indicates that once started, this process tends to proceed for extended periods of time even in the absence of the initiating agent.

With the exception of a more rapid onset, the amiodarone compares quite closely with the bleomycin model. This rapid onset is an asset because it reduces the time needed to induce the disease process, thus allowing investigation of events of pulmonary fibrosis in a shorter time period. Further, with an additional model it is possible to compare the underlying cellular events of the fibrotic process to gain a more detailed understanding of the injury process.

ACKNOWLEDGMENTS

Work in the laboratory of the author was aided by a Grant in Aid from the Stony-Wold Herbert Foundation and by USPHS grant HL15832. Sincere thanks are given to Micheal Doskotz for his technical assistance and to Dr. Jerome O. Cantor for his encouragement throughout these experiments.

REFERENCES

1. **Colby, T. V. and Churg, A.,** Patterns of pulmonary fibrosis, *Pathobiol. Annu.,* 21, 277, 1986.
2. **Huuskonen, M. S., Tossavainen, A., Koskinen, H., Sitting, A., Korhonen, O., Nickels, J., Korhonen, K., and Vaaranen, V.,** Wollastonite exposure and lung fibrosis, *Environ. Res.,* 30, 291, 1983.
3. **Tablan, O. C. and Reyes, M. R.,** Chronic interstitial pulmonary fibrosis following *Mycoplasma pneumoniae* pneumonia, *Am. J. Med.,* 79, 268, 1985.
4. **Chastre, J., Raghu, G., Soler, P., Brun, P., Basset, F., and Gilbert, C.,** Pulmonary fibrosis following pneumonia due to acute Legionnaires disease, *Chest,* 91, 57, 1987.
5. **Trask, C. W., Joannides, T., Harper, P. G., Tobias, J. S., Spiro, S. G., Geddes, D. M., Souhami, R. I., and Beverly, P. C.,** Radiation-induced lung fibrosis after treatment of small cell carcinoma of the lung with very high-dose cyclophosphamide, *Cancer,* 55, 57, 1985.
6. **Adamson, I. Y. R.,** Drug-induced pulmonary fibrosis, *Environ. Health Perspect.,* 55, 25, 1984.
7. **Cantor, J. O., Osman, M., Cerreta, J. M., Suarez, R., Mandl, I., and Turino, G. M.,** Amiodarone-induced pulmonary fibrosis in hamsters, *Exp. Lung Res.,* 6, 1, 1984.
8. **Phan, S. M., Thrall, R. S., and Ward, P. A.,** Bleomycin-induced pulmonary fibrosis in rats: biochemical demonstration of increased rate of collagen synthesis, *Am. Rev. Respir. Dis.,* 121, 501, 1980.
9. **Schoenberger, C. I., Rennard, S. I., Bitterman, P. B., Fukuda, Y., Ferrans, V. J., and Crystal, R. G.,** Paraquat-induced pulmonary fibrosis, *Am. Rev. Respir. Dis.,* 129, 168, 1984.
10. **Shiotsuka, R. N., Yermakoff, J. K., Osheroff, M. R., and Drew, R. T.,** The combination of ozone and silica on the development of pulmonary fibrosis, *J. Toxicol. Environ. Health,* 17, 297, 1986.
11. **Schrier, D. J. and Phan, S. H.,** Modulation of bleomycin-induced pulmonary fibrosis in BALB/c mouse by cyclophosphamide-sensitive T-cells, *Am. J. Pathol.,* 116, 270, 1984.
12. **Scadding, J. G.,** Diffuse pulmonary alveolar fibrosis, *Thorax,* 29, 271, 1974.
13. **Hance, A. J. and Crystal, R. G.,** The connective tissue of the lung, *Am. Rev. Respir. Dis.,* 112, 657, 1975.
14. **Fulmer, J. D. and Crystal, R. G.,** The biochemical basis of pulmonary function, in *The Biochemical Basis of Pulmonary Function,* Crystal, R. G., Ed., Marcel Dekker, New York, 1976, 419.
15. **Crystal, R. G., Fulmer, J. D., Baum, B. J., Bernardo, J., Bradley, K. H., Bruel, S. D., Elson, N. A., Fells, G. A., Ferrans, V. J., Gadek, J. E., Hunningkake, G. W., Kawanami, O., Kelman, J. A., Line, B. R., McDonald, J. A., McLees, B. D., Roberts, W. C., Rosenbery, D. M., Tolstoshev, P., Von Gal, E., and Weinberger, S. E.,** Cells, collagen and idiopathic pulmonary fibrosis, *Lung,* 155, 199, 1978.
16. **Snider, G. L.,** Connective tissue and mechanical behavior of lungs, in *Connective Tissues in Arterial and Pulmonary Disease,* McDonald, T. F. and Chandler, A. B., Eds., Springer-Verlag, New York, 1981, 177.
17. **Rosenquist, T. H.,** Organization of collagen in human pulmonary alveolar wall, *Anat. Rec.,* 200, 447, 1981.
18. **Caton, R. W., Damavskas, J., Twes, B., and Hass, B. M.,** Isolation and study of the elastic tissue network of the lung in three dimensions, *Am. Rev. Respir. Dis.,* 82, 186, 1960.
19. **Starcher, B. C., Kuhn, C., and Overton, J. E.,** Increased elastin and collagen content in the lungs of hamsters receiving an intratracheal injection of bleomycin, *Am. Rev. Respir. Dis.,* 117, 299, 1978.
20. **Goldstein, R. H., Lucey, E. C., Franzblau, C., and Snider, G. L.,** Chronic interstitial pulmonary fibrosis produced in hamsters by endotracheal bleomycin, *Am. Rev. Respir. Dis.,* 117, 289, 1978.
21. **Snider, G. L., Hayes, J. A., and Korthy, A. L.,** Chronic interstitial pulmonary fibrosis produced in hamsters by endotracheal bleomycin, *Am. Rev. Respir. Dis.,* 117, 1099, 1978.
22. **Goldstein, R. H., Lucey, E. C., Franzblau, C., and Snider, G. L.,** Failure of mechanical properties to parallel changes in lung connective tissue composition in bleomycin-induced pulmonary fibrosis in hamsters, *Am. Rev. Respir. Dis.,* 120, 67, 1979.
23. **Clark, J. G., Overton, J. E., Marino, B. A., Uitto, J., and Starcher, B. C.,** Collagen biosynthesis in bleomycin-induced pulmonary fibrosis in hamsters, *J. Lab. Clin. Med.,* 96, 943, 1980.

24. **Cantor, J. O., Osman, M., Mandl, I., and Turino, G. M.,** The sequence of crosslinked elastin synthesis in bleomycin-induced lung fibrosis, *Fed. Proc.*, 41, 323, 1982.
25. **Keller, S., Turino, G. M., and Mandl, I.,** Separation of elastin components by thin layer chromatography and electrophoresis, *Connect. Tissue Res.*, 8, 251, 1981.
26. **Cantor, J. O., Osman, M., Keller, S., Cerreta, J. M., Mandl, I., and Turino, G. M.,** Measurement of cross-linked elastin synthesis in bleomycin-induced pulmonary fibrosis using a highly sensitive assay for desmosine and isodesmosine, *J. Lab. Clin. Med.*, 103, 384, 1984.
27. **Cantor, J. O., Keller, S., Parshley, M. S., Darnule, T. V., Darnule, A. T., Cerreta, J. M., Turino, G. M., and Mandl, I.,** Synthesis of cross-linked elastin by an endothelial cell culture, *Biochem. Biophys. Res. Commun.*, 95, 1381, 1980.
28. **Goldstein, R. A. and Starcher, B. C.,** Elastin peptides in the urine of hamsters following intratracheal injection of elastase, *Am. Rev. Respir. Dis.*, 115, 331, 1977.
29. **Darnule, T. V., Likhit, V., Turino, G. M., and Mandl, I.,** Immune response to peptides produced by enzymatic digestion of microfibrils and elastin of human lung parenchyma, *Connect. Tissue Res.*, 5, 67, 1977.
30. **King, G. S., Mohan, V. S., and Starcher, B. C.,** Radioimmunoassay for desmosine, *Connect. Tissue Res.*, 7, 263, 1980.
31. **Kucich, U., Christner, P., Weinbaum, G., and Rosenbloon, J.,** Immunologic identification of elastin derived peptides in the serum of dogs with experimental emphysema, *Am. Rev. Respir. Dis.*, 122, 461, 1980.
32. **Darnule, T. V., Darnule, A. T., Likhite, V., Turino, G. M., and Mandl, I.,** Antigenic determinants in human elastin peptides, *Connect. Tissue Res.*, 269, 1980.
33. **Harel, S., Janoff, A., Yu, S. Y., Hurewitz, A., and Bergofsky, E. H.,** Desmosine radioimmunoassay for measuring elastin degradation in vivo, *Am. Rev. Respir. Dis.*, 122, 769, 1980.
34. **Cantor, J. O., Bray, B. A., Ryan, S. F., Mandl, I., and Turino, G. M.,** Glycosaminoglycan and collagen synthesis in N-nitroso-N-methylurethane induced pulmonary fibrosis, *Proc. Soc. Exp. Biol. Med.*, 164, 1, 1980.
35. **Cantor, J. O., Cerreta, J. M., Osman, M., Mott, S. H., Mandl, I., and Turino, G. M.,** Autoradiographic study of glycosaminoglycan synthesis in bleomycin-induced pulmonary fibrosis, *Fed. Proc.*, 40, 594, 1981.
36. **Cantor, J. O., Cerreta, J. M., Osman, M., Mott, S. H., Mandl, I., and Turino, G. M.,** Glycosaminoglycan synthesis in bleomycin-induced pulmonary fibrosis. Biochemistry and autoradiography, *Proc. Soc. Exp. Biol. Med.*, 174, 172, 1983.
37. **Rotmensch, H. H., Liron, M., Tupuilske, M., and Laniado, S.,** Possible association of pneumonitis with amiodarone therapy (letter), *Am. Heart J.*, 100, 412, 1980.
38. **Sobol, S. M. and Rakaita, L.,** Pneumonitis and pulmonary fibrosis associated with amiodarone treatment: a possible complication of new antiarrhythmic drug, *Circulation*, 65, 819, 1982.
39. **Marchlinski, F. E., Gansler, T. S., Waxman, H. L., and Josephson, M. E.,** Amiodarone pulmonary toxicity, *Ann. Intern. Med.*, 97, 839, 1982.
40. **Riley, S. A., Williams, S. E., and Cooke, N. J.,** Alveolitis after treatment with amiodarone, *Br. Med. J.*, 284, 161, 1982.
41. **Naccarelli, G. V., Rinkenberger, R. L., Dougherty, A. H., and Giebel, R. A.,** Amiodarone: pharmacology and antiarrhythmic and adverse effects, *Pharmacotherapy*, 5, 298, 1985.
42. **Adams, P. C., Gibson, G. J., Morley, A. R., Wright, A. J., Corris, P. A., Reid, D. S., and Campbell, R. W. F.,** Amiodarone pulmonary toxicity: clinical and subclinical features, *Q. J. Med.*, 229, 449, 1986.
43. **Musto, L.,** Improved iron-hematoxylin stain for elastic fibers, *Stain Technol.*, 52, 173, 1962.
44. **Cantor, J. O., Keller, S., Mandl, I., and Turino, G. M.,** Increased synthesis of elastin in amiodarone-induced pulmonary fibrosis, *J. Lab. Clin. Med.*, 109, 480, 1987.
45. **Sun, X., Davis, W. B., Fukuda, Y., Ferrans, V. J., and Crystal, R. G.,** Experimental polyxin B-induced interstitial lung disease characterized by an accumulation of cytotoic eosinophils in the alveolar structures, *Am. Rev. Resp. Dis.*, 131, 103, 1985.
46. **Baker, J. R., Baker, P. A., Bassett, E. G., and Myers, D. B.,** Collagenolytic activity in peritoneal eosinophils from rats, *Proc. Physiol. Soc.*, 242P, 1976.
47. **Bassett, E. G., Baker, J. R., Baker, P. A., and Myers, D. B.,** Comparison of collagenase activity in eosinophil and neutrophil fractions from rat peritoneal exudates, *Aust. J. Exp. Biol. Med. Sci.*, 54, 459, 1976.
48. **Davis, W. B., Gadek, J. E., Fells, G. A., and Crystal, R. G.,** Role of eosinophils in connective tissue destruction, *Am. Rev. Respir. Dis.*, 123, 55, 1981.
49. **Hibbs, M. S., Mainardi, C. L., and Kang, A. H.,** Type-specific collagen degradation by eosinophils, *Biochem. J.*, 207, 621, 1982.
50. **Cerreta, J. M., Doskotz, M., and Cantor, J. O.,** unpublished data.

THE PARAQUAT MODEL OF LUNG FIBROSIS

Jerold A. Last

INTRODUCTION

Overview

There are many established models in rodents for studying experimental pulmonary fibrosis. Many investigators presently use, as a model of choice, intratracheally instilled bleomycin, with or without supplemental oxygen, for this purpose. The bleomycin model is a simple, convenient animal model to use. However, other animal models continue to be developed and examined in specialized circumstances or for particular purposes, and no rodent model is considered to be an exact analogue of human disease as encountered clinically. The paraquat model is of interest for historical reasons (it was, with intratracheal silica, the first to be exploited for experimental studies of the prefibrotic lung) and continues to be used in several laboratories for the study of lung damage, especially as a model of interstitial lung disease. It is also convenient to use in the sense that intraperitoneal injection requires less technical skill than intratracheal instillation, and animals need not be anesthetized for drug administration.

Authoritative reviews of paraquat have been published previously. Smith and Heath[1] have discussed the chemistry and toxicology of this herbicide. A bibliography of literature references to paraquat toxicity through 1978 was compiled by Imperial Chemical Industries, the manufacturer of paraquat, and is presumably still available with updates to the present through their central toxicology laboratory in Cheshire, England. A symposium volume[2] has also been published.

Paraquat-Induced Lung Damage

Rats — Vijeyaratnam and Corrin[3] have described the histological and ultrastructural changes in rat lungs after administration of 25 to 40 mg/kg of paraquat. The main observations included very early (ultrastructural) changes in the alveolar type I cells (membranous pneumocytes) which could be appreciated within 4 h of paraquat administration. Frank destruction of alveolar epithelial cells (both types I and II) was apparent within less than 1 d of paraquat administration. Severe lung damage, with loss of alveolar structure, occurred within 2 d. Smith and Heath[4] gave rats 25 mg/kg of paraquat. Within 7 d alveolar walls were infiltrated by mononuclear cells (profibroblasts); by 14 d, accumulation of fibroblasts and an increase in parenchymal collagen were evident. Fibrosis was apparent in lungs of rats sampled at 28 and 42 d after administration of paraquat. Similar histopathological changes occurred in lungs of rats administered paraquat orally. Fisher et al.[5] examined rats administered 27 mg/kg of paraquat 3 d earlier by various physiological measurements. They suggested that atelectasis secondary to pulmonary edema caused the observed alterations in lung pressure-volumes curves, which were shifted in the direction of requiring an increased distending pressure to achieve increased lung volumes.

Mice — Mice given an LD_{50} dose of paraquat (20 mg/kg) by intraperitoneal injection were killed 1 to 72 h after injection.[6] Changes consistent with those described in the acute phase of injury in rats (i.e., severe edema and consolidation of the lung; epithelial cell damage) were observed at all time points.

Hamsters — Butler[7] administered 3 to 12 mg/kg of paraquat to hamsters by subcutaneous injection using various dosage and duration regimens between injections. The responses observed were similar to those reported in rats except that lung fibrosis was observed in this study only after a multiple dosage regimen.

Dogs — Kelly et al.[8] reported upon necropsy findings in ten dogs that had accidentally ingested unknown (but lethal) doses of paraquat. Early loss of alveolar epithelial cells (types I and II pneumocytes) with rapid onset of pulmonary fibrosis was consistent with the histopathology described in laboratory rodents studied under controlled circumstances.

Monkeys — Fukuda et al.[9] administered one or two injections of 10 mg/kg of paraquat by the subcutaneous route to adult cynomolgous monkeys (*Macaca fascicularis*). Consistent with observations in other species, they observed damage to and necrosis of alveolar epithelial cells and intra-alveolar edema within 2 to 3 d of paraquat administration. Fibroblast accumulation and early stages of fibrosis were observed focally in the alveolar zone within 1 week after administration of paraquat. By 3 to 4 weeks later, intra-alveolar and septal fibroses were apparent. Obliteration of normal alveolar architecture was apparent (these changes persisted for at least 8 weeks, the last time point examined). It was observed that at least some of the abnormal collagen in the lung parenchyma may have been subject to breakdown by histological criteria, suggesting the possibility that some components of the fibrotic response might be reversible in these animals.

Effects of Paraquat Aerosols on Animals

In early studies, Gage[10] examined the effects of paraquat and diquat aerosols of effective mass median aerodynamic diameters of less than 5 μm on experimental animals. In rats, the lethal concentration (C) × time (T) product was found to be about 6 μg/l × hours (6 mg/m^3 × hours). Guinea pigs and male mice had similar sensitivity. Female mice and rabbits were less sensitive to paraquat aerosols. Dogs were very resistant, tolerating a C × T product of about 25 μg/l (25 mg/m^3) × hours without ill effects. The no-effect level in rat was estimated to be about 0.1 μg/l (0.1 mg/m^3). Popenoe[11] examined the effects of paraquat aerosols on Balb/c mice. Subjective histological examinations demonstrated the characteristic lesions of paraquat that have been associated with administration by other routes in these animals, which were exposed to an aerosol of paraquat dichloride solution of nominal concentration 12 mg/ml for 15 min by nose-only exposure. It is not possible to calculate from this paper actual doses delivered to the lung. Presumably the nose-only exposure protocol would minimize, if not exclude, oral or percutaneous absorption of the paraquat.

Confounding Effects

Rhodes et al.[12] first observed that paraquat seemed to be more toxic when dissolved in water than when dissolved in saline solution. Drew and Gram[13] have also reported that the toxicity of paraquat dissolved in saline was less than when it was dissolved in distilled water and administered to mice either intraperitoneally or subcutaneously. They postulated that this was due to a change in the rate of absorption from the peritoneal cavity or from the subcutaneous space, possibly related to whether the paraquat was oxidized or reduced at the time of injection. In addition, Rhodes et al.[12] have shown that mice injected intraperitoneally with high doses of paraquat have a lower mortality rate when housed under conditions of 10% oxygen than when housed in room air (approximately 20% oxygen). The mice stored under hypoxic conditions rapidly developed pulmonary edema and died when returned to normoxic conditions.

The Effects of Paraquat in Species Other than Rat

In an early study, Murray and Gibson compared the effects of paraquat in rats, guinea pigs, and monkeys.[13] The oral LD_{50} doses were 126, 22, and 50 mg/kg in the three species, respectively. Interestingly, the guinea pigs were resistant to pulmonary fibrosis, while both rats and monkeys developed fibrosis if they survived for as long as 1 to 2 weeks. This study also reports on changes in various other clinical parameters in these three species. Ilett et al.[14] have compared the toxicity of paraquat in rats and rabbits. Rabbits are unusually resistant to paraquat-induced lung damage. However, high doses of paraquat will kill rabbits, usually within the first 24 h of administration. Their apparent resistance may be related to the ability of the rabbit lung to rapidly clear paraquat, rather than to retain it. Rose and co-workers[15] examined the accumulation of paraquat by lungs from several species, using *in vitro* assays. Lung slices from rats, dogs, monkeys, and rabbits all accumulated paraquat from the medium. The apparent K_m (the Michaelis constant), an approximation of the binding constant for the paraquat receptors within the tissue, were very similar for all four species, ranging from 20 to 70 m*M*. The corresponding value for human tissue was reported to be approximately 40 m*M*, almost exactly in the middle of this range. Values of V_{max} (the theoretical maximum velocity of uptake of paraquat by lung slices) ranged from 10 nmol of paraquat per gram of tissue per hour for dogs to 200 to 300 nmol/g/h in rabbits, humans, and rats. The value in monkeys was intermediate, 50 nmol/g/h. There is probably a good correlation between the accumulation of paraquat by the lung of an animal and the subsequent toxicity of paraquat to that animal.[16]

IMPLEMENTING THE MODEL

The Paraquat Model of Pulmonary Fibrosis

Paraquat (1,1′-dimethyl-4,4′-bipyrinidium chloride; methyl viologen) can be conveniently administered by injection. Usually the intraperitoneal or intravenous route is used. Pure paraquat is a white, crystalline substance that is sold by specialty companies as the redox dye methyl viologen. The commercial product (Grammoxone®, Chevron Chemical Company, Richmond, CA or Weedol®, Western Europe), is a 20% aqueous solution of paraquat formulated with stabilizers, spreading agents, and antioxidants. Several dosage regimens have been described for elicitation of pulmonary disease in laboratory animals, using either pure paraquat or diluted commercial material as formulated.

As becomes obvious immediately to workers with this model, there is a tremendous difference in species response among laboratory animals to the effects of paraquat upon the lung (see above), as well as a potential large interanimal variability among otherwise well-matched animals within an experimental grouping. Animal to animal variability tends to be greater at lower doses of paraquat, thereby making dose-response experiments very difficult to perform.

High doses of paraquat are customarily used in the University of California laboratory, usually the LD_{50} dose as determined after 7 d (i.e., 24 mg/kg), to try to suppress the apparent biological variability among rats in response to this agent. The apparent LD_{50} dose in rats seems to differ over a fairly wide range among different laboratories that have reported this value, for reasons that are not completely understood. It has been suggested[12,13] that the solvent used to dissolve the paraquat might affect the experimentally determined LD_{50} dose. Exposure of the material to air, which results in oxidation of the paraquat dye, may also contribute to this apparent variability in LD_{50} dose from laboratory to laboratory. A great deal of additional confusion may result from the practice of expressing paraquat concentrations in terms of the hydrochloride salt, of the free base, or of other parameters. In the University of California laboratory, a dose of 24 mg of paraquat hydrochloride per kilogram body weight has been used for most experiments.

In addition to killing half of the animals injected intraperitoneally after 7 d, this dose causes an approximate 15 to 20% decrease in animal body weight and an approximate 50% increase in lung wet weight 3 d after administration to rats.[17,18] In addition to pulmonary edema, the lungs of animals injected with this dose of paraquat are frankly hemorrhagic over essentially all of their surface area (Table 1).

A useful technique for looking at the nonspecific effects of paraquat is to compare animals

Table 1
EFFECTS OF PARAQUAT ON RATS AT VARIOUS TIMES AFTER ITS ADMINISTRATION

		Toxicologic criteria			Histologic criteria		
Days after paraquat injection	Number of rats evaluated[a]	Rat weight, av percent of initial weight[b]	Percent of lungs hemorrhagic, av by visual estimate[c]	Wet/dry weight ratio of right apical lobe, (mean ± SD)[d]	Masson's trichrome stain[e]	Gomori's stain[e]	Hematoxylin and eosin stain[f]
1	2	90	0	4.68 ± 0.18	0	±	±
2	2	87	35	5.95 ± 0.29[g]	±	+	±
3	2	84	82	7.62 ± 0.55[g]	±	+	+
4	1	74	80	5.74	±	+ +	+ +
5	3	76	88	6.03 ± 0.29[g]	±	+ +	+ +

[a] Two control rats were evaluated in parallel each day.
[b] Percentage of initial weight ranged from 99 to 102% for the control group.
[c] All control lungs were scored as 0% hemorrhagic in this measurement.
[d] Wet/dry weight ratio for control rats was 4.68 ± 0.28 (n = 9, one sample was lost).
[e] Sections were evaluated on a scale of 0 = indistinguishable from normal, ± = marginal subjective increase, + = increased level of staining, and + + = very strong increase in staining.
[f] Sections were evaluated on a scale of 0 = normal, indistinguishable from control; ± = mild edema, cellular infiltration; + = severe edema, cellular infiltration, mild parenchymal destruction; and + + = severe edema, cellular infiltration, severe parenchymal destruction.
[g] Differences significant (Student *t* test) at the $p < 0.01$, level, assuming a normal distribution of the data.

From Greenberg, D. B., Reiser, K. M., and Last, J. A., *Chest,* 74, 421, 1978. With permission.

injected with paraquat with other animals injected with its analog, diquat. Diquat is an analog of paraquat with its quaternary nitrogens in the ortho- rather than the parapositions. It is also active as a herbicide, but it apparently is not specifically accumulated in the lung. Thus, one can compare compounds with very similar chemical and biological properties except for the effects directly upon the lung as a target. Diquat at an equivalent dose to paraquat does not cause either the severe weight loss, pulmonary edema, or lethality associated with paraquat administration.

One of the most obvious effects of paraquat on the lung is to induce severe pulmonary edema with hemorrhage. If animals survive this acute insult, they develop disseminated pulmonary fibrosis. The exact anatomical sites of this fibrosis reflect the extreme destruction of the lung structure associated with paraquat toxicity[9] (Table 1).

Apparently, there is sufficiently severe destruction of the alveolar walls in the early stages of paraquat-induced lung injury that alveoli are literally obliterated and the walls of former alveoli may coalesce to form new structures. In addition to the loss of functional gas exchange units associated with such a formation of "new alveoli" by fusion of components of "old alveoli", one can also appreciate intra-alveolar rather than intraseptal fibrosis in this animal model. Paraquat-induced lung injury is perhaps unique in its severity (commensurate with animal survival) among the various models of lung fibrosis, thereby resulting in an unusual histopathological form of lung fibrosis. This apparent remodeling of alveoli may also be responsible for the concept of "profibroblasts", mononuclear interstitial cells that are appreciated microscopically as appearing within alveoli rather than the interstitium. In fact, as "new alveoli" form by coalescence of the old walls, interstitial cells may appear to be existing free in the airways, even if they are not.

At a constant dose of paraquat, there is a time-dependent increase in lung collagen synthesis that can be appreciated rapidly after administration of this toxicant to rats. At the dose of 24 mg/kg of paraquat that has been used routinely, one can distinguish elevated lung protein synthesis rates and collagen synthesis rates as early as 3 and 2 d after administration of paraquat, respectively[17] (Figures 1 and 2). The apparent collagen synthesis rate increases perceptibly each day after that for at least 5 consecutive days[17] (Figure 2). Using this model, it has also been shown that not only is there more collagen synthesis and accumulation by affected lungs, but the collagen type being made is different than in normal lungs. Specifically, the relative ratio of type I and type III collagen is increased from a normal ratio of about 2:1 to a value of approximately 6:1.[19]

EVALUATING THE MODEL

Determination of Apparent Lung Collagen Synthesis Rate

One can examine apparent collagen synthesis rates using *in vitro* methods in order to quantify the effects upon the lung of paraquat administered *in vivo*. Such *in vitro* methods allow one to study changes in the lung under conditions where the lung tissue is isolated from confounding effects of blood, circulating hormones, and other potential extrapulmonary variables in response to the primary and secondary effects of paraquat intoxication described above. Such *in vitro* culture methods also allow for the accurate determination of the specific activity of the radioactive proline used as a precursor of lung collagen, a necessary determination if rates of synthesis are to be quantified.

For a typical experiment, groups of six or more rats are injected with either paraquat solution or the vehicle used to dissolve the paraquat. To do this, commercial Grammoxone® (20% w/v of paraquat) is diluted with water to give a final concentration suitable for direct injection into rats. In general, stock Grammoxone® (stored in the dark at room temperature) is diluted with distilled water to give a final concentration of 2.9% w/v. For example, 1 ml of stock Grammoxone® diluted with 6 ml of water would give a working solution of 7 ml

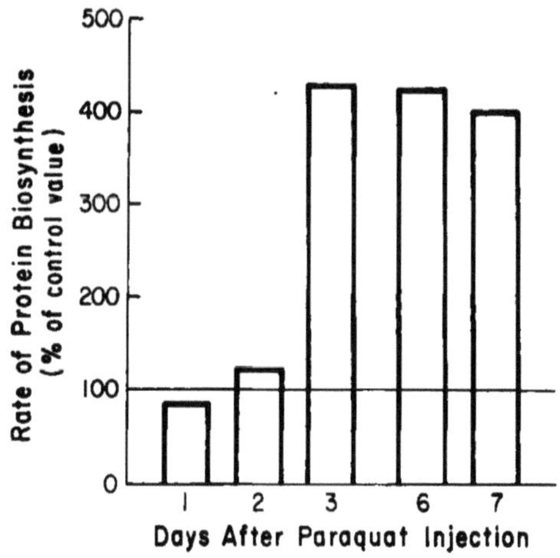

FIGURE 1. Effect of paraquat administration *in vivo* on protein biosynthesis rate measured with cultured lung explants. Lung slices were incubated with 50 μCi of [^{3}H]proline (490 Ci/mmol, generally labeled, New England Nuclear Company) added to the culture medium. At 1, 2, and 3 h after initiating incubation with the isotope, slices were separated from the medium, washed, and homogenized. Total Cl_3CCOOH-precipitable radioactivity was determined on a 20-μl aliquot of the homogenate. The resultant counts per minute of acid-precipitable proline were normalized to the protein content of the incubation vial, and were converted to nanomoles of proline per milligram of protein after determination[12,16] of the specific activity of proline in the acid-soluble fraction of the tissue homogenate. The resultant data were plotted against time of incubation. The best straight line was fitted to the data by a linear regression analysis (least-squares program) to determine the rate of biosynthesis as nanomoles of proline per milligram of protein per hour. The average ratio for these values was then used to calculate the percentage of the control value at each day. Since the control values were similar at all times determined, the average control value was calculated from pooled control data for all days sampled (n = 8). (From Greenberg, D. B., Lyons, S. A., and Last, J. A., *Chest*, 74, 421, 1978. With permission.)

of 2.9% paraquat. Rats are then weighed and injected intraperitoneally with the calculated volume to give a dose of 24 mg/kg. For a 300-g rat, this would correspond to a dose of 0.25 ml of the working solution, conveniently injected with a 1-ml syringe using a 23-gauge needle. Rats free of bacterial or mycoplasmal lung pathogens, obtained from various suppliers including Charles River and Hilltop, have routinely been used for all experiments at the University of California to prevent any confounding effects of intercurrent lung infections.

Rats that have received large doses of paraquat show severe weight loss, about 15% of their preinjection values, within 3 d of receiving this agent (Table 1). Presumably, this is due to failure to eat and drink during this time. They also show other signs of acute distress, including listless behavior, panting and other overt respiratory distress, and pilomotor erection. There seems to be some variability in response of rats within otherwise identical groups. Between 10 and 20% of injected rats will show no symptoms of acute intoxication, as if they did not receive paraquat or it did not become absorbed and reach their lungs. One can deal with this problem by excluding any "nonresponder" animals from a study by the criterion of no overt weight loss even though they received paraquat.

Animals are killed by pentobarbital overdose administered intraperitoneally as specified by the protocol, or when obviously moribund or in obvious respiratory distress. Their chests are opened with a scissors and the lungs are exposed. Lungs are grossly hemorrhagic over their entire surface and are obviously engorged with cells and fluid due to edema. The wet to dry weight ratio of the lung, an index of pulmonary edema, is increased by about 50% 3 d after administration of 24 mg/kg of paraquat to rats (Table 1). The wet to dry weight ratio is easy to determine by weighing the lung (or a lung lobe) before and after drying for 1 to 2 d at 110 to 120°C, and is a precise and reproducible property of the lung. For normal rats, this ratio is 4.7 ± 0.3 (mean ± S.D.), while for animals administered paraquat 3 d

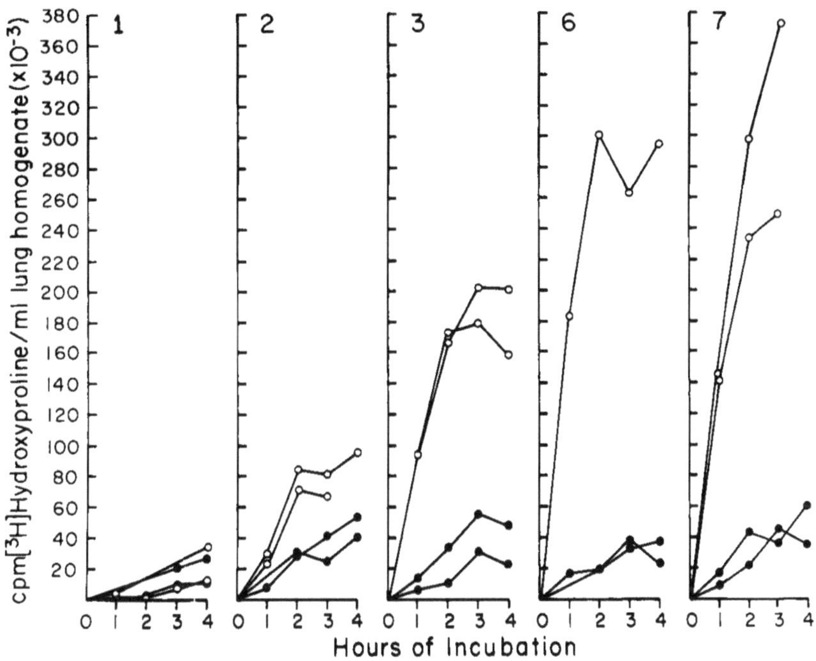

FIGURE 2. Effect of paraquat administration *in vivo* on collagen biosynthesis rate measured with cultured lung explants. Experimental details as in Figure 1. [^{3}H]Hydroxyproline was determined on a 600-μl aliquant of the homogenate. Data are expressed on the basis of nanomoles of hydroxyproline per milligram of lung-slice protein. Numbers at top of panels indicate days after paraquat administration. (From Greenberg, D. B., Reiser, K. M., and Last, J. A., *Chest*, 74, 421, 1978. With permission.)

earlier, ratios of 6 or more are apparent (6.5 ± 0.6, in one series, for example). In affected animals, there is a consistent response (coefficient of variation less than 10%) of this parameter when high doses of paraquat are administered, and the response is proportional to subsequent fibrosis evaluated at later time points.

A large series of papers from the University of California laboratory has shown that there is a correlation between the apparent rate of collagen synthesis as evaluated with minced lung tissue and the subsequent extent of pulmonary fibrosis, defined histologically or as total lung content of hydroxyproline, and ascertained in several different animal models of pulmonary fibrosis. Thus, a lung slice or mince system may be used to quantify the response of the lung to injection of paraquat and thereby the extent of the (pre)fibrotic response of the lung to injury. This type of assay, which is not for everybody to do because it is very technically demanding and tedious to perform, may be done as follows.

Lungs are perfused *in situ* with room-temperature 0.1 *M* NaCl via the pulmonary artery to remove blood trapped in the vasculature. The right apical (cranial) lobe may be tied off or clamped prior to perfusion if wet to dry weight ratio of the lung is to be evaluated. The (remaining) lung is removed from the animal, trimmed free of extraneous tissue and the trachea with scissors, and rinsed in tissue culture medium (usually Dulbecco's modified Eagle's medium (MEM) from Gibco, Grand Island, NY).

The lung lobes are placed on a glass plate well moistened with MEM and sliced with two razor blades into cubes 1 to 2 mm per side. The resultant tissue pieces are thoroughly mixed to blend tissue from all of the lobes and divided randomly into incubation vials (1.8 cm diameter, glass). Such vials are routinely prepared by cutting liquid scintillation sample vials (Wheaton Glass Company, Wheaton, IL) in half with a glass-cutter's wheel. Each vial contains 1.8 ml of MEM containing 100 μg/ml of ascorbic acid and (optionally) 64 μg/ml of β-aminoproprionitrile fumerate (nominally to block cross-linking of newly synthesized collagen, although little difference in results is seen whether or not it is added under culture conditions at the University of California laboratory). Lung tissue (150 to 200 mg wet weight; about 10% of the total lung mince) is added to each of (at least) three vials and preincubated at 37% for 1 h under O_2:CO_2 (95:5) or air (slightly higher incorporation rates are seen with O_2:CO_2, but air is acceptable) with vigorous shaking (Dubnoff® or equivalent metabolic shaker). In control experiments, various individual minced lung lobes have been compared from the same normal or paraquat-treated animal in this assay and the same apparent collagen synthesis rate has been found as when thoroughly mixed tissue pieces from the entire lung are used for the assay.

After approximately 1 h, the equilibration medium is removed and replaced with 0.1 μl of MEM plus ascorbate, (optionally) β-aminoproprionitrile, and 50 μCi of [^{3}H]proline (generally labeled, diluted to a specific activity of about 500 Ci/mol). The three (or more) vials from a single lung are incubated for 1, 2, and 3 h. Incubations are terminated by placing the vial on ice and aspirating off the medium with a Pasteur pipette. The tissue is then quickly washed with a few milliliters of ice-cold phosphate-buffered saline (pH 7.4 to 7.6) and the wash fluid removed by aspiration to remove as much radioactive free proline as possible from the fluid adhering to the tissue pieces.

The washed tissue is suspended in 1.5 ml of 0.5 *M* acetic acid and homogenized with a Polytron Homogenizer® (Brinkman Instruments, Westbury, NY) to give a viscous homogenate. Aliquots of the homogenate may be removed at this point for assay as follows: 100 μl for DNA and for protein content, 20 μl for Cl_3CCOOH-precipitable radioactivity (total protein synthesis), and 1.0 ml for Cl_3CCOOH-precipitable hydroxyproline content (collagen synthesis) and Cl_3CCOOH-soluble radioactivity (specific activity of total tissue proline pool). Total DNA and protein content are determined by colorimetric assays,[20] hydroxyproline by a combined radiometric and colorimetric assay[21] on hydrolyzed tissue, and total protein by a radiometric assay.[22]

Specific activity of the total lung proline pool is quantified for each sample by adjusting the pH of the Cl_3CCOOH-soluble fraction to about 7 (anywhere from pH 3 to 7 is acceptable) with NaOH. The soluble fraction is treated with Permutit® ion-exchange resin (Culligan corporation) to remove basic amino acids that may interfere with the colorimetric assay of proline. An aliquot of the resultant supernate is counted by liquid scintillation, while another aliquot is assayed for its total proline content by the colorimetric method of Troll and Lindsley.[23] Specific activity is calcuated as radioactivity (disintegrations per minute [^{3}H]proline) divided by the amount of proline (nanomoles) in the equivalent volume of lung homogenate. It has been determined in control experiments using paper chromatography and HPLC that essentially all of the soluble radioactivity in such extracts is authentic proline.

A huge amount of data is generated from a single experiment using this protocol, and a great deal of data reduction and computation is required. The University of California laboratory uses a computer for this purpose, and programs have been written in Fortran® for the Burroughs Prime to calculate results, perform linear regression analysis of data (see below), and perform statistical comparisons between groups (see Appendix 1). Details of such computer programs are available upon request. A skilled and experienced technician can work up lungs from up to 18 rats at a time (i.e., 54 incubation vials) and an entire experiment can be analyzed over 3 to 5 working days allowing time for various assays and required hydrolysis steps.

Apparent rates of collagen synthesis are determined as the slopes of lines generated by plotting for each sample (or group of samples) the hours of incubation (0, 1, 2, or 3) vs. the nanomoles of hydroxyproline synthesized per gram of lung tissue protein present in the incubation vials (Figure 2). Normalization to tissue protein allows correction for different amounts of lung tissue present in a given vial, albeit this is not often a large source of variability in a given experiment. Protein synthesis rate (Figure 1) may be quantified as slopes of plots of micromoles of proline incorporated per gram of tissue vs. incubation time.[18] The micromoles or nanomoles of proline are calculated from the observed incorporation of radioactive isotope after correction of *each* sample for the analytically determined specific activity of precursor proline in the lung tissue. The pros and cons of the use of the specific activity of the total lung tissue proline pool as a precursor specific activity are discussed elsewhere.[24] It has been shown that the specific activity of the tissue proline pool decreases dramatically in the hemorrhagic, inflamed, edematous lung after paraquat administration, apparently due to breakdown of tissue protein and influx of unlabeled amino acids. Normal rat lungs have a specific activity of about 1100 ± 110 cpm/ng of proline (mean ± 1 SD n = 6), while lungs from rats administered paraquat 3 d earlier contained proline of specific activity about 650 ± 75 cpm/ng of proline. One would obviously dramatically underestimate collagen synthesis (or protein synthesis) by lungs were such decreases in precursor specific activity not properly taken into account in calculating apparent synthesis rate.

DISCUSSION

A potentially more attractive model to work with using paraquat has recently been described by Selman and co-workers in Mexico City.[25,26] They have given rats intraperitoneal injections of approximately 2.5 to 7.5 mg/kg of body weight at intervals of from 3 d to 1 week between injections. The animals were also exposed to normal air or to 74% oxygen in the inspired air for 2 to 10 weeks. This procedure gives relatively low mortality, causes severe lung damage, and elicits severe and diffuse interstitial fibrosis. The effect of changing variables in usage of this experimental regimen has been described in some detail.[25,26] This variant of the paraquat model would seem to have many of the advantages and few of the disadvantages associated with some of the best models currently used to elicit pulmonary fibrosis in rats.

ACKNOWLEDGMENTS

Several colleagues did much of the experimental work with paraquat from the University of California laboratory, especially Drs. Karen Reiser and Deborah Greenberg and Mrs. Sheila Lyons. This work was supported, in part, by grants from the National Institutes of Health (ES-00628, RR-00169, and HL-07013).

APPENDIX I

```
C
C
C     SUBROUTINE HPRO
C
C
C   This is (hopefully) the last of a series of three
C   programs written to analysis Hydroxproline assay data.
C
C
C
C       Written by: Tracy Jones
C                   06/07/84
C       Modified:   Steve Fisher 12-19-84
C       Modified:   Steve Fisher  6-27-86
C         At: The California Primate Research Center
C             Data Sevices
C             Davis, Ca.
C
C
C
      INTEGER*2
     *   MPNT(25,3),                      /* MISSING POINT FLAG
     *   DGRP,                            /* NUMBER OF DARRENS T-TEST RUNS
     *   DGFREE(25),                      /* DEGREES OF FREEDOM (DARRENS T-TEST)
     *   DATITL(25,10),DBTITL(25,10),     /* TITLES OF GROUPS (DARRENS T-TEST)
     *   DFA(25),DFB(25),                 /* DEGREES OF FREEDOM (T-TEST)
     *   DLNGA(25,25),DLNGB(25,25),       /* LUNG NUMBERS FOR EACH GRP (D'S TT)
     *   DNUMA(25),DNUMB(25),             /* NUMER OF LUNGS (DTT)
     *   ATITL(25,10),BTITL(25,10),       /* TITLES OF GROUPS (T-TEST)
     *   ASSCOM(15),                      /* ASSAY COMMENT / KILL DATE
     *   IGRP,                            /* NUMBER OF T-TEST RUNS (GROUPS)
     *   TINC,                            /* TIME INCREMENT (ALWAYS 3)
     *   NLUNG,                           /* NUMBER OF LUNGS ASSAYED
     *   PRLNC,                           /* NUMBER OF PROLINE POINTS IN STND
     *   PROTC,                           /* NUMBER OF PROTEIN POINTS IN STND
     *   HYPRC,                           /* NUMBER OF HYPRO   POINTS IN STND
     *   I, J, CMMD,                      /* INDEXES, AND THE COMMAND VAR
     *   TCMMD,                           /* TTEST COMMAND VAR
     *   FILNAM(16),                      /* DATA FILE NAME
     *   NMLNG,                           /* LENGTH OF FILENAME
     *   XYZZY1                           /* LAST I*2 (NO COMMA)
C
      INTEGER*4
     *       HCPM(25,3),                  /* HPRO CPM
     *       TCAPPT(25,3),                /* TCA PPT CPM
     *       TTCA(25,3),                  /* TOTAL TCA PROTIEN/ML
     *       CHPRML(25,3),                /* CPM HYPRO/ML
     *       PCPM(25,3)                   /* PROLINE CPM
C
      REAL
     *       NMHPRO(25,3),                /* NUMBER OF MOLES HYPRO RECOVERED
     *       UGPRL(25,3),                 /* UG PROTIEN/ML
     *       CPRLNM(25,3),                /* SPECIFIC ACTIVITY (cpm/nmole proline)
     *       TVALUE(25),                  /* TVALUE FOR DARRENS T-TEST
     *       DMNA(25),DMNB(25),           /* DDT - MEANS FOR A&B
     *       DSDA(25),DSDB(25),           /* DDT - SD'S FOR A&B
     *       SLP1A(25),SLP1B(25),         /* SLOPE (T-TEST)
     *       SLP2A(25),SLP2B(25),
     *       INT1A(25),INT1B(25),         /* INTERCEPT (T-TEST)
     *       INT2A(25),INT2B(25),
     *       R1A(25),R2A(25),             /* REGRESSION (T-TEST)
     *       R1B(25),R2B(25),
     *       TVA(25),TVB(25),             /* T VALUE (T-TEST)
     *       UGPROT(25,3),                /* UG PROTIEN
     *       MGPROT(25,3),                /* MG PROTIEN/ML
     *       UMPLPT(25,4),                /* UMOL PROLINE/G PROTIEN
     *       UMI(25),UMS(25),UMR(25),     /* INTERCEPT,SLOPE,R↑2 (uM prol/g Protein
     *       NMHPPT(25,4),                /* NMOLE HYPRO/G PROTIEN
     *       NMI(25),NMS(25),NMR(25),     /* INTERCEPT,SLOPE,R↑2 (nM HPRO/g Protein
     *       LF(25,3),                    /* LUNG FRACTION
     *       HPRLNG(25,3),                /* HPRO/LUNG
     *       AHPRO(25,3),                 /* A560 HPRO.
     *       PRLNI,PRLNS,PRLN2,           /* INTERCEPT,SLOP,R↑2 (PROLINE)
     *       PROTI,PROTS,PROT2,           /* INTERCEPT,SLOP,R↑2 (PROTEIN)
     *       HYPRI,HYPRS,HYPR2,           /* INTERCEPT,SLOP,R↑2 (HYPRO)
     *       VOL(25,3),                   /* HPRO SAMPLE VOLUME
     *       APROL(25,3),                 /* A515 PROLINE
     *       APROT(25,3),                 /* A750 PROTEIN
     *       PRLNX(25),                   /* STANDARD CURVE (PROLINE)
     *       PRLNY(25),                   /* STANDARD CURVE (PROLINE)
```

```
     *      PROTX(25),                  /* STANDARD CURVE (PROTEIN)
     *      PROTY(25),                  /* STANDARD CURVE (PROTEIN)
     *      HYPRY(25),                  /* STANDARD CURVE (HYPRO)
     *      HYPRX(25),                  /* STANDARD CURVE (HYPRO)
     *      VOLDFL,                     /* DEFAULT VOLUME FOR ASSAY
     *      NORMSL(25),                 /* NORMALIZED HPRO SLOPE (TO DARRENS TT)
     *      XYZZY4                      /* LAST REAL
C
      LOGICAL
     *          DATAIN,                 /* ASSAY DATA READ IN OR NOT FLAG
     *          USRINP,                 /* USER NEEDS TO INTERACT W/ SUBROUT.?
     *          CNTLDN,                 /* CONTROL DATA SUB. DONE - NMLZD DATA?
     *          XYZZY5                  /* LAST LOGICAL
C
C
C---- INITIALIZE STUFF
C
      CALL INIT(PRLNX,PRLNY)
      CALL INIT(PROTX,PROTY)
      CALL INIT(HYPRX,HYPRY)
      NLUNG=0
      DATAIN=.FALSE.                    /* WE DONT HAVE THE DATA YET
      CNTLDN=.FALSE.                    /* WE HAVENT DONE CONTROL DATA YET
      PRLNC=0                           /* SET NUMBER OF STANDARD CURVE
      PROTC=0                           /*      POINTS TO 0 FOR ALL THREE:
      HYPRC=0                           /*      PROLINE, PROTEIN, HYPRO
      DO 12 I=1,25
         NMHPPT(I,4)=0.
         UMPLPT(I,4)=0.
         NORMSL(I)=0.0
         DO 11 J=1,3
            MPNT(I,J)=1                 /* ASSUME HAVE NO POINTS AT BEGINNING
11       CONTINUE
12    CONTINUE
C
C---  INITALIZE ALL THE ARRAYS ETC DEALING WITH THE TTEST
C
      DGRP=0
      IGRP=0
      DO 112 I=1,25
          TVALUE(I) = 0.0               /* DARRENS T-TEST INIT VALUES
          DGFREE(I) = 0
        R1A(I)=0.                       /* TTEST VALUES =0
        R1B(I)=0.
        R2A(I)=0.
        R2B(I)=0.
        INT1A(I)=0.
        INT1B(I)=0.
        INT2A(I)=0.
        INT2B(I)=0.
        SLP1A(I)=0.
        SLP1B(I)=0.
        SLP2A(I)=0.
        SLP2B(I)=0.
        TVA(I)=0.
        TVB(I)=0.
        DFA(I)=0
        DFB(I)=0
112   CONTINUE
C
C
1     CONTINUE                    /* TOP OF THE WORLD
      CALL TNOU ('                                          ',40)
      CALL TNOU (' [1] Exit program                         ',40)
      CALL TNOU (' [2] Read Muse Assay Data File            ',40)
      CALL TNOU (' [3] View Raw Data on Screen              ',40)
      CALL TNOU (' [4] HPRO and Proline regressions         ',40)
      CALL TNOU (' [5] T-test Procedures                    ',40)
      CALL TNOU (' [6] Print Hydroxy Proline Results        ',40)
      CALL TNOU (' [7] Select, Store & Print Control Data ',40)
      CALL TNOU (' [8] Read Old-Style Data                  ',40)
      CALL TNOU ('                                          ',40)
      CALL TNOUA(' Enter Command:   ',17)
      READ(1,10,ERR=1)CMMD
10    FORMAT(I2)
C
C---  BRANCH TO WHERE THE USER WANTS TO GO
C
      GO TO (900,200,300,400,500,600,700,800),CMMD
         CALL TNOU ('Invalid command.  Must be between 1 and 8',41)
         GO TO 1                        /* TRY AGAIN
C
```

```
200     CONTINUE                              /* READ DATA FROM DISK
C
        CALL READDT(MPNT,NLUNG,TINC,AHPRO,VOL,HCPM,  /* READ IN THE
     *             TCAPPT,APROL,PCPM,APROT,          /* JUNK FROM DISK
     *             PRLNC,PRLNX,PRLNY,
     *             PROTC,PROTX,PROTY,
     *             HYPRC,HYPRX,HYPRY,ASSCOM,VOLDFL,FILNAM,NMLNG)
        IF (NLUNG .GT. 0) GO TO 210
           CALL TNOU('Returning to main menu.',23)
           GO TO 1
210     CONTINUE
        CALL REGRSS(PRLNY,PRLNX,PRLNC,PRLN2,PRLNI,PRLNS)
        CALL REGRSS(PROTY,PROTX,PROTC,PROT2,PROTI,PROTS)
        CALL REGRSS(HYPRY,HYPRX,HYPRC,HYPR2,HYPRI,HYPRS)
        CALL MCALC (MPNT,NLUNG,TINC,AHPRO,VOL,HCPM,TCAPPT,
     *             APROL,PCPM,APROT,PRLNS,PRLNI,PROTS,PROTI,
     *             HYPRS,HYPRI,NMHPRO,CHPRML,TTCA,UGPRL,CPRLNM,
     *             UGPROT,MGPROT,UMPLPT,NMHPPT,LF,HPRLNG)
        DATAIN=.TRUE.
        USRINP=.FALSE.                        /* SET USER INTERACT FLAG TO FALSE
        CNTLDN=.FALSE.                        /* RESET THE CONTROL DATA DONE
        CALL HPREG(MPNT,NLUNG,TINC,NMHPPT,NMI,NMS,NMR,
     *                             UMPLPT,UMI,UMS,UMR,USRINP)
C
        GO TO 1
C
C
300     CONTINUE
        IF(DATAIN) GO TO 310
           CALL TNOU ('Not enough data to enter this section',37)
           GO TO 1
310     CONTINUE
        WRITE(1,301)
301     FORMAT('LUNG TIME  A560   VOL.  HCPM   TCPM   A515   PCPM   ',
     *        ' A750 ')
        WRITE(1,302)
302     FORMAT('---- ---- ------ ---- ------ ------ ------ ------   ',
     *        '------')
        DO 380 I=1,NLUNG
            DO 340 J=1,TINC
                WRITE(1,303)I, J, AHPRO(I,J), VOL(I,J), HCPM(I,J),
     *                     TCAPPT(I,J), APROL(I,J), PCPM(I,J), APROT(I,J)
303             FORMAT(1X,I2,3X,I2,2X,F6.1,2X,F4.1,2X,I5,2X,I5,
     *                 2X,F6.1,2X,I5,2X,F6.1)
340         CONTINUE
            WRITE(1,304)
304         FORMAT(1X)
            IF (MOD(I,5) .NE. 0) GO TO 380
                WRITE(1,305)
305             FORMAT('--More--')
                READ(1,306)
306             FORMAT(1X)
                WRITE(1,301)
                WRITE(1,302)
380     CONTINUE
C
        GO TO 1
C
C
C
400     CONTINUE                              /* HYPRO AND PROLINE REGRESSIONS
        IF(DATAIN) GO TO 410
           CALL TNOU ('Not enough data to enter this section',37)
           GO TO 1
410     CONTINUE
        USRINP=.TRUE.                         /* SET USER INTERACT FLAG TO TRUE
        CALL HPREG(MPNT,NLUNG,TINC,NMHPPT,NMI,NMS,NMR,
     *                             UMPLPT,UMI,UMS,UMR,USRINP)
        GO TO 1
C
C
C
500     CONTINUE
        IF(DATAIN) GO TO 510
           CALL TNOU ('Not enough data to enter this section',37)
           GO TO 1
510     CONTINUE
        CALL TNOU (' [1] TTest on HPRO Slopes                  ',40)
        CALL TNOU (' [2] Old-Style TTest.  (On nmHPPT)         ',40)
        CALL TNOU ('                                           ',40)
        CALL TNOUA(' Enter TTest:  ',15)
        READ(1,511,ERR=1)TCMMD
```

```
  511   FORMAT(I2)
        GO TO (520,540),TCMMD
        GO TO 1
C
  520   CONTINUE
        CALL DTEST(NMS    ,NLUNG,DGRP,TVALUE,DGFREE,DATITL,DBTITL,
     *              DLNGA,DLNGB,DNUMA,DNUMB,DMNA,DMNB,DSDA,DSDB)
        GO TO 1
C
  540   CONTINUE
        CALL TTEST(R1A,R1B,R2A,R2B,
     *             INT1A,INT1B,INT2A,INT2B,
     *             SLP1A,SLP1B,SLP2A,SLP2B,
     *             TVA,TVB,
     *             DFA,DFB,
     *             NMHPPT,UMPLPT,MPNT,ATITL,BTITL,IGRP,NLUNG)
        GO TO 1
C
C
  600   CONTINUE
        IF(DATAIN) GO TO 610
           CALL TNOU ('Not enough data to enter this section',37)
           GO TO 1
  610   CONTINUE
        CALL SPLDAT (MPNT,NLUNG,AHPRO,VOL,HCPM,
     *                TCAPPT,APROL,PCPM,APROT,
     *                PRLNY,PRLNX,PRLNC,PRLN2,PRLNI,PRLNS,
     *                PROTY,PROTX,PROTC,PROT2,PROTI,PROTS,
     *                HYPRY,HYPRX,HYPRC,HYPR2,HYPRI,HYPRS,
     *                NMHPRO,CHPRML,TTCA,UGPRL,CPRLNM,
     *                UGPROT,MGPROT,UMPLPT,NMHPPT,
     *                LF,HPRLNG,NMI,NMS,NMR,
     *                            UMI,UMS,UMR,
     *                SLP1A,SLP1B,SLP2A,SLP2B,
     *                INT1A,INT1B,INT2A,INT2B,
     *                R1A,R2A,R1B,R2B,TVA,TVB,DFA,DFB,
     *                ATITL,BTITL,IGRP,TINC,ASSCOM,
     *                DGRP,TVALUE,DGFREE,DATITL,DBTITL,
     *                DLNGA,DLNGB,DNUMA,DNUMB,DMNA,DMNB,DSDA,DSDB)
C
        GO TO 1
C
C
  700   CONTINUE
        IF(DATAIN) GO TO 710
           CALL TNOU ('Not enough data to enter this section',37)
           GO TO 1
  710   CONTINUE
        CALL CNTLDT(NLUNG, NMS, NORMSL)
        CNTLDN = .TRUE.
C
        GO TO 1
C
C
  800   CONTINUE                              /* READ DATA FROM DISK OLD FORMAT
C
        CALL RDOLDT(MPNT,NLUNG,TINC,AHPRO,VOL,HCPM,   /* READ IN THE
     *              TCAPPT,APROL,PCPM,APROT,          /* JUNK FROM DISK
     *              PRLNC,PRLNX,PRLNY,
     *              PROTC,PROTX,PROTY,
     *              HYPRC,HYPRX,HYPRY,ASSCOM,VOLDFL,FILNAM,NMLNG)
        CALL REGRSS(PRLNY,PRLNX,PRLNC,PRLN2,PRLNI,PRLNS)
        CALL REGRSS(PROTY,PROTX,PROTC,PROT2,PROTI,PROTS)
        CALL REGRSS(HYPRY,HYPRX,HYPRC,HYPR2,HYPRI,HYPRS)
        CALL MCALC (MPNT,NLUNG,TINC,AHPRO,VOL,HCPM,TCAPPT,
     *              APROL,PCPM,APROT,PRLNS,PRLNI,PROTS,PROTI,
     *              HYPRS,HYPRI,NMHPRO,CHPRML,TTCA,UGPRL,CPRLNM,
     *              UGPROT,MGPROT,UMPLPT,NMHPPT,LF,HPRLNG)
        DATAIN=.TRUE.
        USRINP=.FALSE.                        /* SET USER INTERACT FLAG TO FALSE
        CALL HPREG(MPNT,NLUNG,TINC,NMHPPT,NMI,NMS,NMR,
     *                            UMPLPT,UMI,UMS,UMR,USRINP)
C
        GO TO 1
C
C
  900   CONTINUE
        IF(DATAIN) GO TO 910
           GO TO 920
  910   CONTINUE
        CALL STORDT(MPNT,NLUNG,TINC,AHPRO,VOL,HCPM,
     *              TCAPPT,APROL,PCPM,APROT,
```

```
         *          PRLNC,PRLNX,PRLNY,
         *          PROTC,PROTX,PROTY,
         *          HYPRC,HYPRX,HYPRY,ASSCOM,VOLDFL,FILNAM,NMLNG)
C
920   CONTINUE
      STOP
      END
```

REFERENCES

1. **Smith, P. and Heath, D.,** Paraquat, *Crit. Rev. Toxicol.*, 4, 411, 1976.
2. **Autor, A. P.,** *Biochemical Mechanisms of Paraquat Toxicity*, Academic Press, New York, 1977.
3. **Vijeyaratnam, G. S., and Corrin, B.,** Experimental paraquat poisoning: a histological and electron-optical study of the changes in the lung, *J. Pathol.*, 103, 123, 1971.
4. **Smith, P. and Heath, D.,** The ultrastructure and time sequence of the early stages of paraquat lung in rats, *J. Pathol.*, 114, 177, 1973.
5. **Fisher, H. K., Clements, J. A., and Wright, R. R.,** *J. Appl. Physiol.*, 35, 268, 1973.
6. **Etherton, J. E. and Gresham, G. A.,** Early bronchiolar damage follwoing paraquat poisoning in mice, *J. Pathol.*, 128, 21, 1979.
7. **Butler, C.,** Pulmonary interstitial fibrosis from paraquat in the hamster, *Arch. Pathol.*, 99, 503, 1975.
8. **Kelly, D. F., Morgan, D. G., Darke, P. G. G., Gibbs, C., Pearson, H., and Weaver, B. M. Q.,** Pathology of acute respiratory distress in the dog associated with paraquat poisoning, *J. Comp. Pathol.*, 88, 275, 1978.
9. **Fukuda, Y., Ferrans, V. J., Schoenberger, C. I., Rennard, S. I., and Crystal, R. G.,** Patterns of pulmonary structural remodeling following experimental paraquat toxicity: the morphogenesis of intraalveolar fibrosis, *Am. J. Pathol.*, 118, 452, 1985.
10. **Gage, J. C.,** Toxicity of paraquat and diquat aerosols generated by a size-selective cyclone: effect of particle size distribution, *Br. J. Ind. Med.*, 125, 304, 1968.
11. **Popenoe, D.,** Effects of paraquat aerosol on mouse lung, *Arch. Pathol. Lab. Med.*, 103, 331, 1979.
12. **Rhodes, M. L., Zavala, D. C., and Brown, D.,** Hypoxic protection in paraquat poisoning, *Lab. Invest.*, 35, 496, 1976.
13. **Drew, R. and Gram, T. E.,** Vehicle alteration of paraquat lethality in mice, *Toxicol. Appl. Pharmacol.*, 48, 479, 1979.
13. **Murray, R. E. and Gibson, J. E.,** A comparative study of paraquat intoxication in rats, guinea pigs and monkeys, *Exp. Mol. Pathol.*, 17, 317, 1972.
14. **Ilett, K. F., Stripp, B., Menard, R. H., Reid, W. D., and Gillette, J. R.,** Studies on the mechanism of the lung toxicity of paraquat: comparison of tissue distribution and some biochemical parameters in rats and rabbits, *Toxicol. Appl. Pharmacol.*, 28, 216, 1974.
15. **Rose, M. S., Lock, E. A., Smith, L. L., and Wyatt, I.,** Paraquat accumulation: tissue and species specificity, *Biochem. Pharmacol.*, 25, 419, 1976.
16. **Sharp, C. W., Ottolenghi, A., and Posner, H. S.,** Correlation of paraquat toxicity with tissue concentrations and weight loss of the rat, *Toxicol. Appl. Pharmacol.*, 22, 241, 1972.
17. **Greenberg, D. B., Reiser, K. M., and Last, J. A.,** Correlation of biochemical and morphologic manifestations of acute pulmonary fibrosis in rats administered paraquat, *Chest*, 74, 421, 1978.
18. **Greenberg, D. B., Lyons, S. A., and Last, J. A.,** Paraquat-induced changes in the rate of collagen biosynthesis by rat lung explants, *J. Lab. Clin. Med.*, 92, 1033, 1978.
19. **Reiser, K. M. and Last, J. A.,** Lung fibrosis in rats: changes in lung collagen in several experimental models, *Am. Rev. Respir. Dis.*, 123, 58, 1981.
20. **Shatkin, A. J.,** Colorimetric reactions for DNA, RNA, and protein determinations, in *Fundamental Techniques in Virology*, Habel, K. and Salzman, N. P., Eds., Academic Press, New York, 1969, 231.
21. **Juva, K. and Prockop, D. J.,** Modified procedure for the assay of H^3- or C^{14}-labelled hydroxyproline, *Anal. Biochem.*, 15, 77, 1986.
22. **Modolell, J.,** The S-30 system from *Escherichia coli*, in *Protein Biosynthesis in Bacterial Systems*, Last, J. A. and Laskin, A. I., Eds., Marcel Dekker, New York, 1971, 1.
23. **Troll, W. and Lindsley, J.,** The colorimetric determination of proline in tissue and body fluids, *J. Biol. Chem.*, 215, 655, 1955.
24. **Last, J. A.,** Measuring collagen synthesis, *Am. Rev. Respir. Dis.*, 124, 346, 1981.
25. **Selman, M., Montano, M., Montfort, I., and Perez-Tamayo, R.,** The duration of the pulmonary paraquat toxicity — enhancement effect of O_2 in the rat, *Exp. Mol. Pathol.*, 43, 375, 1985.
26. **Selman, M., Montano, M., Montfort, I., and Perez-Tamayo, R.,** A new model of diffuse interstitial pulmonary fibrosis in the rat, *Exp. Mol. Pathol.*, 43, 388, 1985.

RADIATION-INDUCED PULMONARY FIBROSIS

William F. Ward and Yoon T. Kim

INTRODUCTION

Animal models of radiation-induced lung injury are relevant to three distinct human populations: (1) thoracic radiotherapy patients, (2) radiation workers exposed occupationally or accidentally, and (3) victims of atomic weapons exposures. Of these three populations, the first is foremost in terms of numbers of individuals exposed, the magnitude of radiation doses, and the frequency of pulmonary complications. Approximately 150,000 new cases of lung cancer were detected in the United States in 1987, and approximately 90% of these will be fatal.[1] Lung cancer is now the leading single cause of cancer mortality among both men and women in the U.S.[1] In these patients, pneumonitis and progressive interstitial lung fibrosis are major dose-limiting sequelae of aggressive radiotherapy. In addition to lung cancer, radiation pneumotoxicity is an important consideration in the treatment planning of some patients with breast cancer, certain lymphomas, certain pediatric tumors, and in patients receiving whole-body radiation prior to bone marrow transplantation.[2] Previous attempts to ameliorate radiation pneumonitis and fibrosis using antibiotics, anticoagulants, cytostatics, and glucocorticoids have proven inadequate, and lung tolerance remains a primary dose-limiting factor in the radiotherapy of thoracic and mediastinal disease.[2] Compounding the problem is the fact that several common chemotherapy agents (e.g., bleomycin, cyclophosphamide, and actinomycin D) reduce the radiation tolerance of the lung in patients receiving combined modality therapy.[3,4] In the absence of an effective treatment for radiation-induced pulmonary injury, therefore, the dose limitations imposed by normal lung tolerance presently preclude curative radiotherapy for all but the smallest or most responsive thoracic tumors. Clearly, methods of increasing lung tolerance to ionizing radiation would be of considerable clinical significance. As a result, many animal models are directed toward a more complete understanding of the mechanisms, early indicators, and possible modification of radiation pneumotoxicity in cancer patients. These models contribute the bulk of the data considered in this chapter. Lung damage produced by photon radiations, X-rays, and gamma rays is emphasized because these are the conventional modalities in radiation oncology and radiobiology. Particulate ionizing radiations such as neutrons and heavy ions are an interesting variation on the experimental theme, but rarely are used clinically and are not considered here. Internal radioisotopes, the model for inhalation toxicology in radiation workers and atomic weapons victims, also are not discussed.

As described below, there are certain similarities between radiation- and chemical-induced lung injury with respect to cellular and perhaps even molecular mechanisms of damage. However, there are at least three important differences as well. First, X- and gamma radiation are penetrating and uniformly ionizing photon beams. They deposit the same amount of energy and presumably produce the same amount of molecular damage in all cells of the lung, regardless of anatomic location. In contrast, blood-borne or inhaled chemotoxins impact, at least initially, on selected subpopulations of lung cells. Second, radiation-induced molecular injury is virtually instantaneous, although cellular events triggered by the initial damage may develop or perpetuate for decades. Thus, pharmacological processes such as uptake, distribution, metabolism, and excretion, important in an analysis of chemical toxins, are less relevant in the external radiation model. Third, external radiation can be localized to a single lung or a portion of one lung by simple shielding techniques in order to study tissue volume effects, compensatory reactions, cellular migration, etc. Blood-borne chemotoxins, in contrast, inevitably damage both lungs and may produce confounding systemic effects.

Table 1
SOURCES OF IONIZING RADIATION

Modality	Advantages	Disadvantages
Orthovoltage X-rays (180—300 kVp)	Relatively inexpensive and simple machines to operate Reduced shielding requirement (mm of lead) simplifies construction of radiation jigs for localized exposures	No skin sparing No longer used clinically
Megavoltage X-rays (4—18 MeV)	Skin sparing Used clinically	Expensive and complex machines to operate Greater shielding requirements (several in. of lead) complicates construction of radiation jigs for localized exposure
Gamma rays (^{137}Cs; ^{60}Co)	Relatively inexpensive and simple machines to operate Skin sparing Used clinically	Greater shielding requirements, as for megavoltage X-rays

Despite intensive laboratory investigation and despite an obvious clinical need, the pathogenesis of radiation-induced plumonary injury has yet to be clarified. Considerable gains can be made simply by applying the techniques developed and information obtained in other models of pulmonary injury to the radiation model. In the past decade there have been several reviews of radiation-induced pulmonary injury in general[5,6] and connective tissue reactions in particular.[7-9] As a consequence, recent contributions to this subject are emphasized in this chapter.

MODELS OF RADIATION-INDUCED PULMONARY INJURY

While the radiation model of lung injury offers some interesting advantages over chemical models, it also suffers some major disadvantages. X-ray machines and gamma-ray units are expensive, they occupy an entire room, and they require extensive shielding. Furthermore, it is neither prudent nor legal to operate a source of ionizing radiation without the supervision of a radiation or medical physicist. A trained professional in this field is essential for accurate calibration and dosimetry of the radiation unit and for compliance with federal, state, and local safety regulations. It is beyond the scope of this chapter to review the physical and regulatory aspects of the radiation model. Professional expertise is required, either in-house or at least on a consultant basis. Radiation dosimetry in the lung, with its intimate association of air spaces and soft tissue surrounded by bone, is a complex special case.

Several sources of ionizing photons are available for experimental purposes. These are listed in Table 1 along with a notation of the advantages and disadvantages of each. X-rays typically are identified by the energy of the electrons used to generate them. Thus, orthovoltage X-rays are generated by 180 to 300 keV electrons impinging on a tungston target and megavoltage X-rays by 4 to 18 MeV electrons on a copper target. Gamma rays typically are identified by the radioactive isotope from which they are emitted (e.g., ^{60}Co, ^{137}Cs). Both X-rays and gamma rays are electromagnetic radiations, differing only in the process by which they are produced. Both modalities deposit relatively little average energy per unit track lengh (<3 keV/μm) and both have approximately equivalent biological effectiveness. Particulate radiations, in contrast, deposit considerably more energy per unit of track length (10 to 100+ keV/μm), and exhibit greater biological effectiveness relative to photons. At the molecular level, X-ray and gamma-ray damage acutally is produced by secondary electrons ejected from the initial target atoms at very high kinetic energies. Megavoltage X-rays and gamma rays are sufficiently energetic that a full complement of secondary electrons does not build up until the photons have penetrated 1 to 2 mm below the skin surface. This phenomenon results in a reduction in undesired skin damage and is one of the reasons why these modalities have replaced orthovoltage X-rays in the radiotherapy of deep-seated tumors. On the other hand, the highly energetic photons also require more extensive shielding, which complicates construction of radiation jigs for localized lung exposures.

External radiation exposures are quantitated in units of Grays (Gy). One Gray is defined as that amount of ionizing radiation which deposits one joule of energy per kilogram of target materal (e.g., soft tissue). The Gray has replaced the rad as the preferred unit of absorbed radiation dose. The rad is defined as that quantity of radiation which deposits 100 ergs of energy per gram of soft tissue. Conveniently, 1 Gy = 100 rads and 1 cGy = 1 rad. Radiation energy deposited in target atoms produces damage by both direct and indirect mechanisms. Direct damage, as the name implies, results from a direct, physical energy exchange between the incident photon (or its associated secondary electrons) and one of the electrons of target atom. The latter absorbs sufficient energy to be ejected from its nuclear orbit, often with high kinetic energy. This process creates an ion pair consisting of a positively charged target atom and a negatively charged ejected electron. Photons with sufficient energy to create ion pairs are called ionizing radiations. Direct radiation damage is rapid (10^{-14} s) and random. That is, although DNA is regarded as the critical target molecule for radiation kill in most cells, ionizing photons do not seek out DNA for direct damage. The latter occurs randomly, in direct proportion to the contribution of DNA to the total cellular target volume ($< 1\%$). Thus, if 80% of the cell volume is occupied by water, approximately 80% of the direct radiation damage will occur in water molecules. This, in fact, forms the basis of the second and more common mechanism of radiation-induced cytocidal damage, i.e., indirect damage. Indirect damage occurs when products of the radiolysis of water react chemically with the target molecule. These products include many ions and radicals, with half-lives in cytoplasm of approximately 10^{-6} s. Thus, they can diffuse short distances (10 to 30 Å) before reacting to achieve chemical equilibrium. Compared to direct damage, indirect radiation damage is slow and nonrandom in that the ions and radicals exhibit differential reaction kinetics with the many atoms in the vicinity. The products of the radiolysis of water include both oxidizing and reducing species. However, on balance, indirect radiation damage represents chemical oxidation. The hydroxyl radical ($OH\cdot$) in particular is regarded as a primary agent of indirect radiation damage, thus providing a common molecular pathway for lung damage induced by radiation, chemicals, and inflammatory effector cells.

There are four basic techniques of lung irradiation, each relevant to a specific category of radiotherapy patient and each with characteristic advantages and disadvantages (Table 2). All of the techniques except whole-body irradiation require some degree of shielding with an attendant increase in the complexity of the dosimetry. The position of the shield for an upper hemibody port may be located by an external anatomic landmark (e.g., the tip of the xiphoid process), but whole- and hemithorax ports require radiographic verification (Figure 1). Depending on port selection, radiation dose, and autopsy time, one also must consider possible confounding effects of radiation esophagitis, myelitis, cardiopathy, and hypothyroidism on the lung reactions. Anesthesia is an additional variable in some instances, particularly when the location of the irradiated volume must be precise.

MECHANISMS OF RADIATION-INDUCED LUNG INJURY

While the emphasis of this chapter is fibrosis, it is necessary to consider the early inflammatory phase as well as the chronic fibrotic phase of radiation pneumotoxicity. The following discussion represents a distillation of several excellent reviews of radiation lung

Table 2
PULMONARY IRRADIATION TECHNIQUES

Technique	Advantages	Disadvantages
Whole-body (WB)	Simplicity No shielding required No anesthesia required Model for bone marrow transplantation patients and for accidental or military exposures	Doses limited to <8 Gy by acute bone marrow lethality Systemic radiation reaction (bone marrow, immune system, liver, etc.) may influence lung reaction
Upper hemibody (UHB)	Simple shielding May avoid anesthesia Can deliver higher doses than WB Model for some lymphoma and other patients treated paliatively for disseminated disease	Slightly more complex than WB Other thoracic organs (thyroid, spinal cord, thymus, esophagus, and heart) in treatment field may influence lung reactions
Whole-thorax (WT)	Can deliver higher doses than UHB Model for patients receiving bilateral lung irradiation and for fatal radiation pneumonitis	More complex shielding May require anesthesia, especially with megavoltage X-rays and gamma rays Other thoracic organs in treatment field may influence lung reactions
Hemithorax (HT) or partial thorax	Can deliver higher doses than WT Model for patients receiving localized lung irradiation Can avoid thyroid, esophagus, spinal cord, and most of heart	Most complex shielding May require anesthesia Cardiac output shunted to shielded lung

injury[2,5,6,10] to which the interested reader is referred for a more complete account. References for more recent information can be found in the legend to Table 3.

A prevailing opinion in radiation pathology holds that radiation-induced organ dysfunction results largely, if not solely, from cell kill, although irradiated cells typically do not die until and unless they attempt mitosis. Supporting this hypothesis is the clinical and experimental experience that organs which react quickly to radiation (skin, mucosa, gut, bone marrow, and testis) are those with rapidly proliferating stem cells, while organs which are slow to express radiation dysfunction (lung, central nervous system, heart, and kidney) are those whose cell populations turn over more slowly. In the authors' opinion, this hypothesis is overly influenced by the *in vitro* colony formation assay of radiation cytotoxicity and ignores lessions from cell physiology and other models of lung injury. The spectrum of cellular responses to insult is broader than simply death. In fact, even in the radiation model, pulmonary cell dysfunction occurs in the absence or at least in advance of a cytocidal effect. Almost immediately after moderate to high radiation doses to the lung there is evidence of perivascular and interstitial edema, a release of surfactant from type II pneumocytes into the airspaces, a reduction in the number of macrophages recovered by bronchoalveolar lavage (BAL), and a decrease in basal laminar proteoglycans. These reactions persist for up to 1 month after irradiation without evidence of significant cell kill or inflammatory cell infiltration and may be considered an early exudative phase of radiation pneumotoxicity. Additional features of this early phase may include vascular hyperemia or congestion, endothelial blebbing, microthrombus formation, and the appearance of proteinaceous fluid in the airspaces. These reactions typically are asymptomatic, and some authors describe this interval as a latent period.

Starting approximately 1 month after irradiation and developing for the next several months are the signs and symptoms of the acute inflammatory phase or pneumonitis. These include the accumulation of histiocytes, granulocytes, plasma cells, mast cells, and fibroblasts in

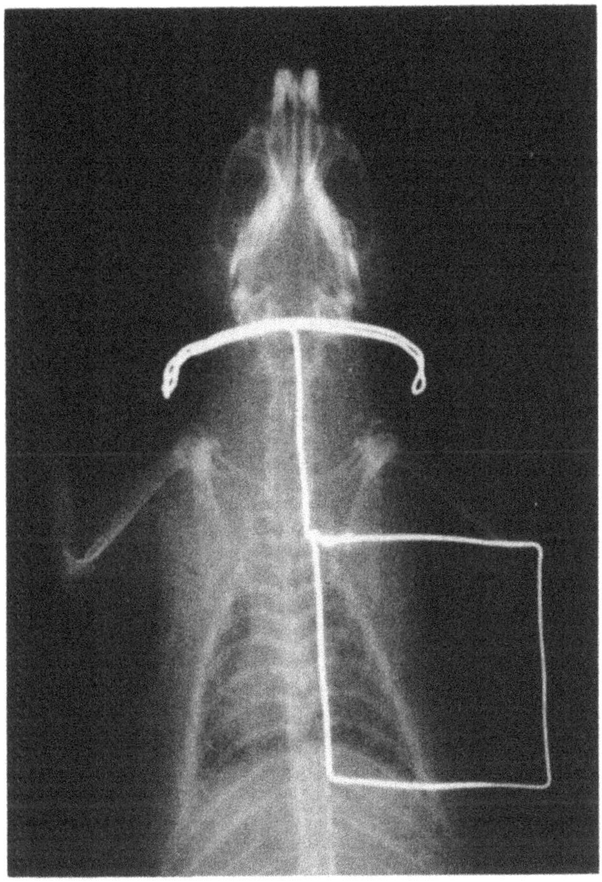

FIGURE 1. Radiographic verification of the location of a 3.5-cm² right hemithorax port in a rat. The proximal margin of the port is located 2 mm to the right of midline in order to avoid radiation esophagitis and myelitis as confounding factors.

the interstitium and a gradual increase in the number of intra-alveolar cells. In the radiation model of pneumonitis, unlike some of the chemical pneumotoxins, the vast majority of cells in the airspace are macrophages, with very few polymorphonuclear leukocytes or lymphocytes. At this time, a modest degree of degeneration may be observed in endothelial cells and in alveolar and bronchial epithelial cells. Desquammated type II cells may appear in the airspaces, where they are difficult to distinguish from foamy macrophages engorged with lamellar bodies. Persistence of fibrin-rich exudative fluid may result in interstitial and intra-alveolar fibrin and hyaline membranes, particularly in species (humans and dogs) with low intrinsic tissue plasminogen activator activity. Alveolar septa are thickened due to persistent edema and cellular infiltration and the airspaces contain macrophages, cellular debris, and proteinaceous fluid. At this time the individual will exhibit mild dyspnea and a productive cough. If the radiation regimen and treatment volume were conservative, these reactions will regress spontaneously. If treatment was more aggressive, however, the pneumonitis becomes clinically significant and requires management. The barrier to gas exchange increases, arterial perfusion defects appear, and degeneration and proliferative repair of the capillary endothelium may partially obstruct the fine vasculature. Hyperplasia and desquam-

Table 3
ENDPOINTS OF RADIATION-INDUCED LUNG INJURY

Endpoint	Ref.
Anatomic	
Histopathology	
Qualitative	1—5
Morphometric	6—9
Ultrastructure	
Transmission EM	3,10—18
Scanning EM	3,7,19
Radiography	6,20—22
Computed tomography	23,24
Microangiography	5
Cytologic	
^{3}H-Thymidine incorporation	11,25,26
Cultured cell lines	27,28
Flow cytometry	61
Functional	
Survival (LD_{50}/60—180; LD_{50}/180—360)	5,29,30
Arterial gases	31—33
Respiration rate	2,30
CO-diffusion	21,34
Compliance	5,21,35,36
Endothelial function	37—42
Perfusion	6,22,40,43—46
Surfactant metabolism	14,47—49
Vascular permeability	28,50—52
Ventilation	21,22,43,45
Biochemical	
Collagen polymorphism	
Gel electrophoresis	53
Immunohistochemical or immunofluorescent staining	54
Connective tissue enzymes	53
Extracellular matrix metabolism	55—57
Hydroxyproline content and concentration	25,35,38,39, 50,58,59
^{3}H-Proline incorporation	60

REFERENCES

1. **Travis,** *Int. J. Radiat. Oncol. Biol. Phys.*, 6, 345, 1980.
2. **Travis et al.,** *Radiat. Res.*, 84, 133, 1980.
3. **Travis et al.,** *Int. J. Radiat. Oncol. Biol. Phys.*, 2, 475, 1977.
4. **Jennings and Arden,** *Arch. Pathol.*, 71, 437, 1961.
5. **Kurohara and Casarett,** *Radiat. Res.*, 52, 263, 1972.
6. **Ward et al.,** *Radiology*, 131, 751, 1979.
7. **Penney et al.,** *Scan. Electron Microsc.*, 1, 221, 1986.
8. **Woodruff et al.,** *Am. J. Pathol.*, 82, 287, 1976.
9. **Rhoads et al.,** *Radiat. Res.*, 88, 266, 1981.
10. **Watanabe et al.,** *Lab. Invest.*, 31, 55, 1974.
11. **Adamson et al.,** *Am. J. Pathol.*, 58, 481, 1970.
12. **Port and Ward,** *Radiat. Res.*, 92, 61, 1982.
13. **Phillips,** *Radiology*, 87, 49, 1966.
14. **Penney and Rubin,** *Int. J. Radiat. Oncol. Biol. Phys.*, 2, 1123, 1977.
15. **Leroy et al.,** *Lab. Invest.*, 15, 1544, 1966.
16. **Maisin,** *Radiat. Res.*, 44, 545, 1970.
17. **Moosavi et al.,** *Int. J. Radiat. Oncol. Biol. PHys.*, 2, 921, 1977.
18. **Faulkner and Connolly,** *Lab. Invest.*, 28, 545, 1973.
19. **Penney et al.,** *Scan. Electron Microsc.*, 1, 131, 1982.

Table 3 (continued)
ENDPOINTS OF RADIATION-INDUCED LUNG INJURY

20. **DeHalleux,** *Strahlentherapie,* 139, 196, 1970.
21. **Teates,** *J. Appl. Physiol.,* 20, 628, 1965.
22. **Bradley et al.,** *Int. J. Radiat. Oncol. Biol. Phys.,* 7, 1055, 1981.
23. **Miller et al.,** *Int. J. Radiat. Oncol. Biol. Phys.,* 12, 1971, 1986.
24. **El-Khatib et al.,** *Int. J. Radiat. Oncol. Biol. Phys.,* 9, 853, 1983.
25. **Adamson and Bowden,** *Am. J. Pathol.,* 112, 224, 1983.
26. **Coultas et al.,** *Radiat. Res.,* 85, 516, 1981.
27. **Guichard et al.,** *Int. J. Radiat. Oncol. Biol. Phys.,* 6, 441, 1980.
28. **Friedman et al.,** *J. Cell. Physiol.,* 129, 237, 1986.
29. **Ward et al.,** *Radiat. Res.,* 90, 321, 1982.
30. **Travis and Down,** *Radiat. Res.,* 87, 166, 1981.
31. **Siemann et al.,** *Radiat. Res.,* 81, 303, 1980.
32. **Siemann and Hill,** *Radiat. Res.,* 93, 560, 1983.
33. **Schreiner et al.,** *Am. Rev. Respir. Dis.,* 99, 205, 1968.
34. **Rappaport et al.,** *Radiat. Res.,* 93, 254, 1983.
35. **Dubrawsky et al.,** *Radiat. Res.,* 73, 111, 1978.
36. **Shrivastava et al.,** *Radiology,* 112, 439, 1974.
37. **Ts'ao and Ward,** *Radiat. Res.,* 103, 393, 1985.
38. **Dancewicz et al.,** *Radiat. Res.,* 67, 482, 1976.
39. **Ward et al.,** *Int. J. Radiat. Oncol. Biol. Phys.,* 11, 1985, 1985.
40. **Heinz et al.,** *Prog. Biochem. Pharmacol.,* 20, 74, 1985.
41. **Ward et al.,** *Radiat. Res.,* 98, 397, 1984.
42. **Ts'ao et al.,** *Radiat. Res.,* 96, 301, 1983.
43. **Vieras et al.,** *Radiology,* 147, 839, 1983.
44. **Teates,** *Am. J. Roentgenol.,* 102, 875, 1968.
45. **Moustafa and Hoperwell,** *in Radiobiological Research and Radiotherapy,* Vol. 1, International Atomic Energy Agency, Vienna, 1977, 75.
46. **Ward,** *Radiology,* 139, 201, 1981.
47. **Penney et al.,** *Virchows Arch.,* 37, 327, 1981.
48. **Shapiro et al.,** *Int. J. Radiat. Oncol. Biol. Phys.,* 10, 375, 1984.
49. **Gross,** *J. Lab. Clin. Med.,* 93, 627, 1979.
50. **Law,** *Radiat. Res.,* 103, 60, 1985.
51. **Evans et al.,** *Radiat. Res.,* 107, 262, 1986.
52. **Gross,** *J. Lab. Clin. Med.,* 95, 19, 1980.
53. **Walklin and Law,** *Br. J. Cancer,* 53 (Suppl. 7) 368, 1986.
54. **Miller et al.,** *Radiat. Res.,* 105, 76, 1986.
55. **Penney and Rosenkrans,** *Radiat. Res.,* 99, 410, 1984.
56. **Rosenkrans and Penney,** *Radiat. Res.,* 109, 127, 1987.
57. **Rosenkrans and Penney,** *Int. J. Radiat. Oncol. Biol. Phys.,* 11, 1629, 1985.
58. **Law et al.,** *Radiat. Res.,* 65, 60, 1976.
59. **Ward et al.,** *Radiology,* 146, 533, 1983.
60. **Pickrell et al.,** *Radiat. Res.,* 62, 133, 1975.
61. **Sablonniere et al.,** *Int. J. Radiat. Biol.,* 44, 575, 1983.

mation of atypical type II epithelial cells becomes prominent and an acute inflammatory reaction with secondary infection may develop. Dyspnea and cyanosis progress rapidly and the individual may succumb to acute radiation pneumonitis.

Survivors of acute radiation lung injury develop pulmonary fibrosis after several months. The mechanism of radiation-induced lung fibrosis is not clear, but some of the cells thought to participate in this process (fibroblasts, macrophages, and mast cells) begin to accumulate in the interstitium at about 1 month postirradiation. Increased interstitial collagen and elastin first are observed between 1 and 2 months after exposure using electron microscopy, hydroxyproline content, and collagen anabolic enzyme activities as endpoints (Table 3). Fibrotic foci appear initially in the subpleural region and adjacent to major airways and vessels. With

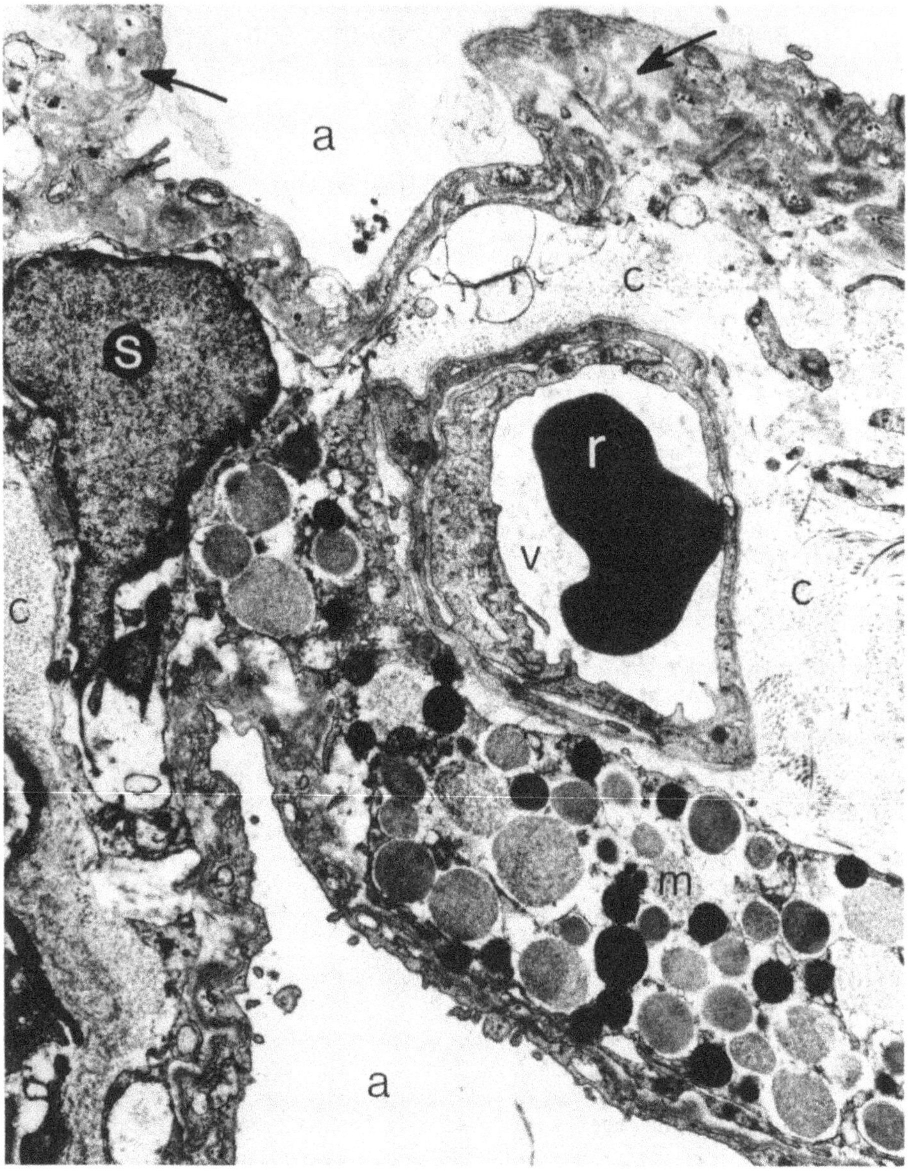

FIGURE 2. Interstitial lung fibrosis in a rat sacrificed 6 months after a single dose of 25 Gy of ^{60}Co gamma rays to the right hemithorax. A blood vessel (v) is partially surrounded by collagen fibers (c). Note the intimate association between a mast cell (m) and a septal cell (s). Excessive, convoluted epithelial basement membrane also is present (arrows). a = alveolus and r = erythrocyte. (Uranyl acetate and lead citrate; magnification × 3500.)

time, these foci enlarge, become confluent, and produce fairly large areas of diffuse interstitial lung fibrosis. Curiously, mast cells figure prominently in most descriptions of radiation-induced lung fibrosis in rats (Figure 2), whereas in mice, mast cells rarely are mentioned. In most other respects, however, these two species exhibit comparable radiation pulmonary histopathology. Fibrosis may progress to atelectasis, scarring, and atrophy of the irradiated volume, although necrosis is rare. Progressive distortion of the alveolar architecture, decreased compliance, and fibrotic obliteration of the microcirculation renders the irradiated tissue afunctional. Chronic fibrotic reactions are considered irreversible and lead essentially to a radiation pneumonectomy.

Radiographic examination reveals increased vascular and diffuse interstitial markings, traction or tenting of the pericardial and diaphragmatic surfaces, zones of pleural thickening and calcification, partial immobilization of the diaphragm, and compensatory emphysema. Retraction and atrophy of the scarred parenchyma may cause displacement of the mediastinal structures toward the involved side.

Functional impairment is, of course, the critical issue for thoracic radiotherapy patients whose treatment has been successful. The severity of radiation-induced respiratory dysfunction is influenced by many factors, including radiation regimen, treatment volume, age, concomitant chemotherapy, preexisting or secondary disease, infection, and tumor regression or regrowth. The frequency of symptomatic dysfunction usually is low due to the large functional reserve of the lung and to compensatory hypertrophy and voluntary restrictions in activity. However, when the treatment volume is large and is treated aggressively, functional impairment is common. In these individuals, pulmonary function tests reveal decreased compliance, decreased vital capacity, increased vascular resistance, decreased arterial perfusion, decreased gas diffusion, decreased ventilation, increased respiration rate, and arterial hypoxemia. In extreme cases the individual may become a respiratory cripple, experiencing resting dyspnea, cyanosis, cor pulmonale, and even fibrotic lung death.

Morphologically, the pneumonitis and interstitial fibrosis induced by ionizing radiation are not pathognomonic; rather they resemble the reactions to many other lung insults, at least in their general features. There is some evidence to suggest that radiation pneumonitis and fibrosis are independent phenomena, perhaps with different pathways of development and different critical target cells. For example, in irradiated rodents there is a dissociation between pneumonitis and fibrosis with respect to the effect of radiation dose, dose fractionation, and the action of dose-modifying agents. Furthermore, in thoracic radiotherapy patients there is not a strong correlation between the severity of pneumonitis and the severity of fibrosis. In fact, lung fibrosis can develop in patients who do not exhibit symptoms or signs of pneumonitis. On the other hand, the belief is emerging from the study of interstitial pulmonary fibrosis of other known and unknown etiology that inflammation (alveolitis) usually, if not always, precedes fibrosis.[11] This alveolitis may be chronic and subclinical, but can be identified by careful evaluation such as by BAL in asymptomatic children at risk for the development of familial interstitial pulmonary fibrosis. Lung fibrosis, perhaps including radiation fibrosis, may thus be less a primary process than a secondary reactive or repair phenomenon, initiated and modulated by prior or concurrent inflammation.

The pathogenesis of radiation-induced pulmonary injury, however, may be somewhat unique in that the edema, the activation of macrophages, the influx of inflammatory cells and fibroblasts, and the accumulation of collagen all develop on a substratum of radiation-induced parenchymal cell kill. Thus, the inflammatory effector cells whose influx is triggered by early radiation reactions produce parenchymal cell injury superimposed on chronic radiation-induced cell kill. To the extent that the inflammatory reactions themselves are cytotoxic or mitogenic, there may be premature expression of occult radiation lethality as parenchymal cells proliferate on an accelerated timetable. It is possible, however, that the radiation dose may be sufficiently low to preclude significant inflammation, leaving only the gradual expression of radiation cytotoxicity. This insidious radiation-induced parenchymal atrophy, accompanied by an increase in fibroconnective tissue, and a reduction in functional fine vasculature resembles spontaneous aging reactions. In this case, the effect of low doses of radiation to the lung may be virtually indistinguishable from premature aging.

There has been a tendency among radiation pathologists to evaluate lung injury in terms of a single target cell, or perhaps one target cell for pneumonitis and a second for fibrosis.

With more than 40 resident cell types, however, the pathophysiology of radiation pneumotoxicity almost certainly is complex, particularly since radiation damages all cells equally and instantly. It seems reasonable to assume that each cell reacts in some way to the radiation insult. These reactions in many cases may be subtle, may include dysfunction as well as death, may occur at different times, and may escape detection in the absence of a suitable assay; nevertheless they occur. Furthermore, the reaction of each cell will influence its microenvironment and the reactions of its neighbors. Evidence exists to firmly implicate at least three lung cells in the pathogenesis of radiation lung injury, and future research should add to this number. In identifying important radiation targets and pathways of damage, it is useful to determine which cells react quickly e.g., within 24 h. These are logical candidates for the initial triggering of subsequent reactions. Likewise, it is helpful to know which pneumocyte radiation reactions also occur *in vitro,* since these may be regarded as possible direct effects *in situ.* In the radiation model of lung injury, the same three cell types test positively by both the "early" and the "*in vitro*" criteria: vascular endothelial cells, type II epithelial cells, and alveolar macrophages. These are interesting targets because they all exhibit features of both stem cells and mature functional cells. It thus may be appropriate to briefly summarize the radiobiology of each of these cell types.

Endothelial cells — It was once accepted as dogma that vascular damage was the major, if not the sole cause of radiation fibrosis in normal tissues.[2] This dogma presently is under challenge, particularly in the case of the lung.[10] Nevertheless, sufficient data are available to demonstrate clearly that the vascular endothelial cell is one factor in the pathogenesis of radiation pneumotoxicity. Within 24 h after high single doses of radiation to the thorax, there is a significant increase in the leakage of radiolabeled plasma proteins into the lung interstitium[12] and a significant but transient decrease in lung prostacyclin production.[13] Thus, there is almost immediate perturbation of both the barrier and the metabolic functions of the lung endothelium. This early increase in vascular permeability, well documented by ultrastructural criteria, results in interstitial and intra-alveolar edema and presumably in extravasation of fibrinogen, immunoglobulins, complement, enzymes, growth factors, etc. These may contribute to the interstitial fibrin formation, macrophage activation, and extracellular matrix disorganization typical of early inflammatory processes. The transient reduction of PGI_2 production[13] might perturb the hemodynamic and thrombogenic status of the irradiated lung. Law[14] has proposed that interstitial fibrin resulting from the radiation-induced increase in vascular permeability persists due to defective fibrinolytic activity after irradiation[15,16] and serves as a chronic stimulus to fibroblast migration, proliferation, and collagen synthesis. Most of the early radiation responses observed in the pulmonary endothelium *in situ* also have been reported in cultured endothelial monolayers irradiated *in vitro.* These include increased permeability,[17] decreased PGI_2 production,[18] and decreased release of plasminogen activator.[19]

Type II epithelial cells — Perhaps the earliest radiation reaction in rodent lung is the release of surfactant from type II pneumocytes into the airspaces.[20] Furthermore, the concentration of surfactant in BAL fluid during the first few days after irradiation correlates fairly closely with the risk of fatal pneumonitis months later.[21] It has been suggested that this early release of surfactant contributes to tissue oncotic changes that quickly result in increased vascular permeability. Type II pneumocytes irradiated *in vitro* also release surfactant in a dose-dependent manner[22] demonstrating that this cell, like the endothelial cell, is capable of responding directly to the radiation insult.

Alveolar macrophages — The number of macrophages recoverable by BAL decreases in the first few days after thoracic irradiation[23,24] and their phagocytic function is impaired.[23,25] Injured or activated macrophages are known to release OH·, elastase, and other proteases[11] which might account for the early changes in basal laminar glycosaminoglycans known to occur in irradiated mouse lung.[26,27] Activated macrophages also produce leukocyte chemotaxins, fibroblast growth factors, and fibronectin and are thought to mediate other models of lung inflammation and fibrosis.[11] It has recently been reported that soluble fibronectin levels in mouse lung increase significantly at 13 weeks after 5 to 13 Gy, preceding the onset of fibrosis.[28] Alveolar macrophages also react to radiation *in vitro,* with a dose-dependent loss of colony-forming ability in the small subpopulation of these cells with proliferative capability.[29] Thus, the macrophage also appears to be a direct target of radiation damage and may play as crucial a role in radiation pneumotoxicity as it appears to play in other models of lung injury.[11]

ENDPOINTS OF RADIATION-INDUCED LUNG INJURY

Given the complexity of pulmonary pathophysiology, one would prefer multiple rather than single endpoints of response. Yet when multiple endpoints are utilized, they may not correlate well. Thus, a lung with increased hydroxyproline content may not appear fibrotic by light microscopy or radiography, or a lung with decreased compliance may not necessarily contain an excessive amount of hydroxyproline. This lack of correlation among endpoints, while disconcerting, probably should not be surprising. Endpoints differ in radiosensitivity and time of onset. They may be interrelated in complex and even inverse ways, or may be directed at different functions of a multifunctional organ. Perhaps the most frequently discussed example is the lack of correlation between hydroxyproline content and concentration in the irradiated lung. In order for these two indices of fibrosis to correlate with one another, changes in wet weight, dry weight, protein, or DNA content (or other denominator of concentration) must parallel changes in hydroxyproline content. Sometimes this occurs (Table 3),[25,35,39] but more frequently it does not. It is now apparent, however, that hydroxyproline is not a particularly sensitive marker of radiation-induced lung fibrosis. Furthermore, it is subject to false negatives. That is, a lung with "normal" hydroxyproline content may suffer the consequences of abnormal connective tissue metabolism, if there are abnormalities in collagen location, polymorphism or turnover, or in extracellular matrix metabolism. A more detailed and sophisticated analysis of connective tissue metabolism, underway for several years in other models of lung fibrosis, now is required in the radiation model.

Somewhat unique to the radiation model of lung injury is the frequent use of lethality in mice irradiated to the whole-thorax as an endpoint (Table 3). This protocol has been employed to study the effect of virtually every conceivable modifier of radiation response, including genetic strain, age, sex, radiation dose-rate and quality, repair, chemotherapy, sensitizers, protectors, anesthetics, and oxygen. The single, whole-thorax radiation dose required to kill half of the animals at risk (LD_{50}) is determined for deaths occurring between 60 and 180 d after irradiation (pneumonitis) or between 180 and 360 d after irradiation(fibrosis). Deaths occurring during the first 60 d after whole-thorax exposure are the result of radiation esophagitis and are eliminated from an analysis of pneumotoxicity. This model clearly illustrates the delayed onset of radiation pneumonitis compared to that induced by most chemical insults.

In general, the methodology employed to evaluate radiation-induced lung injury has been traditional rather than state of the art. To our knowledge, there has been only a single application of the isolated perfused lung preparation (Table 3)[40] and of flow cytometry (Table 3)[61] to this problem. The authors are not aware of any published efforts to apply current methods of molecular genetics or immunology to questions in radiation pneumotoxicity. Likewise, phenomena such as growth factors, cytokines, cellular interactions, and angiogenesis remain to be explored in this model. Radiation biologists and oncologists should exploit lessons learned in other models of lung injury because, in terms of the size of the relevant patient population, the radiation model is among the most important.

EARLY INDICATORS OF RADIATION-INDUCED LUNG DAMAGE

Predictors of lung damage would have universal clinical appeal. The perfect indicator should be (1) sensitive (steep dose-response curve), (2) accurate (low incidence of false positives and negatives), (3) specific (correlated with known events in pulmonary pathophysiology), (4) early in onset, (5) predictive of the severity of late damage, (6) minimally invasive, (7) serially applicable, and (8) inexpensive. Ideally, the indicator of choice should discriminate between radiation pneumotoxicity, tumor regression or recurrence, and superimposed infections.

Recent literature on minimally invasive markers of radiation-induced lung damage is reviewed in Table 4. Of these markers, the early indicators, i.e., those which develop during the first month of fractionated radiation, would be of prime clinical interest. A reliable early indicator would allow the radiotherapist to interrupt or shorten the original treatment plan, if lung tolerance were in danger of being exceeded in an unexpectedly sensitive patient. Advanced warning is particularly important for pulmonary radiation damage, since symptoms of the acute inflammatory phase of the radiation reaction do not occur until weeks or months after completion of the intended course of radiotherapy. In this respect, the lung differs from several early responding normal tissues (skin, mucosa, bone marrow, and gut) whose acute reactions develop during radiotherapy and whose severity can be controlled, if necessary, by a short break in therapy. Indicators of lung damage which develop during the second month after the start of radiotherapy might be too late to influence the treatment plan, but sufficiently early to select patients for modifier therapy (Table 5). Early intervention presumably would be more effective.

While the number of minimally invasive markers of radiation pneumotoxicity is large (Table 4), none is perfect. Some (e.g., decrements in alveolar macrophage function) occur early but are relatively insensitive. Others (e.g., CO-diffusion) are sufficiently sensitive but occur too late. Because of a delayed onset and/or low sensitivity, none of the reported blood tests appears useful. The standard clinical marker of lung damage, the chest radiograph, is neither early nor particularly sensitive. Computed tomography offers greater sensitivity than radiography but is expensive. The sensitivity of pulmonary function tests typically is compromised by the considerable reserve capacity of the lung. Most functional deficits also are relatively late to develop. Pulmonary arterial perfusion scans may represent the best combination of speed of onset and sensitivity among the standard clinical tests. However, radionuclide scans identify regional defects in function and may be less reliable in cases of uniform, whole-lung radiation damage. Experimentally, the most promising of the present indicators appear to be surfactant concentration in BAL fluid and thromboxane (TXA_2) concentration in the urine (Table 4). Both responses occur during the first 24 h after single doses of radiation and both are reasonably sensitive. Unfortunately, neither has been evaluated in a fractionated radiation setting. Therefore, there is room for improvement. Radiation oncologists and biologists should consider some of the recent and more sophisticated indicators identified in other models of lung injury. These include flow cytometric analysis of alveolar cell populations, BAL enzymes, and procollagen III peptide levels in blood, urine, or BAL fluid, among others.

MODIFICATION OF RADIATION-INDUCED LUNG INJURY

Since radiation-induced pulmonary injury presently is refractive to management, the lung remains a major dose-limiting organ to the radiotherapist. Strategies to increase pulmonary radiation tolerance would therefore be welcome. At least 12 distinct classes of modifying agents have been reported to ameliorate radiation reactions in lung; five of these have been tested in thoracic radiotherapy patients (Table 5). However, controlled clinical trials in this

Table 4
EARLY OR MINIMALLY INVASIVE MARKERS OF RADIATION-INDUCED LUNG DAMAGE

Indicator	Early	Sensitive	Ref.
Bronchoalveolar lavage fluid			
Cytology	***	**	1—6
Macrophage function	***	*	4—6
Protein	**	**	7,8
ACE	*	0	8
TXA_2	*	**	9
Surfactant	****	***	10—13
Adrenergic receptor	?	?	13
Blood			
ACE	0	*	8,14
von Willebrand factor	0	0	15,16
pO_2	0	*	17—20
CO-diffusion	0	***	21,22
Urine			
TXA_2	****	***	23
Hydroxyproline	*	*	24
Radiographs			
Chest radiographs	0	*	21,25—27
Computed tomography	*	**	28,29
Radionuclide scans			
Perfusion	**	***	26,27,30—32
Ventilation	**	**	21,27,31,32
Lung function tests			
Respiration rate	*	**	33—35
Other	*	**	17,18,21

Note: Early:**** = <1 day, *** = <1 week, ** = <1 month, * = <2 months, and 0 = >2 months. Sensitive: *** = very, ** = moderately, * = slightly, and 0 = not.

REFERENCES

1. **Moyer and Riley,** *Radiat. Res.*, 39, 716, 1969.
2. **Gross,** *Radiat. Res.*, 72, 325, 1977.
3. **Gross and Balis,** *Lab. Invest.*, 39, 381, 1978.
4. **Peel and Coggle,** *Radiat. Res.*, 81, 10, 1980.
5. **Sablonniere et al.,** *Int. J. Radiat. Biol.*, 44, 575, 1983.
6. **Ts'ao and Ward,** *Radiat. Res.*, 103, 393, 1985.
7. **Gross,** *J. Lab. Clin. Med.*, 95, 19, 1980.
8. **Ward et al.,** *Int. J. Radiat. Oncol. Biol. Phys.*, 11, 1985, 1985.
9. **Ward and Hinz,** in *Prostaglandin and Lipid Metabolism in Radiation Injury*, Walden and Hughes, Eds., Plenum Press, 1987, in press.
10. **Gross,** *J. Lab. Clin. Med.*, 93, 627, 1979.
11. **Shapiro et al.,** *Int. J. Radiat. Oncol. Biol. Phys.*, 8, 879, 1982.
12. **Shapiro et al.,** *Int. J. Radiat. Oncol. Biol. Phys.*, 10, 375, 1984.
13. **Rubin et al.,** *Int. J. Radiat. Oncol. Biol. Phys.*, 12, 469, 1986.
14. **Mansfield et al.,** *J. Clin. Oncol.*, 2, 452, 1984.
15. **Sporn et al.,** *Blood*, 64, 567, 1984.
16. **Fajardo et al.,** *Int. J. Radiat. Oncol. Biol. Phys.*, 12, 107, 1986.
17. **Schreiner et al.,** *Am. Rev. Respir. Dis.*, 99, 205, 1968.
18. **Michaelson and Schreiner,** *Radiat. Res.*, 47, 168, 1971.
19. **Siemann et al.,** *Radiat. Res.*, 81, 303, 1980.
20. **Siemann and Hill,** *Radiat. Res.*, 93, 560, 1983.
21. **Teates,** *J. Appl. Physiol.*, 20, 628, 1965.
22. **Rappaport et al.,** *Radiat. Res.*, 93, 254, 1983.

Table 4 (continued)
EARLY OR MINIMALLY INVASIVE MARKERS OF RADIATION-INDUCED LUNG DAMAGE

23. **Schneidkraut et al.,** *J. Appl. Physiol.*, 61, 1264, 1986.
24. **Pickrell et al.,** *Chest*, 69, 311, 1976.
25. **DeHalleux,** *Stahlentherapie*, 139, 196, 1970.
26. **Ward et al.,** *Radiology*, 131, 751, 1979.
27. **Bradley et al.,** *Int. J. Radiat. Oncol. Biol. Phys.*, 7, 1055, 1981.
28. **Miller et al.,** *Int. J. Radiat. Oncol. Biol. Phys.*, 12, 1971, 1986.
29. **El—Khatib et al.,** *Int. J. Radiat. Oncol. Biol. Phys.*, 9, 853, 1983.
30. **Teates,** *Am. J. Roentgenol.*, 102, 875, 1968.
31. **Moustafa and Hopewell,** in *Radiobiological Research and Radiotherapy*, Vol. 1, International Atomic Energy Agency, Vienna, 1977, 75.
32. **Vieras et al.,** *Radiology*, 147, 839, 1983.
33. **Travis et al.,** *Radiat. Res.*, 84, 133, 1980.
34. **Travis and Down,** *Radiat. Res.*, 87, 166, 1981.
35. **Parkins et al.,** *Br. J. Radiol.*, 58, 225, 1985.

area have been few and their results generally disappointing. Only cortisone and its derivatives are used regularly in the treatment of radiation pneumonitis, with opinion divided regarding their efficacy.[2] The chronic fibrotic phase of radiation lung injury is not amenable to any known intervention, and its frequency and severity are controlled solely by limitations on the radiation dose and treatment volume. These limitations also restrict the probability of local tumor control. Research on modifiers of radiation lung injury is significant from at least two perspectives. First, there is the practical clinical need to increase the radiation tolerance of the lung. Second, specific modifiers can be used to interdict or influence putative steps in the pathogenesis of radiation pneumotoxicity and thereby contribute to a clearer understanding of the molecular and cellular mechanisms of damge. Thus, anti-inflammatory agents and inhibitors of collagen metabolism are heavily represented on lists of successful modifiers of lung injury in general and of radiation pneumotoxicity in particular (Table 5). A thorough review of the early literature on this subject is available,[2] and recent work is emphasized in the present chapter.

The 12 classes of modifying agents include 1 radioprotector and 11 therapeutic agents (Table 5). Radioprotectors must be present during irradiation in order to be effective. These agents usually are administered 15 to 30 min prior to exposure in order to reach a protective intracellular concentration. Most radioprotectors are sulfhydryl or antioxidant compounds and are thought to reduce the initial yield of radiation damage via mechanisms such as radical scavenging, hydrogen donation, or chemical reduction reactions. Radiotherapeutic agents, in contrast, are administered after irradiation and influence the biological expression rather than the initial yield of damage.

While the number of successful modifiers appears promising, there are several problems which prejudice their clinical application. For some of these agents (e.g., antibiotics, anticoagulants, cytostatics, and glucocorticoids), negative results have been obtained in other studies. Most of these agents produce side effects in humans and some (e.g., BAPN, colchicine) are unacceptably toxic. To the authors' knowledge, only three of the modifying agents (WR-2721, angiotensin converting enzyme [ACE] inhibitors, and penicillamine) have been studied in sufficient detail to yield a quantitative estimate of their efficacy, and these have been evaluated only in rodent models. Modifier effectiveness typically is expressed in terms of the dose-reduction factor (DRF) i.e., the ratio of radiation doses in the presence and absence of modifier, required to produce a comparable level of damage. True dose-reducing agents shift the dose-response curve toward higher doses by a constant fraction.

Table 5
MODIFIERS OF RADIATION-INDUCED LUNG INJURY

Modifier	Endpoint	Species	Ref.
Protectors			
WR-2721	Lethality	Mouse	1—3
	Respiration rate		
	Hydroxyproline content		
	DNA content		
Therapeutics			
ACE inhibitors	Endothelial dysfunction	Rat	4,5
	Hydroxyproline content		
Antibiotics	Compliance	Rat	6
Anticoagulants	Histopathology	Rabbit	7
	Histopathology	Human	8
Beta-aminopropionitrile	Hydroxyproline content	Rat	9
Colchicine	Hydroxyproline concentration	Rat	10
	Compliance		
Cytostatics	Histopathology	Dog	11
	Fibrinolytic activity		
Dextran sulfate	Radiographic	Human	12
Glucocorticoids	Compliance	Dog	6
	Respiratory symptoms	Human	13
	Lethality	Mouse	14,15
	Vascular permeability	Rat	16
Oxyphenbutazone	Radiographic	Human	17
Penicillamine	Lethality	Mouse	18
	Histopathology	Rat	19,20
	Ultrastructure	Rat	19,20
	Arterial perfusion	Rat	19,21
	Hydroxyproline content and concentration	Rat	22,23
	Endothelial dysfunction	Rat	5,24,25
Triiodothyronine	Breathing capacity	Human	26
	Compliance	Dog	27

REFERENCES

1. **Travis et al.,** *Int. J. Radiat. Oncol. Biol. Phys.*, 10, 243, 1984.
2. **Travis et al.,** *Radiat. Res.*, 103, 219, 1985.
3. **Travis and DeLuca,** *Int. J. Radiat. Oncol. Biol. Phys.*, 11, 521, 1985.
4. **Ward et al.,** presented at 8th Int. Cong. Rad. Res., Edinburgh, Scotland, July 19 to 26, 1987.
5. **Ward and Hinz,** Prostaglandin and lipid metabolism, in *Radiation Injury*, Walden, T. L. and Hughes, H. N., Eds., Plenum Press, New York, 1987, 147.
6. **Moss et al.,** *Radiology*, 75, 50, 1960.
7. **Boys and Harris,** *Am. J. Roentgenol.*, 50, 1, 1943.
8. **Macht and Pearlberg,** *Am. J. Roentgenol.*, 63, 335, 1950.
9. **Percarpio and Fischer,** *Radiology*, 121, 737, 1976.
10. **Dubrawsky et al.,** *Radiat. Res.*, 73, 111, 1978.
11. **Fleming et al.,** *J. Nucl. Med.*, 3, 341, 1962.
12. **Ohno et al.,** *Nippon Kyobu Shik. Gakkai Zasshi*, 16, 756, 1978.
13. **Karg,** *Strahlentherapie*, 142, 490, 1971.
14. **Phillips et al.,** *Cancer*, 35, 1678, 1975.
15. **Gross,** *J. Clin. Invest.*, 66, 504, 1980.
16. **Evans et al.,** *Int. J. Radiat. Oncol. Biol. Phys.*, 13, 563, 1987.
17. **Klein et al.,** *Strahlentherapie*, 144, 421, 1972.
18. **Ward et al.,** *Radiat. Res.*, 90, 321, 1982.
19. **Ward et al.,** *Radiology*, 131, 751, 1979.
20. **Port and Ward,** *Radiat. Res.*, 92, 61, 1982.
21. **Ward,** *Radiology*, 139, 201, 1981.

Table 5 (continued)
MODIFIERS OF RADIATION-INDUCED LUNG INJURY

22. **Ward et al.**, *Radiology*, 146, 533, 1983.
23. **Ward and Kim**, in *Handbook of Animal Models of Pulmonary Disease*, Cantor, J. O., Ed., CRC Press, Boca Raton, FL, 1989.
24. **Ward et al.**, *Radiat. Res.*, 98, 397, 1984.
25. **Ward et al.**, *Int. J. Radiat. Oncol. Biol. Phys.*, 13, 1505, 1987.
26. **Glicksman et al.**, *Radiology*, 73, 178, 1959.
27. **Tyree et al.**, *Radiat. Res.*, 28, 30, 1966.

Thus, they do not change the slope of the response curve substantially and their effect is largely independent of radiation dose. Other modifiers, in contrast, act primarily to reduce the slope of the dose-response curve. In this case their efficacy is quantitated in terms of the slope ratio (SR) i.e., the ratio of the response curve slopes in the absence and presence of modifier. The effectiveness of slope-modifying agents is directly proportional to radiation dose. Another problem with the current literature is that only one of the modifiers (WR-2721) has been quantitated in the clinically relevant model of radiation lung injury, i.e., a fractionated dose regimen. The others have been tested against large, single radiation doses. Additional work is clearly required if the clinical objective of increased radiation tolerance of the lung is to be realized.

The Northwestern University Medical School laboratory has evaluated several potential modifiers of radiation injury in rat lung, three of which have exhibited significant therapeutic activity. These are D-penicillamine (β,β-dimethylcysteine), an inhibitor of collagen cross-linking, and the ACE inhibitors Captopril® and CL242817 (Table 5). Adult male rats were sacrificed 1 to 12 months after a range of single doses (0 to 30 Gy) of ^{60}Co gamma rays to the right hemithorax (Figure 1). Half of each dose group consumed control feed continuously after irradiation and half consumed feed containing one of the modifiers. Initial studies evaluated the effect of penicillamine on radiation pneumotoxicity in rats sacrificed up to 12 months after a single dose of 25 Gy to the right hemithorax. These data demonstrated that penicillamine delayed the onset, reduced the peak severity, and/or accelerated recovery from radiation-induced pulmonary histopathology, hypoperfusion, endothelial dysfunction, and hydroxyproline accumulation (Table 5). Furthermore, penicillamine was effective at a regimen which appeared to be free of significant side effects for up to 12 months of continuous drug administration. From these temporal studies, two times at which to quantitate the therapeutic effect were selected: 2 months (early fibrosis) and 6 months (peak fibrosis) after irradiation. All three agents significantly ameliorated pulmonary endothelial dysfunction and hydroxyproline accumulation at 2 months postirradiation (Table 5). DRF values ranged from 1.2 to 2.1, with SRs of 1.1 to 5. The ACE inhibitors, perhaps because they were tested at a higher concentration, exhibited slightly higher DRF and SR values than did penicillamine. The side effects of all three agents were minor. Only the penicillamine experiment has been completed at 6 months postirradiation. These data demonstrate that (1) radiation-induced lung fibrosis progresses significantly between the second and sixth months in all animals, (2) penicillamine delays the onset of fibrosis at 2 months and reduces the peak severity of fibrosis at 6 months, and (3) penicillamine is more effective as a slope modifier than a dose modifier, particularly at 6 months postirradiation (Figure 3). Although encouraging, these data cannot be interpreted as evidence of an increase in pulmonary radiation tolerance in the clinical sense until the studies are repeated in animals exposed to multiple small fractions of thoracic irradiation. The mechanism of therapeutic action of these agents is not clear at present. Perhaps it is only a coincidence that all three are thiol or thioacetate compounds. Nevertheless, these data raise the very interesting and potentially significant possibility of

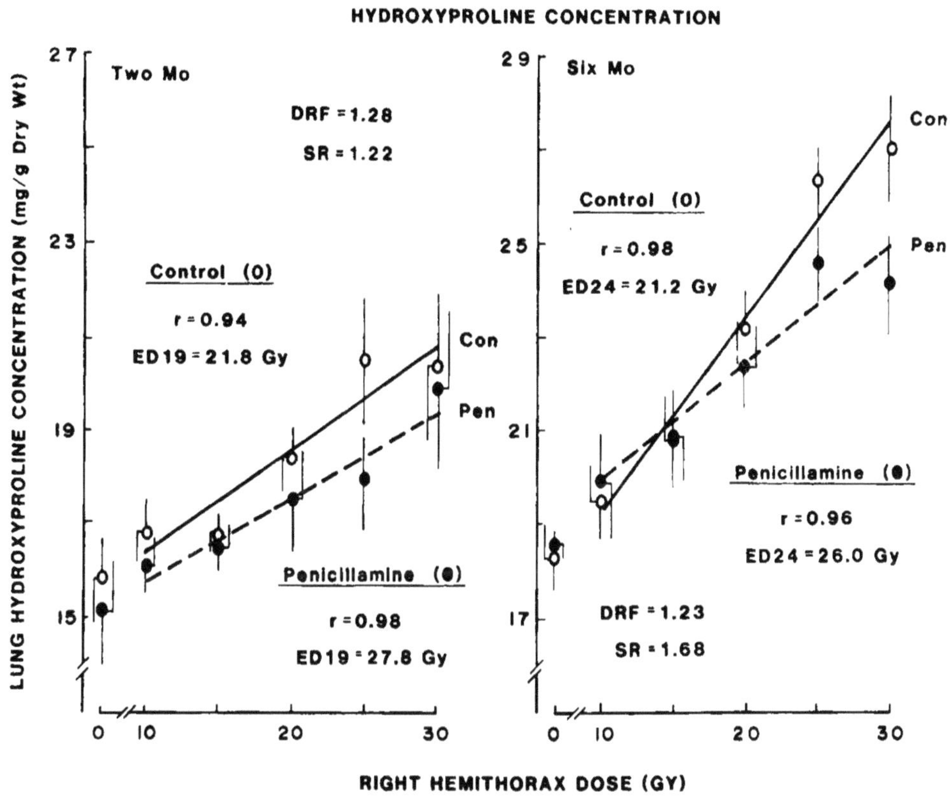

FIGURE 3. Hydroxyproline concentration (milligrams per gram dry weight) in the right lung of rats sacrificed 2 (left panel) or 6 (right panel) months after a range of single doses of ^{60}Co gamma rays to the right hemithorax. Animals consumed either control feed (open circles, solid line) continuously after irradiation or feed containing D-penicillamine, 25 mg/kg/d (solid circles, broken line). Mean ± SEM; n = 6 to 8. Lines were fitted by linear regression analysis (r = linear correlation coefficient). The penicillamine dose-reduction factor (DRF) was defined as the ratio of isoeffective doses (ED) at the level of response noted, and as the ratio of response curve slopes (SR).

a novel application for penicillamine and ACE inhibitors as injury modifying agents in irradiated lung. The need is great because ionizing radiation, generally unknown as an experimental model of lung injury, is a significant pneumotoxin to a large and rapidly growing patient population.

REFERENCES

1. *Cancer Facts and Figures,* American Cancer Society, New York, 1988.
2. **Rubin, P. and Casarett, G. W.,** Respiratory system, in *Clinical Radiation Pathology,* Vol. 1, W.B. Saunders, Philadelphia, 1968, 423.
3. **Phillips, T. L., Wharam, M. D., and Margolis, L. W.,** Modification of radiation injury to normal tissues by chemotherapeutic agents, *Cancer,* 35, 1678, 1975.
4. **Rubin, P.,** The Franz Buschke lecture: late effects of chemotherapy and radiation therapy: a new hypothesis, *Int. J. Radiat. Oncol. Biol. Phys.,* 10, 5, 1984.
5. **Gross, N. J.,** Pulmonary effects of radiation therapy, *Ann. Intern. Med.,* 86, 81, 1977.
6. **Gross, N. J.,** The pathogenesis of radiation-induced lung damage, *Lung,* 159, 115, 1981.
7. **Gerber, G. B. and Altman, K. I.,** Biochemical effects of ionizing radiation on connective tissue, *Nukleonika,* 25, 725, 1980.

8. **Pickrell, J. A.,** *Lung Connective Tissue: Location, Metabolism, and Response to Injury*, CRC Press, Boca Raton, FL, 1981.
9. **Altman, K. I. and Gerber, G. B.,** The effect of ionizing radiations on connective tissue, *Adv. Radiat. Biol.*, 10, 237, 1983.
10. **Van den Brenk, H. A. S.,** Radiation effects on the pulmonary system, in *Pathology of Irradiation*, Berdjis, C. C., Ed., Williams and Wilkins, Baltimore, 1971, 569.
11. **Crystal, R. G., Bitterman, P. B., Rennard, S. I., Hance, A. J., and Keogh, B. A.,** Interstitial lung diseases of unknown cause: disorders characterized by chronic inflammation of the lower respiratory tract, *N. Engl. J. Med.*, 310, 154, 1984.
12. **Evans, M. L., Graham, M. M., Mahler, P. A., and Rasey, J. S.,** Changes in vascular permeability following thorax irradiation in the rat, *Radiat. Res.*, 107, 262, 1986.
13. **Ts'ao, C., Ward, W. F., and Port, C. D.,** Radiation injury in rat lung. I. Prostacyclin (PGI_2) production, arterial perfusion, and ultrastructure, *Radiat. Res.*, 96, 284, 1983.
14. **Law, M. P.,** Radiation-induced vascular injury and its relation to late effects in normal tissues, *Adv. Radiat. Biol.*, 9, 37, 1981.
15. **Dancewicz, A. M., Mazanowska, A., and Gerber, G. B.,** Late biochemical changes in the rat lung after hemithoracic irradiation, *Radiat. Res.*, 67, 482, 1976.
16. **Ts'ao, C., Ward, W. F., and Port, C. D.,** Radiation injury in rat lung. III. Plasminogen activator and fibrinolytic inhibitor activities, *Radiat. Res.*, 96, 301, 1983.
17. **Friedman, M., Ryan, U. S., Davenport, W. C., Chaney, E. L., Strickland, D. L., and Kwock, L.,** Reversible alterations in cultured pulmonary artery endothelial cell monolayer morphology and albumin permeability induced by ionizing radiation, *J. Cell. Physiol.*, 129, 237, 1986.
18. **Eldor, A., Vlodavsky, I., Hyam, E., Atzmon, R., and Fuks, Z.,** The effect of radiation on prostacyclin (PGI_2) production by cultured endothelial cells, *Prostaglandins*, 25, 263, 1983.
19. **Ts'ao, C. and Ward, W. F.,** Acute radiation effects on the content and release of plasminogen activator activity in cultured aortic endothelial cells, *Radiat. Res.*, 101, 394, 1985.
20. **Rubin, P., Shapiro, D. L., Finkelstein, J. N., and Penney, D. P.,** The early release of surfactant following lung irradiation of alveolar type II cells, *Int. J. Radiat. Oncol. Biol. Phys.*, 6, 75, 1980.
21. **Penney, D. P., Shapiro, D. L., Rubin, P., Finkelstein, J., and Siemann, D. W.,** Effects of radiation on the mouse lung and potential induction of radiation pneumonitis, *Virchows Arch. B*, 37, 327, 1981.
22. **Shapiro, D. L., Finkelstein, J. N., Rubin, P., Penney, D. P., and Siemann, D. W.,** Radiation-induced secretion of surfactant from cell cultures of type II pneumocytes: an in vitro model of radiation toxicity, *Int. J. Radiat. Oncol. Biol. Phys.*, 10, 375, 1984.
23. **Gross, N. J. and Balis, J. V.,** Functional, biochemical, and morphologic changes in alveolar macrophages following thoracic x-irradiation, *Lab. Invest.*, 39, 381, 1978.
24. **Peel, D. M. and Coggle, J. E.,** The effect of x-irradiation on alveolar macrophages in mice, *Radiat. Res.*, 81, 10, 1980.
25. **Sablonniere, B., Nicolas, J., Neveux, Y., and Drouet, J.,** Effect of whole-body irradiation on phagocytic activity of rat alveolar macrophages, *Int. J. Radiat. Biol.*, 44, 575, 1983.
26. **Penney, D. P. and Rosenkrans, W. A., Jr.,** Cell-cell matrix interactions in induced lung injury. I. The effects of x-irradiation on basal laminar proteoglycans, *Radiat. Res.*, 99, 410, 1984.
27. **Rosenkrans, W. A., Jr. and Penney, D. P.,** Cell-cell matrix interactions in induced lung injury. II. X-irradiation mediated changes in specific basal laminar glycosaminoglycans, *Int. J. Radiat. Oncol. Biol. Phys.*, 11, 1629, 1985.
28. **Rosenkrans, W. A., Jr. and Penney, D. P.,** Cell-cell matrix interactions in induced lung injury. IV. Quantitative alterations in pulmonary fibronectin and laminin following x-irradiation, *Radiat. Res.*, 109, 127, 1987.
29. **Lin, H., Kuhn, C., and Chen, D.,** Radiosensitivity of pulmonary alveolar macrophage colony-forming cells, *Radiat. Res.*, 89, 283, 1982.

AN ANIMAL MODEL OF ASBESTOS-INDUCED INTERSTITIAL LUNG DISEASE

Arnold R. Brody and Lila H. Overby

INTRODUCTION

Inhaled asbestos fibers are known to cause debilitating and life threatening lung disease in exposed individuals.[1-4] There are two major types of asbestos, serpentine and amphibole,[1-4] and both induce a fibrogenic condition leading to a "stiff" or noncompliant lung and concomitant shortness of breath.[1-4] In addition, inhaled asbestos appears to be associated with an increased risk of lung cancer,[1-4] but many of the epidemiological findings have been confounded by a history of cigarette smoking. This chapter considers only nonneoplastic asbestos-induced lung disease.

Since the original observations of Cooke in 1924,[5] evidence for the role of asbestos in causing lung fibrosis has grown until this view is now undisputed. Pathologists have learned that chronic exposure to asbestos results in a diffuse thickening of the alveolar walls through deposition of connective tissue components.[1-4] Although the lesion extends diffusely along the alveolar walls, it tends to be more severe in alveolar regions located at the junctions of bronchioles and alveolar ducts (Figure 1A). The degree of cellularity of the lesions varies from a rather bland acellular fibrosis to a mixed inflammatory exudate of polymorphonuclear leukocytes, lymphocytes, and macrophages.[1-4] As with any chronic process studied at one point in time in fixed tissue, it is impossible to know which cells have contributed to the pathogenesis of the disease and which are simply secondary to the process. Similar to other inhalation-induced lung diseases, the basic pathobiological mechanisms of asbestosis (i.e., the pulmonary fibrosis caused by asbestos) are difficult to understand because individuals have been exposed for varying periods of time to different types of dust (asbestiform or not), and there are a multitude of postexposure time periods as well as susceptibilities to infections and personal decisions such as cigarette smoking and residential setting. These, of course, are the major reasons for attempting to establish an animal model of the disease process.

The first animal model demonstrating the fibrogenic effects of asbestos was developed by King et al.[6] in 1946. Their study, published in *Thorax,* preceded by 5 years the classic studies of Vorwald et al.[7] who showed that long asbestos fibers (>20 μM) induced more fibrosis than short fibers (<5 μM). However, this concept has been challenged by original animal studies by Holt et al.[8] using fibers less than 5 μM in length in animal models with rats and guinea pigs. Instillation and inhalation models evolved slowly until the study of Wagner and associates in 1974.[9] Using the large steel exposure chambers and dependable fiber-generating system of Timbrell and Skidmore,[10] Wagner and associates[9] designed a large inhalation experiment which conclusively demonstrated the pathogenesis of asbestosis and the development of pulmonary neoplasms. They showed that the rat is an excellent model system for studying asbestosis because the lesions which developed appeared precisely as they do in man and proved to be progressive, i.e., the fibrosis continued even after cessation of exposure. This issue of progression remains controversial today and is considered below in more detail.

In the earlier models described above, several potential mechanisms of asbestos-induced lung injury were offered. Generally, the thinking was that cells were impaled by sharp fibers or that macrophages which phagocytized the fibers released enzymes into alveolar spaces. Others believed that air spaces were obliterated by the organization of fibrin and plasma released into alveoli subsequent to asbestos-induced injury. However, none of these studies

trations ranging from 4 to 15 mg/m^3. Nose-only exposures can be carried out for 0.5 to 1 h and insure that large amounts of the respirable mass of fibers will be inhaled.[20-22] The disadvantages of this procedure include stress-related effects such as rapid, shallow breathing and the increased handling and chamber space required for each animal. Despite these problems, nose-only methods are recommended for brief exposure to high concentrations of fibers. The animals cannot shield their noses and remain clean for rapid handling immediately after exposure. For longer than 1-h duration, whole-body methods are better.

Whole-Body Exposure

This method is less labor intensive and allows more animals to be exposed in a single chamber. All these studies, except the initial ones on deposition and distribution of fibers,[20-22] were accomplished with whole-body exposures. Although unrestrained animals maintained singly in open cages will attempt to hide their noses in their fur, it has been shown that large numbers of fibers reach the alveolar levels of the lung.[23,24] It is also known that the initial deposition pattern of the fibers on alveolar surfaces is identical for both exposure methods.[21,23] Sufficient fiber deposition to cause interstitial pulmonary inflammation and fibrosis is achieved by 1 h of nose-only[25] or 3 to 5 h of whole-body exposure[23] at respirable dust concentrations ranging from 4 to 15 mg/m^3 (Figure 1B).

Animals exposed to room air in the chambers serve as controls for all the time points studied.

Lung Fixation, Embedding, and Dissection

To study the distribution of inhaled particles and consequent cell responses, it is essential that the tissue be fixed appropriately for light and electron microscopy. Determining the initial deposition pattern is accomplished by inoculating exposed animals intraperitoneally with sodium pentabarbital (1 ml of 1% solution) immediately after cessation of exposure. To prevent artifactual redistribution of inhaled particles during tissue preparation, fixative must flow through the vasculature, not into the airways.[20,21] This is done by exposing and clamping the trachea of an anesthetized animal just below the larynx. The chest wall is then cut on both sides of the sternum which is pulled anteriorly to expose the heart still beating rythmically. At this time, a 19-gauge hypodermic needle is inserted into the right ventricle, up to the entrance of the pulmonary artery. Just as the renal artery is cut to provide an outflow tract, a 1% saline solution is released into the right ventricle at a pressure of 20 cm of water. Blood clears from the vasculature and the lungs should appear light pink to white within 30 s. At this time, the fluid flow is switched to fixative, preferably a glutaraldehyde or glutaraldehyde-paraformaldehyde combination for 10 to 15 min at 20 cm H_2O pressure. With the trachea still clamped, the tracheo-bronchial tree and lungs are carefully dissected from the chest cavity and immersed in fresh fixative for at least 18 h. To study the initial deposition pattern of inhaled asbestos fibers, fixative was introduced into the vasculature of the lungs by 4.5 min after cessation of a 1-h exposure.[20,21] For routine studies of lung anatomy or morphometry at varying time points postexposure, the fixative can simply be introduced into the trachea at a pressure of 10 cm H_2O.[23]

Tissues are prepared for light microscopy as well as scanning and transmission electron microscopy. It is important to monitor the lungs of all animals, sham- and dust-exposed, by light microscopy of paraffin-embedded tissue. Routine histopathology of haematoxylin-eosin stained lung tissue will alert the investigator to the presence of infections and inflammation not caused by the experimental exposure.

Scanning electron microscopy is the method of choice for studying alveolar and cell surfaces at high resolution (Figure 2).[20,21,26] Tissues fixed as described above are dissected into 1-cm^3 blocks and dehydrated through a series of ethyl alcohols. The blocks are then critical-point dried in liquid CO_2 and mounted in carbon paint on carbon disks.[20,21] To demonstrate the appropriate anatomic regions and inhaled particles at the alveolar level, the

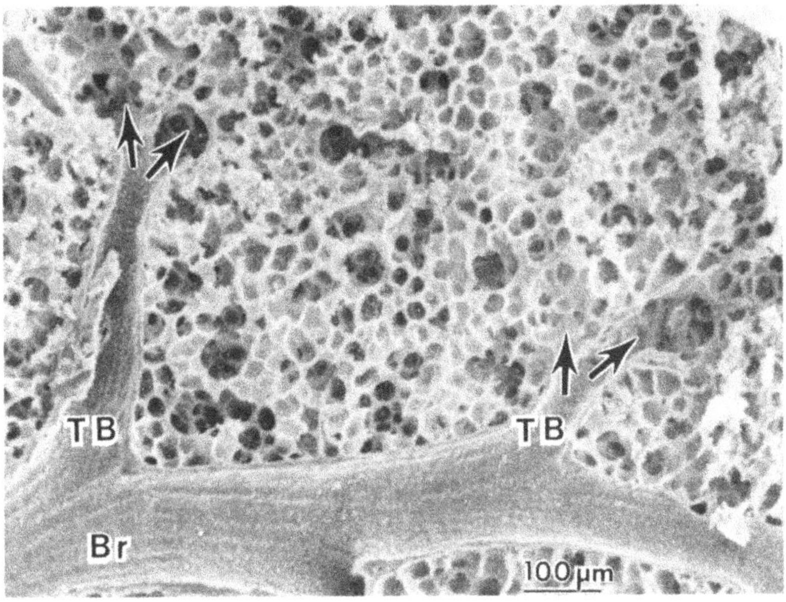

A

FIGURE 2. (A) Lower-power scanning electron micrograph (SEM) of rat lung parenchyma dissected to reveal bronchioles (Br), terminal bronchioles (TB), and their branching alveolar ducts (arrows). (B) SEM of a terminal bronchiole (TB) and its alveolar ducts exhibiting three bifurcations. The first (1) is closest to the bronchiole and the second (2) and third (3) are respectively more distal. The majority of fibers small enough to pass through the bronchiole is deposited on the bifurcations.[21] The first bifurcation (rectangle) is enlarged in (C) to demonstrate deposited asbestos fibers. (From Brody, A. R. and Roe, M. W., *Am. Rev. Respir. Dis.*, 128, 724, 1983. With permission.) (C) Higher magnification of the first bifurcation illustrated in (B).The rat from which this was dissected had been exposed to chrysotile asbestos for 1 h. Note the large numbers of fibers (arrowheads) littering the bifurcation surface, particularly the anterior edge. Few fibers are observed on more distal alveolar surfaces. (From Brody, A. R. and Roe, M. W., *AM. Rev. Respir. Dis.*, 128, 724, 1983. With permission.)

critical-point-dried tissue blocks are dissected further. This is accomplished by observing a block under a dissecting microscope at magnification × 4 and, using a new razor blade, thinly shaving the block face, thus revealing a number of terminal bronchioles and their alveolar ducts (Figure 2). The blocks are then coated with a thin (250 Å) layer of gold in a sputter-coater and can be stored in a dessicator for extended periods.

For transmission electron microscopy, two methods can be used. First, the fixed tissue blocks can be embedded in plastic for ultrathin sectioning of randomly oriented tissues.[25] Second, the fixed tissue can be embedded and then the plastic blocks warmed so that slabs of tissue approximately 1 mm thick can be sliced off with a razor blade and dissected further.[23,25] This technique allows the investigator to select the specific anatomic locations which may be the site of focal inflammation after exposure (Figure 3). As has been shown[23,25] and as discussed in some detail below, asbestos-induced lung injury is initiated as a focal event at bifurcations of the alveolar ducts. Using a random sectioning technique will cause the investigator to completely miss this phenomenon.

EVALUATING THE MODEL AND KEY EXPERIMENTAL FINDINGS

There is one major question which should be asked first in any model of disease. This is

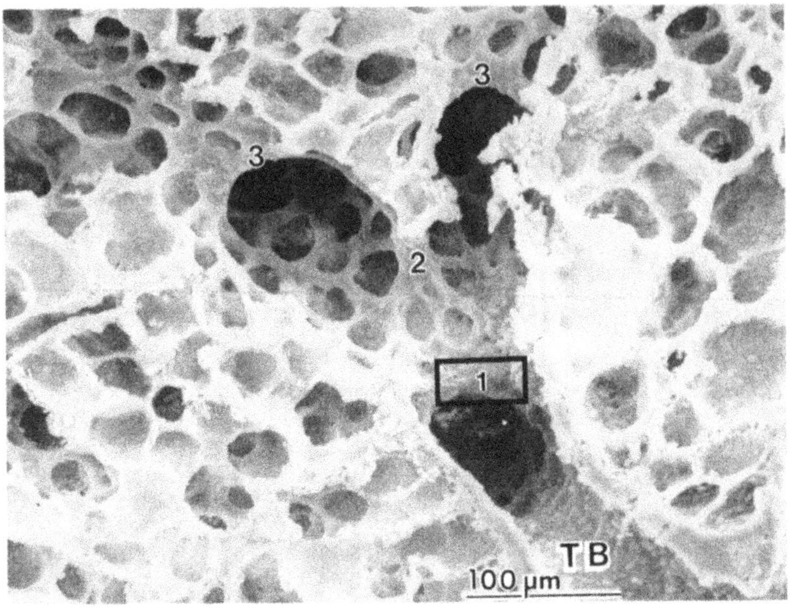

FIGURE 2B.

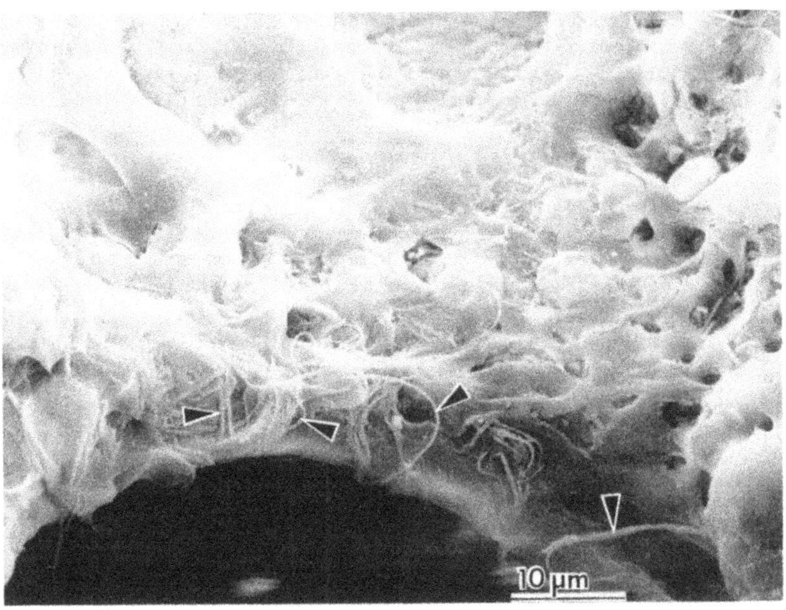

FIGURE 2C.

whether or not the animals develop a condition which is comparable to the human disease. If so, the door obviously is open to work out the basic biological mechanisms of the disease process. Using light microscopy and ultrastructural morphometry, rats and mice have been shown to provide such a model for the study of asbestosis.[9,13,19,20,23-25]

As described earlier, Wagner et al.[9] established that rats develop a progressive interstitial fibrosis from asbestos exposure. It was asked whether or not it would be possible to establish

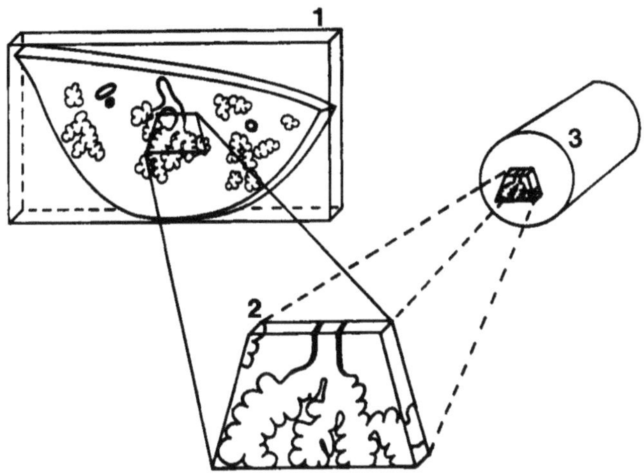

FIGURE 3. Diagram of method for isolating alveolar duct bifurcations for transmission electron microscopy. A thick slice of lung tissue is embedded in plastic (1). A terminal bronchiole and its alveolar ducts are recognized under a dissecting microscope; this region (2) is cut out of the large plastic block and glued to a blank plastic cylinder (3) for thin-sectioning. (From Warheit, D. B., Chang, L. Y., Hill, L. H., Hook, G. E., Crapo, J. D., and Brody, A. R., *Am. Rev. Respir. Dis.*, 129, 301, 1984. With permission.)

the original deposition sites of fibers inhaled during the first 1 h of exposure and then sort out the very first cellular responses to the particles. Fortunately, it was feasible to fix the lungs of the exposed animals rapidly enough after exposure to reveal the fibers on alveolar surfaces[20,21] (see previous section). Surprisingly, the fibers small enough to pass through the conducting airways and reach the alveolar level were concentrated at the first and second bifurcations of the alveolar ducts[20,21] (Figure 2). Although particles of this small size (~10 μM long and 0.2 μM diameter) would be predicted to distribute evenly by diffusion on alveolar surfaces,[27] there appeared to be sufficient flow of air to cause interception of fibers at the alveolar duct bifurcations where about 80% of the deposited fibers were observed immediately after exposure.[21] Electron microscopy of vascular-perfused tissue is the only method currently available to make such a quantitative observation in this model. The significance of this finding becomes apparent when one attempts to study the clearance of the fibers and subsequent cellular events.

The next question was where the inhaled particles go, which cells respond, and over what time frame. The first attempts to section the lung randomly and study particle distribution proved fruitless because of the focal nature of the original deposition pattern. When the alveolar duct bifurcations were dissected out and sectioned (see above), a series of fascinating pathobiological events could be deduced. Below is outlined a chronological series of events which shed some light on the pathogenesis of asbestosis as revealed in our model.

The first detectable cellular response to the presence of asbestos is the active uptake of fibers by type I epithelial cells.[20] This takes place during the first hour of exposure and continues as long as there are fibers on the alveolar surface. This event is significant because epithelial uptake appears to provide the major route for asbestos fibers to reach the lung interstitium. Fibers have been observed in alveolar-capillary basement membranes and interstitial connective tissue within the first 1 to 5 h after a 1-h exposure.[20] Although the mechanistic details have not yet been firmly established, it appears that asbestos fibers are translocated through the alveolar epithelium by an actin-containing microfilament system[28,29] similar to that which transports secretory granules and other organelles.[30] It has been estimated from lung digestion studies that approximately 20% of the fibers which reach the alveolar level are translocated to the interstitium[22] and many are retained in the connective tissue region for at least 3 months postexposure.[31] There is little doubt that a significant number of fibers will be entrained in the lung for the lifetime of the animal. It is proposed that the asbestos fibers in the lung interstitium play a major role in disease progression postexposure.

During the 1- to 5-h exposure periods employed in the model, asbestos fibers accumulate on the alveolar duct bifurcation surfaces as discussed above.[19,23] It is during this exposure period that the fifth component of complement (i.e., C5) in the alveolar lining layer is cleaved to form C5a.[19,32] C5a is a well-known chemotactic factor which attracts both polymorphonuclear leukocytes and macrophages *in vitro* and *in vivo*.[19,32] In a variety of normal animals, substantial amounts of C5 can be recovered from the lung. In the model, it was shown that a 3-h exposure to chrysotile asbestos depleted all detectable C5 and induced the production of a complement-dependent chemotactic factor.[19,32] Biochemical studies established the identity of the factor as C5a, which most likely was a product of the asbestos-induced alternative pathway of complement activation.[19] The major source of alveolar complement appeared to be from serum, as determined in exposured and unexposed rats treated with cobra venom factor which depletes all circulating complement.[19,32] Alveolar C5 returned to normal levels by 2 weeks after a 3-h exposure to chrysotile asbestos.[32]

Within 12 to 24 h after a 3-h asbestos exposure, increased numbers of alveolar macrophages accumulated at alveolar duct bifurcations (Figure 4)[23] where the fibers originally were deposited.[21] There were no increases in total numbers of macrophages per lung, but the percentages of bifurcations with clusters of macrophages increased from virtually none in the sham and zero-time animals to over 90% by 48 h postexposure.[23] Using complement-deficient animals and biochemical techniques, it was shown that the macrophage accumulation was dependent upon activation of C5 to produce C5a on alveolar surfaces as discussed above. Thus, alveolar macrophages were concentrated on bifurcation surfaces by 48 h after asbestos exposure of 1, 3, or 5 h. Interestingly enough, macrophage numbers returned to normal by 1 week postexposure. After learning the kinetics of the macrophage response, it was possible to design experiments in which these cells could be recovered from the lung by lavage and studied *in vitro*.[23] Studies of macrophage biology *in vitro* form a major component of work with the model,[33,34,35] but these studies will be reviewed in a different forum.

Epithelial cells, macrophages, and fibroblasts all are exposed to asbestos fibers in the first few hours and days postexposure.[24] Since this brief exposure results in an interstitial lesion in which there are increased volumes and numbers of cells,[25] the earliest evidence of proliferation in animals injected with tritiated thymidine (^{3}H-TdR) was studied. Immediately after a 5-h exposure, there was no evidence of ^{3}H-TdR incorporation by any cells of the bronchiolar-alveolar regions.[24,36] The first evidence of ^{3}H-TdR uptake by pulmonary cells was seen at 19 h postexposure. The numbers of labeled cells climbed to between 8 and 18% by 24 to 30 h after exposure and remained significanty elevated through a 48-h period (Figure 5). The percentage of labeled cells returned to normal by 1 week postexposure and remained at this level throughout the 1-month period studied. At no time did labeled cells of unexposed or iron-exposed animals go higher than ~1%.[36] These data are consistent with morphometric findings which show increased numbers of epithelial and interstitial cells at alveolar duct bifurcations after a 1-h exposure to chrysotile asbestos.[25] Electron microscopy correlated with the autoradiography showed that the epithelial cells proliferating were Clara cells in the airways and type II cells at the bifurcations. Both interstitial fibroblasts and macrophages also had incorporated ^{3}H-TdR to a highly significant degree.[36]

This model of asbestos exposure is exceptionally good for evaluating the mass of inhaled dust which is deposited in the lung and retained over time postexposure.[22,31] It is so useful because the investigator knows the respirable mass of aerosolized asbestos, the ventilatory

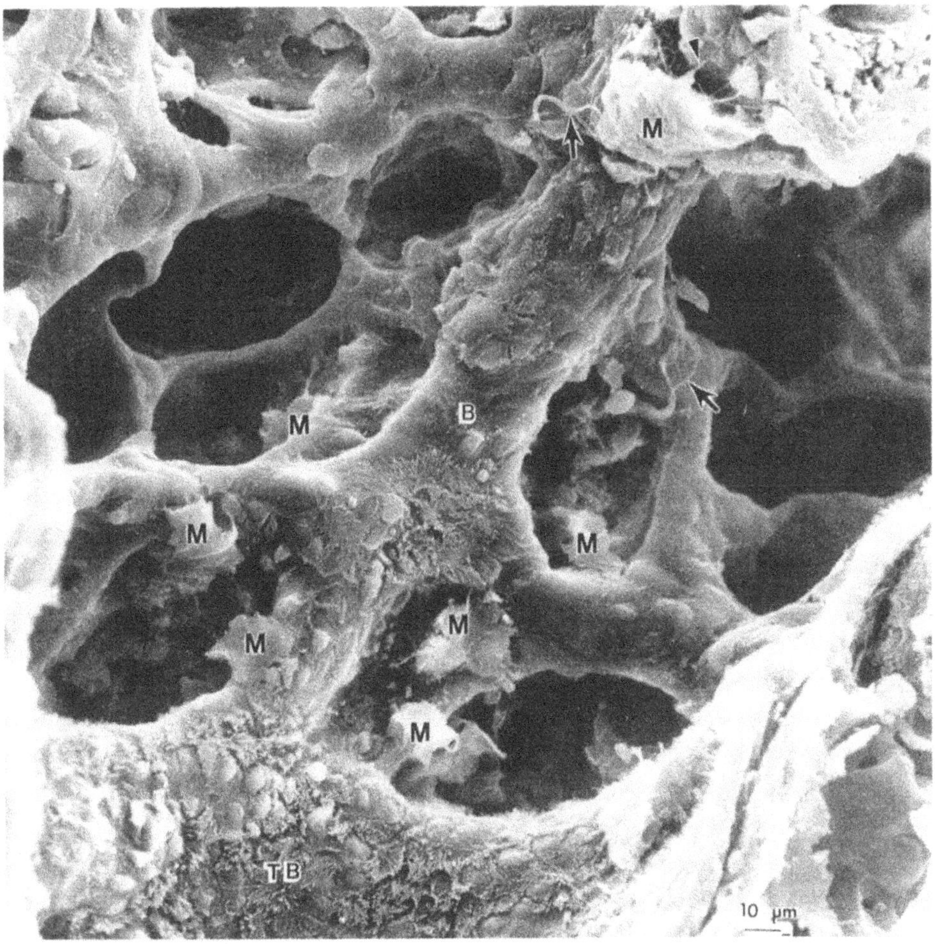

FIGURE 4. SEM of lung tissue from a rat 48 h after a 3-h exposure to chrysotile asbestos. Macrophages (M) have accumulated on the first alveolar duct bifurcation (B) and in adjacent alveolar spaces. Some of the cells exhibit phagocytized asbestos fibers (arrows). Ciliated and nonciliated cells of the terminal bronchiole (TB) are observed.

rate of the animal, and the unit mass-density of the fiber type inhaled.[22,31] In addition, because the duration of exposure is precise, when measurements of lung asbestos burden are made immediately after exposure, one has a meaningful number to which all clearance data can be compared postexposure.[22,31] Such experiments have been carried out with both chrysotile[22] and crocidolite asbestos.[31] It has been learned that approximately 20% of the respirable fraction was deposited in the lung, and 20% of this amount was still present in the lungs 1 month after exposure. At this 1-month time, significantly longer fibers were found in the lung compared to earlier time points. The chrysotile fibers also were thinner at 1 month postexposure,[22] but the crocidolite fibers exhibited no changes in diameter.[31] These two separate experiments support the concept that smaller fibers are cleared from the lung preferentially and that chrysotile fibers are likely to fragment longitudinally as well as in cross-section.

CONCLUSIONS AND FUTURE DIRECTIONS

It is believed that the animal model of asbestosis described herein is most useful for

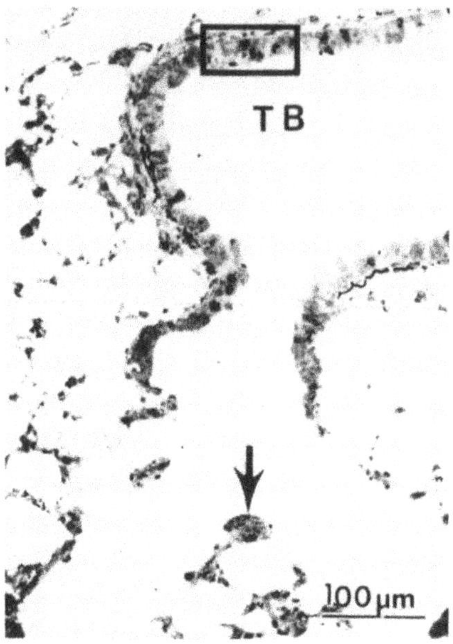

A

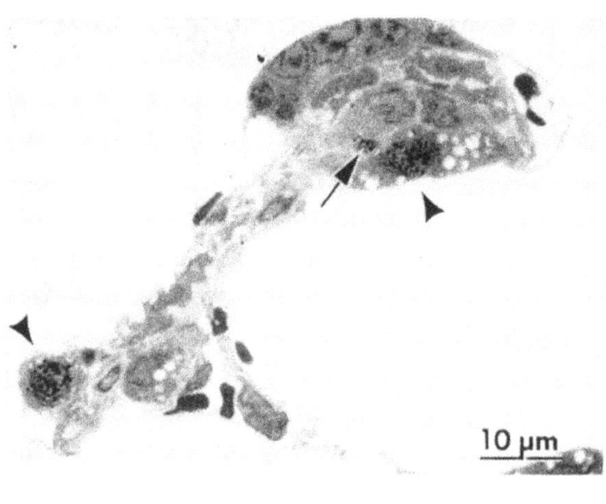

B

FIGURE 5. Light micrographs of lung tissue from a rat exposed to chrysotile asbestos for 5 h and then inoculated with tritiated thymidine 2 h before sacrifice at 24 h postexposure. (A) Low magnification showing the terminal bronchiole (TB) and first alveolar duct bifurcation (arrow). (B) Higher magnification of the bifurcation illustrating interstitial cell (arrow) and epithelial cells (arrowheads) labeled with ^{3}H-TdR. (C) Portion of the bronchiole (rectangle in Figure 1A) exhibiting numerous labeled epithelial cells (arrowheads), a submucosal interstitial cell (arrow), and macrophages (M) in the airway. (From Brody, A. R., *Chest,* 89, 155, 1986. With permission.)

studying the earliest cellular events after exposure. The target cells have been identified and categorized as to the nature of their response *in vivo.*[37] This paves the way for making intelligent decisions on which cells to separate from the lung. Once the appropriate cells can be maintained in culture, studies using biochemical and modern cell biological techniques will provide an understanding of the mechanisms leading to the asbestos-induced lung

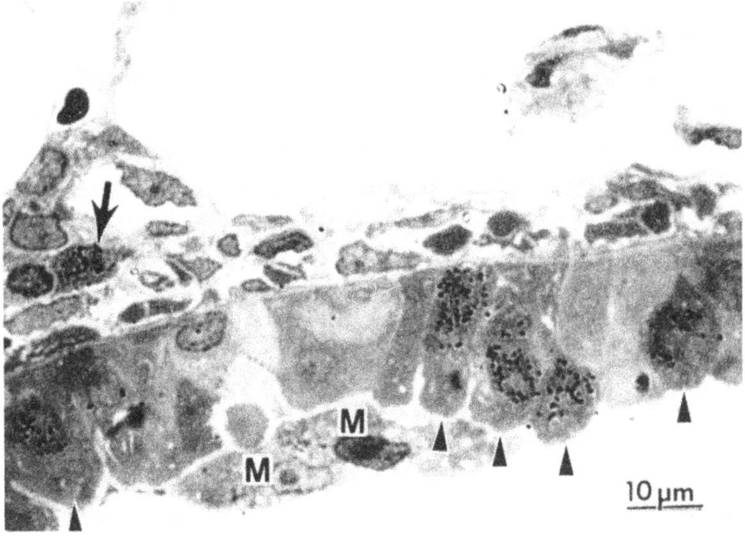

FIGURE 5C.

fibrosis. Thus far, results using this model have suggested that *in vitro* studies on alveolar and interstitial macrophages as well as on type II epithelial cells and fibroblasts should be carried out. It was most interesting to learn that pulmonary macrophages produce a platelet-derived, growth factor-like protein which stimulates lung fibroblasts to proliferate *in vitro*.[38,39] In addition, the macrophages produce the whole range of arachidonic acid metabolites from both the cyclooxygenase and lipoxygenase pathways.[33,34] It is proposed that the growth factors and the inflammatory metabolites are mediators of asbestos-induced interstitial disease, and studies are ongoing to test this hypothesis.

The future for *in vivo* studies with the model appears bright. Plans to carry out sequential 5-h exposures on 3 consecutive days have been made. This will allow us to ask what happens to lung macrophages which have already responded after the first day of exposure and then are exposed again; i.e., "hit on the head" by newly inhaled fibers as they reside on the duct bifurcations. Will more macrophages be attracted? Will the cells already there exhibit increased activation? Will this result in increased or sustained proliferation of the surrounding bronchiolar-alveolar epithelial and interstitial cells; if so, is the severity or time course of the progressive lesion altered? Animals exposed to multiple doses of fibers provide even better models of human exposure because individuals are likely to be exposed repeatedly during construction or demolition of homes and buildings. Before it is possible to understand the pathobiological events associated with multiple exposures, it will be necessary to establish these events after single brief exposures. It is obvious that many interesting and provocative questions remain concerning the biological mechanisms of asbestos-induced lung fibrosis. It is hoped that this model will aid in elucidating many of them.

ACKNOWLEDGMENTS

The authors are grateful for continuing collaboration with Drs. David B. Warheit, Ling Yi Chang, James D. Crapo, and Victor Roggli. They also are indebted to the highly cooperative individuals from Northrop Services, Inc., who operate the NIEHS Inhalation Exposure Facility under contract from the National Toxicology Program.

REFERENCES

1. **Selikoff, I. J. and Lee, D. H. K.,** *Asbestos and Disease*, Academic Press, New York, 1978.
2. **Becklake, M. R.,** Asbestos-related diseases of the lungs and pleura, *Am. Rev. Respir. Dis.*, 126, 187, 1982.
2. **Becklake, M. R.,** Exposure to asbestos and human disease, *N. Engl. J. Med.*, 306, 1480, 1982.
4. **Craighead, J. E. and Mossman, B. T.,** The pathogenesis of asbestos-associated diseases, *N. Engl. J. Med.*, 306, 1446, 1982.
5. **Cooke, W. E.,** Fibrosis of the lungs due to the inhalation of asbestos dust, *Br. Med. J.*, II, 147, 1924.
6. **King, E. J., Clegg, J. W., and Rae, V. M.,** The effect of asbestos and of aluminum on the lungs of rabbits, *Thorax*, 1, 188, 1946.
7. **Vorwald, A. J., Durkan, T. M., and Pratt, P. C.,** Experimental studies of asbestosis, *Arch. Ind. Hyg. Occup. Med.*, 3, 1, 1951.
8. **Holt, P. F., Mills, J., and Young, D. K.,** Experimental asbestosis with four types of fibers: importance of small particles, *Ann. N.Y. Acad. Sci.*, 132, 87, 1965.
9. **Wagner, J. C., Berry, G., Skidmore, J. W., and Timbrell, V.,** The effect of inhalation of asbestos in rats, *Br. J. Cancer*, 29, 252, 1974.
10. **Timbrell, V.,** The inhalation of fibrous dusts, *Ann. N.Y. Acad. Sci.*, 132, 255, 1965.
11. **Begin, R., Rola-Pleszczynski, M., Masse, S., et al.,** Asbestos induced injury in the sheep model: the initial alveolitis, *Environ. Res.*, 30, 195, 1983.
12. **Kagan, E., Oghiso, Y., and Hartmann, D. P.,** Enhanced release of a chemoattractant for alveolar macrophages following asbestos inhalation, *Am. Rev. Respir. Dis.*, 128, 680, 1983.
13. **Bozelka, B. E.,** A murine model of asbestosis, *Am. J. Pathol.*, 112, 326, 1983.
14. **Begin, R., Cantin, A., Drapeau, G., Lamoureux, G., Boctor, M., Masse, S., Rola-Pleszczynski, M.,** Pulmonary uptake of Gallium-67 in asbestos-exposed humans and sheep, *Am. Rev. Respir. Dis.*, 127, 623, 1983.
15. **Lemaire, I., Beaudoin, H., Moore, S., and Grondin, D.,** Alveolar macrophage stimulation of lung fibroblast growth in asbestos-induced pulmonary fibrosis, *Am. J. Pathol.*, 122, 205, 1986.
16. **Brain, J. D., Knudson, D. E., Sorokin, S. P., and Davis, M. A.,** Pulmonary distribution of particles given by intratracheal instillation or by aerosol inhalation, *Environ. Res.*, 11, 13, 1976.
17. **Begin, R., Masse, S., and Bureau, M. A.,** Morphologic features and function of the airways in early abestosis in the sheep model, *Am. Rev. Respir. Dis.*, 126, 870, 1982.
18. **Adamson, I. Y. R. and Bowden, D. H.,** Crocidolite-induced pulmonary fibrosis in mice; cytokinetic and biochemical studies, *Am. J. Pathol.*, 122, 621, 1986.
19. **Warheit, D. B., George, G., Hill, L. H., Snyderman, R., and Brody, A. R.,** Inhaled asbestos activates a complement-dependent chemoattractant for macrophages, *Lab. Invest.*, 52, 505, 1985.
20. **Brody, A. R., Hill, L. H., Adkins, B., Jr., O'Connor, R. W.,** Chrysotile asbestos inhalation in rats: deposition pattern and reaction of alveolar epithelium and pulmonary macrophages, *Am. Rev. Respir. Dis.*, 123, 670, 1981.
21. **Brody, A. R. and Roe, M. W.,** Deposition pattern of inorganic particles at the alveolar level in the lungs of rats and mice, *Am. Rev. Respir. Dis.*, 128, 724, 1983.
22. **Roggli, V. and Brody, A. R.,** Changes in numbers and dimensions of chrysotile asbestos fibers in lungs of rats following short-term exposure, *Exp. Lung Res.*, 7, 133, 1984.
23. **Warheit, D. B., Chang, L. Y., Hill, L. H., Hook, G. E., Crapo, J. D., and Brody, A. R.,** Pulmonary macrophage accumulation and asbestos-induced lesions at sites of fiber deposition, *Am. Rev. Respir. Dis.*, 129, 301, 1984.
24. **Brody, A. R.,** Pulmonary cell interactions with asbestos fibers *in vivo* and *in vitro*, *Chest*, 89, 155, 1986.
25. **Chang, L. Y., Hill, L. H., Brody, A. R., and Crapo, J. D.,** Early morphologic alterations induced in the lung by acute exposure to asbestos, *Am. J. Pathol.*, 131, 156, 1988.
26. **Brody, A. R., George, G., and Hill, L. H.,** Interactions of chrysotile and crocidolite asbestos with red blood cell membranes: chrysotile binds to sialic acid, *Lab. Invest.*, 49, 468, 1983.
27. **Lippmann, M., Yeates, D. B., and Albert, R. E.,** Deposition retention and clearance of inhaled particles, *Br. J. Ind. Med.*, 37, 337, 1980.
28. **Brody, A. R., Hill, L. H., and Adler, K. B.,** Actin-containing microfilaments of pulmonary epithelial cells provide a mechanism for translocating asbestos to the interstitium, *Chest*, 83, 11, 1983.
29. **Brody, A. R., Hill, L. H., Hesterberg, T. W., Barrett, J. C., and Adler, K. B.,** Intracellular transport of inorganic particles, in *The Cytoskeleton, A Target For Toxic Agents*, Clarkson, T. W., Sager, P. R., and Syverson, T. L., Eds., Plenum Press, New York, 1986, 221.
30. **Adler, K. B., Brody, A. R., and Craighead, J. E.,** Studies on the mechanism of mucin secretion by cells of the porcine tracheal epithelium, *Proc. Soc. Exp. Biol. Med.*, 166, 37, 1981.
31. **Roggli, V. L., George, M. H., and Brody, A. R.,** Clearance and dimensional changes of crocidolite asbestos fibers isolated from lungs of rats following short-term exposure, *Environ. Res.*, 42, 94, 1987.

32. **Warheit, D. B., Hill, L. H., George, G., and Brody, A. R.,** Time course of chemotactic factor generation and the corresponding macrophage response to asbestos inhalation, *Am. Rev. Respir. Dis.*, 134, 128, 1986.
33. **Kouzan, S., Gallagher, J., Eling, T. E., and Brody, A. R.,** Particle binding to sialic acid residues on macrophage plasma membranes stimulates arachidonic acid metabolism, *Lab. Invest.*, 53, 320, 1985.
34. **Kouzan, S., Brody, A. R., Nettesheim, P., and Eling, T. E.,** Production of arachidonic acid metabolites by macrophages exposed *in vitro* to asbestos, carbonyl iron particles or calcium ionophore, *Am. Rev. Respir. Dis.*, 131, 624, 1985.
35. **Gallagher, J. E., George, G., and Brody, A. R.,** Sialic acid mediates the initial binding of positively-charged inorganic particles to alveolar macrophage membranes, *Am. Rev. Respir. Dis.*, 135, 1345, 1987.
36. **Brody, A. R. and Overby, L. H.,** Incorporation of tritiated thymidine by epithelial and interstitial cells in bronchiolar-alveolar regions of asbestos-exposed rats, *Am. J. Pathol.*, 134, 133, 1989.
37. **Brody, A. R. and Hill, L. H.,** The target cells for inhaled mineral dusts, in *In Vitro Effects of Mineral Dusts*, (NATO ASI Series), Vol. G3, Beck, E. G. and Bignon, J., Eds., Springer-Verlag, Berlin, 1985, 55.
38. **Bauman, M. D., Jetten, A. M., and Brody, A. R.,** Biologic and biochemical characterization of a macrophage-derived growth factor for rat lung fibroblasts, *Chest*, 91, 15, 1987.
39. **Bonner, J. C., Hoffman, M., Brody, A. R.,** α-Macroglobulin secreted by alveolar macrophages serves as a binding protein for a macrophage-derived homologue of platelet-derived growth factor, *Am. J. Respir. Cell Mol. Biol.*, in press.

SILICA-INDUCED PULMONARY FIBROSIS

Ian Y. R. Adamson

INTRODUCTION

The induction of pulmonary fibrosis by the inhalation of silica has been known for many years. Although silicosis may be regarded as purely an occupational lung disease, the list of potential exposure victims has greatly increased from the first stone cutters or miners to include those working with substances such as abrasives, fillers, and glass.[1] The element silicon makes up over 25% of the crust of the earth, with silicon dioxide or free silica being its most abundant substance. The structure $(SiO_2)x$ is a tetrahedron joined by the oxygen atoms to give a silica crystal, the most common of which is quartz. Other related structures that incorporate iron, aluminum, and magnesium make up various silicates of which asbestos fibers form one group. Natural pure silica is closely associated with certain rock formations including granite, so that miners and quarrymen are frequently victims of a fibrotic lung disease that is distinct from a simple pneumoconcosis which is characterized by accumulations of carbon pigment in the lung without fibrosis or emphysema.

The toxic nature of the actual silica component of inhaled mixed dusts is often masked by other factors such as cigarette smoking or pulmonary infections, especially tuberculosis, which complicate the pulmonary pathology. However, it is well established that a nodular fibrotic reaction of the lung may follow inhalation of small crystalline particles of silica that lie in the respirable range, i.e., <3 μm. Although most inhaled particles of this size are eliminated by microciliary clearance, many are deposited in bronchiolar and alveolar regions of the lung. Repeated occupational exposure increases the total dust bruden and eventually produces clinical symptoms of lung disease such as shortness of breath. By radiologic examination, small nodular opaque areas are seen that correspond to discrete pulmonary granulomas (Figure 1A). These nodules are well circumscribed, often found in peribronchiolar regions, and are usually composed of collagen with closely associated macrophages and fibroblasts at the periphery. The interstitial macrophages frequently contain silica particles among carbonaceous cytoplasmic material. Silica crystals can be distinguished in mixed pulmonary pigments by examining lung sections under polarized light. Crystals are usually more evident in fibrotic granulomas, but silica is also found in free alveolar macrophages and some diffuse fibrosis may be found in alveolar walls. With time, the fibrotic process develops further to increase the size of the nodules, destroying alveolar tissue and often compressing or eliminating bronchiolar lumina. Pleural fibrosis may also be produced and the combined effects of the fibrotic reactions reduce pulmonary function.[1,2] Since free silica particles penetrate lung tissue and can be cleared either in the free state or within interstitial macrophages to the lymphatics, fibrosis in hilar lymph nodes may also be detected in the long term.

The pathogenesis of silica-induced fibrosis cannot be determined solely from these observations on human disease, although the involvement of the alveolar macrophage in the mechanism of injury seems clear. This cell type is the major defender of the alveolar surface due to its phagocytic function and rapid clearance of most inhaled contaminants. However, elimination of particles from the alveoli may be inhibited if there is any reduction in these macrophagic functions. For example, the system may be overloaded by a heavy inhaled burden that necessitates an adaptive response with the recruitment of extra phagocytes to the alveoli from blood and interstitium.[3,4] In the case of silica, this sequence seems to occur but there is incomplete clearance of particles. It has been suggested that silica is toxic to alveolar macrophages and that death of this cell type after phagocytosis releases the non-

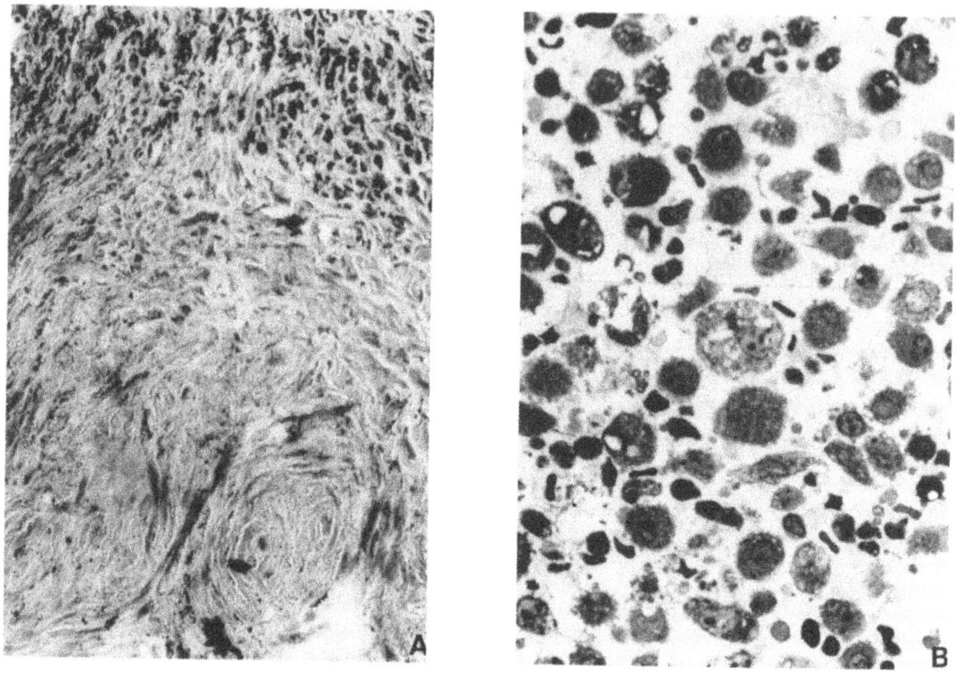

FIGURE 1. (A) Silicotic granuloma in human lung showing a dense fibrotic area with fibroblasts and pigmented macrophages at the periphery. (Magnification × 80.) (B) Cells lavaged from the lung are PMN and macrophages, many of which show vacuolated lysosomes. (Magnification × 400.)

digestible silica for further phagocytosis by other macrophages.[5] Cell lysis also results in release of lysosomal enzymes into lung tissue where further cell injury can occur. These hypotheses were based initially on the morphology of silica-exposed alveolar macrophages in human lung. These cells were often swollen, vacuolated with disrupted lysosomal membranes (Figure 1B). A cytotoxic action for silica on macrophages was the basis of the first experimental studies on silicosis.

Exposure of animals to inhaled silica was carried out initially to determine whether the fibrotic reaction seen in humans could be reproduced and the sequence of cellular events studied. It is now well recognized that inhalation studies require such careful monitoring of particle dose, size, flow characteristics, and proper animal housing that few laboratories can carry out satisfactory experiments.[6] Although long-term exposure of animals to inhaled silica did produce fibrotic nodules, there was some dispute over the question of infection in stressed animals and whether silicosis is seen in pathogen-free rats. Small rodents are particularly prone to respiratory infections such as mycoplasma pneumonia, and the incidence seems to increase with age so that the results of long-term exposures (1 to 2 years) were suspect without proper screening of the animals for bacteria or viruses. Inhalation studies have largely been used in toxicologic studies to establish permissible airborne levels for industrial standards. In most cases, the pulmonary disease seen in this more natural exposure route matches that produced by the more convenient experimental procedure of intratracheal instillation. The major advantages of this method of delivering particles to the lung are that a measured dose can be administered and that a precise time zero is known for the temporal relationships of cell injury and inflammation through to fibrosis. Efforts can be concentrated on studying the pathogenesis of silicosis rather than on establishing minimal injury levels for extrapolation to human exposure.

SILICA ADMINISTRATION

Choice of Animal

Silicotic nodules have been produced in a variety of animals such as mice,[7,8] rats,[9,10] guinea pigs,[11] and rabbits.[12] The pulmonary reactions appear similar and dose dependent; for example, fibrosis is produced after injecting 5 to 50 mg to rats, 1 mg to mice, or 400 mg to rabbits. Usually, the major factors in determining the species to use for an *in vivo* study are (1) the end points being used, e.g., for physiologic studies, a larger animal is required[13] and (2) the frequency of time points under study and the number of animals per group, both of which considerably increase the costs, especially with large animals. There do not appear to be strain differences in lung reaction, although pathogen-free animals may show less fibrosis. It is possible that compromising macrophages with silica may make animals or humans more susceptible to an infectious agent whose pathology may be superimposed on the pure silica reaction.

For studies using a combination of morphology, biochemistry, and cell kinetics, the mouse has proved to be suitable and gives enough tissue for many parameters of study. However, physiologic measurements are not possible and alveolar macrophages recovered by lavage may not be present in sufficient numbers of metabolic studies unless cells from several animals are pooled. If this type of study is required, it is suggested that rats be used. One normal rat can give 2 to 4 $\times$ 10^6 cells when the lung is lavaged four times, a tenfold increase over the number obtained from one mouse lung.

Sample Preparation

Pulmonary fibrosis can be induced by injecting various forms of crystalline silica, the most commonly used being quartz. A standard sample used for many years came from South Africa (Dowson and Dobson quartz from the Pneumoconiosis Research Unit, Johannesburg) and is pure sample with mean particle diameter of approximately 0.3 μm, well within the respirable range. More recently, a product called MinUsil® (Pennsylvania Glass Sand Company, Pittsburgh) has been used. Both of these samples have been used in the University of Manitoba laboratory with comparable results at an injection level of 10 mg to rats and 1 or 2 mg to mice. These doses gave measurable fibrosis starting at 4 to 6 weeks after a single injection.

The main consideration in preparing the sample for intratracheal injection is the liquid volume that can reasonably be injected to the lungs of small animals. From early trials using sterile water injections, examining animal recovery, and quantitating fluid and inflammatory cells in subsequent lung lavage, it was determined that the optimal level for injected fluid should be 0.1 ml to mice and 0.5 to 1 ml to rats depending on size. For the injections to mice described below, a stock silica suspension of 200 mg in 10 ml was made up. This gives a dose of 2 mg silica in 0.1 ml sterile water when injected to mice. The stock solution was thoroughly shaken and dispersed further using ultrasonic agitation just prior to drawing into a syringe for injection.

Silica Injection

These techniques have been developed in the laboratory of the author using mice and have been successfully used in rats with the appropriate increase in anesthetic dose. The key to successful intratracheal injections lies in the initial phase of anesthesia while administering the particles. As a general rule, ether is never used when pulmonary changes are to be examined, particularly when studying ultrastructure or lipid biochemistry. Sodium pentobarbital is a preferred anaesthetic because with careful dose control, recovery is faster than after ether exposure. When using mice, the stock solution usually supplied at 50 mg/ml is diluted 1:4 with sterile water and 0.08 to 0.12 ml is then injected to a 25-g mouse, i.e., about 40 mg/kg body weight. This same approximate dose level is used for rats and is injected by the intraperitoneal route.

These are definite strain differences when considering the response of mice to this level of pentobarbital and it is best to judge the correct dosage by the level of anesthesia. The animal should not be deeply anesthetized and should not show shallow breathing. If this is the case, the injected particulate suspension will lie in the major airways without being deeply inhaled into the lung and can cause blockage. Frequently these animals die. Best results are obtained when the mouse is mildly anesthetized, still reacts to the touch, and even gives mild resistance to handling. The animal can be attached to a board by making slipknots of suture thread to pass over each extremity and attaching the other end of the thread to a permanent hook or pin in the board. When pinning the mouse out in this way, it is also useful to hold the head back for easy access to the trachea. This can be accomplished by using a slipknot of thread around the front teeth, pulling straight back over the head, and attaching to the board. The animal may squirm somewhat at this procedure if the correct level of pentobarbital is used.

At this stage the hair around the tracheal injection site is shaved or, easier with mice, clipped close to the skin over a region at least 1 cm long. The area is swabbed with alcohol or iodine which dries quickly and an incision about 1 cm in length is then made along the line of the trachea using a scalpel. Using fine forceps, the tissue around the trachea, including the various salivary glands, is separated to expose the trachea. The tip of a pair of micro-tip hemostatic forceps in the locked position can be inserted under the trachea and left in place. This gives a solid backing underneath and produces a slightly elevated region of the trachea which can be readily accessed from the top or "head" end of the animal for injection.

Using a 1-ml syringe, draw up 0.4 ml air and then 0.1 ml of the sterile silica suspension. A small needle 0.5 in. long of 26-g gauge (no less than 23 g for mice) is attached and the needle is inserted into the trachea just above the elevation made by the forceps. When injecting the suspension, it is important to inject all the air in the syringe to ensure good dispersion of particles to distal lung. The animal will often take deep breaths at this point which also helps to distribute the silica. The needle, syringe, and forceps can be quickly removed and the skin wound sewn with two stitches. It has been found that a curved needle is much easier to handle when using mice. The animal can be removed from the table by releasing the slipknots and placed in a plastic box in a well heated and ventilated area to recover from the anesthesia, which usually takes 15 to 45 min.

The procedure takes some practice to get the correct anesthetic dose and particle delivery in a short time with good particle dispersal and animal survival. The injections usually result in some mortality in the first 24 h. The animal mortality rate was initially 10 to 20% at the author's laboratory, and it is now about 5%. It was found to be particularly useful to use a particle such as colloidal carbon for trial experiments. On gross examination of the lungs, the nature of this black preparation makes it easy to see the location of the injection material, from a few minutes to a few days after intratracheal injection.

Controls for these experiments are usually mice injected in a similar manner with the same volume of sterile water. In order to distinguish effects of silica from the nonspecific pulmonary response to an inert particle, control injections of colloidal carbon or polystyrene latex can be used. In this way, effects of fibrogenic and nonfibrogenic particles of similar size can be compared.

Animal Sacrifice

Various studies on the fibrotic process have focused on different stages of the pulmonary reaction, ranging from animal sacrifice in the first few days when initial largely inflammatory events are to be studied to killing weeks and months after injection to investigate the fibrotic phase of the response. The development of fibrosis or at least of impaired pulmonary function in living animals can be assessed using larger animals in which pulmonary function tests can be carried out.[13] Many other types of investigation are possible in small animals where frequent times of sacrifice allow the reaction to silica to be followed serially. The procedure now described for mice gives great potential for studying such parameters as morphology, biochemistry, and cell kinetics on single animals using both whole lung tissue as well as cells and fluid recovered by bronchoalveolar lavage.

To obtain an estimate of overall cell proliferation at any particular time and identify which cell type in the lung is dividing, mice are injected with tritiated thymidine 1 h before death. A dose of 2 μCi/g body weight of ^3H thymidine (sp.act. 2 Ci/mmol) is injected intraperitoneally.[14] In studies of cell turnover, it is advisable to use the same sex of animal for the experiments and, because there is a diurnal variation in DNA synthesis, the time of day for animal sacrifice should be kept constant.[14] Thymidine is rapidly taken up by cells in DNA synthesis which can be subsequently identified and quantified on autoradiographs.

Bronchoalveolar Lavage

Mice are killed by intraperitoneal injection of an overdose of sodium pentobarbital (0.1 ml of 50 mg/ml solution). As soon as the animal is completely anesthetized, it is pinned out as before with the head held back. The trachea is again exposed and usually there is no sign of the silica injection site. Below the likely site of the original injection, a small incision is made to allow a narrow plastic tube to be inserted and tied in place by suture thread around the trachea. This intratracheal tube should be at least 2 in. long and a catheter of 26 G with a syringe connection at the other end is preferred. The skin is then peeled back from the upper body, an incision made into the peritoneal cavity, and a hole made in the diaphragm which allows the lungs to collapse. A syringe is then connected to the end of the intratracheal tube and normal saline is injected into the lung. A volume of 1 ml will inflate the lung to about the normal volume. The diaphragm can be cut to allow direct viewing of the lungs to make sure that inflation occurs with no leakage. Leaving the chest wall in place ensures that overdistention of the lung does not occur. The saline is then slowly withdrawn and usually a volume of approximately 0.8 ml is recovered. The syringe can be detached, this volume transferred to a centrifuge tube, and an equivalent volume of fresh saline (i.e., 0.8 ml) can be drawn into the same syringe, which is reattached to the tracheal tube, and injected. This procedure can be repeated several times. Experience has shown that most cells are recovered in four such lung lavages; further washing increases the risk of blood vessel rupture and contamination of lavage fluid with blood cells and plasma without significantly increasing the cellular yield.

This procedure has recently been modified to carry out split lung lavage. Using this method, the tracheal tube is inserted, the main bronchus is exposed, and the branch leading to the left lung is clamped. The lavage procedure is followed as above using 0.5 ml saline. Because of the size of the animal, this procedure is easier on rats but with practice it can be done on mice. The lavage fluids are pooled per animal and the volume noted. The cellular content of a sample is counted using a hemocytometer and samples are spun using a cytocentrifuge to prepare slides for staining and making differential counts of cells present in the lavage.[7] The remaining sample can be spun at 1000 rpm to give a separation of cells and fluid.

A variety of studies can be carried out on this material. The cells can be resuspended to separate a pure population of alveolar macrophages, usually by their property of attachment to glass or plastic. These cells have been used for various metabolic studies. After short-term culture, their secreted products have been examined particularly for growth-promoting activity for fibroblasts.[15,16] The lavage fluid itself is also of value. Pooled samples from four washes in mice can give more than 3 ml fluid, an adequate amount to measure chemotactic activity for inflammatory cells as well as total protein and a variety of enzymes that may be indicative of lung injury.[7]

Lung Tissue Preparation

Samples of lung may be used for biochemical analysis. This can be done on tissue taken after lavage, taking care to express results as dry weight since there is usually some fluid retained in alveoli which makes lung wet weight unreliable. It has been found that the split lung lavage procedure is valuable because the clamped right lung can be tied off and excised intact for biochemical studies without the complication of a lavage procedure. The tissue can be weighed, both wet and after lyophilizing and various measurements of collagen biosynthesis and phospholipid metabolism can be carried out.

Returning to the lavaged lung, immediately after withdrawing the final saline wash, a syringe containing 2% buffered glutaraldehyde fixative is attached to the tracheal tube and 0.5 ml for a split lung or 1.0 ml for whole lung is instilled to reinflate the lung. A variation in fixation will be necessary if morphometric measurements are to be made since the lungs must be inflated at a constant pressure. This is done by placing a reservoir of fixative at a constant height, usually 20 cm above the animal, and slowly infusing fixative into the lungs before ligation. This ensures comparable inflation among the experimental groups and controls and allows comparison of measurements of air space or interstitial volume, for example, which can be used as indices of lung disease.

After instilling fixative, the lung is removed, immersed in the same fixative for 1 h, and then cut for further processing. For light microscopy on paraffin or plastic sections, slices of each lobe (approximately 3 mm thick) are cut, fixed for 2 more hours in glutaraldehyde, and postfixed in formalin. For detailed light microscopy, cellular identification is required and this necessitates the use of plastic sections which can be cut at 1 μm or less. Sections of tissue embedded in glycol methacrylate, for example, can be stained for regular light microscopy or used unstained for autoradiography.[4,7,8] Sections mounted on slides are dipped briefly in a photographic emulsion such as Kodak® NTB2 in total darkness and allowed to remain in the dark for 2 weeks to allow the emulsion to be exposed by the radioactive molecules. The slides are subsequently developed photographically before staining.[14] Cells in DNA synthesis are marked by black silver grains over the nuclei; these cells can be identified and counted. From the glutaraldehyde-fixed tissue, lung samples are also taken and postfixed in osmic acid prior to embedding in a harder plastic to allow cutting thin sections for electron microscopy.[8]

ASSESSMENT OF LESIONS

Bronchoalveolar Lavage

The most rapid indication that there has been a pulmonary reaction to silica is the presence of an inflammatory response which can be quantitated by serial lavage studies. It has been established that inhalation or instillation of a low particulate dose is handled by resident macrophages which phagocytize the particles prior to clearance from the lung by the mucociliary route. As the deposition load increases, a stimulus is generated for chemotaxis of new cells, both macrophages and polymorphonuclear leukocytes (PMN), to the alveoli as part of an inflammatory reaction.[4,17] In this respect, the pulmonary reaction to silica is not different from that seen in response to a nonfibrogenic particle.[3] The number of alveolar macrophages and PMN recovered after injecting 2 mg of silica or carbon to mice is shown in Figure 2. In the acute phase, both cell types increase in number, the larger peak being found after carbon since, in a 2-mg dose, more carbon particles are delivered due to their smaller size (0.03 μm diameter compared to 0.3 μm for silica). The PMN response is immediate. With carbon, however, these cells are not found in lavage after 1 week, whereas after silica approximately 1×10^5 PMN are always recovered up to 20 weeks, indicating an ongoing inflammatory response. The number of alveolar macrophages rises sharply after silica to about four times normal but returns to normal values at least by 4 to 6 weeks.

Although the numbers of recovered macrophages may not be different from normal several weeks after silica, there is evidence that the function of this cell type may not be normal.

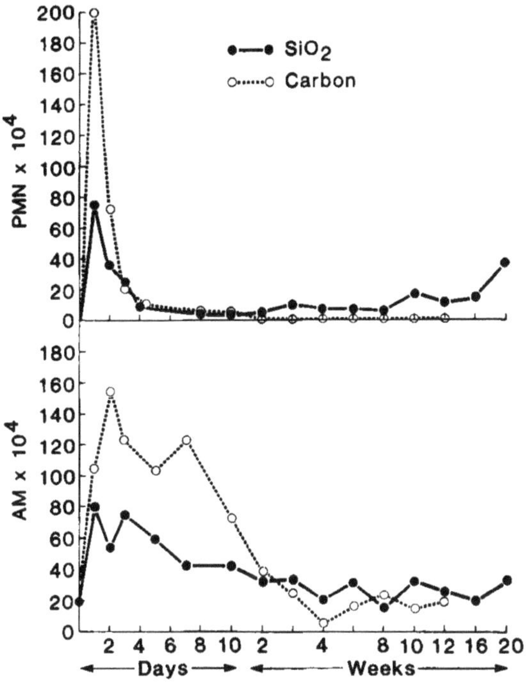

FIGURE 2. Numbers of PMN and alveolar macrophages (AM) in lavage up to 20 weeks after injecting 2 mg carbon or silica to mice.

In several studies, cells recovered by lavage have been incubated in serum-free media and the supernatants tested in fibroblast proliferation assays. It has been shown that macrophages recovered shortly after silica injection to rat lung did not release stimulatory factors, while supernatants of cells lavaged from lungs 6 weeks after silica secreted a growth factor for fibroblasts.[15] The cellular component of the bronchoalveolar lavage provides ample material for similar studies during the development of fibrosis.

Information on lung injury can also be obtained from biochemical analysis of the lavage fluid. Elevated protein levels are indicative of leakage at the air blood barrier. During an inflammatory stimulus, PMN emigration is usually accompanied by an increase in alveolar protein. This has been found after carbon and silica, although after silica the value may be increased further due to some injury to the epithelial surface. In addition, the PMN in the alveoli tend to degranulate and increased levels of lysosomal enzymes such as glucosaminidase and glucuronidase are detected by lavage (Figure 3).[7] It should be remembered that many such observations in the early phase of silica-induced lung injury are features of a general inflammatory response and are not specific for silica. Similarly, the interaction of silica and macrophages release a chemotaxin for PMN and macrophages, a finding which also applies to particle-macrophage reactions in general. However, one feature of lavage studies that suggests continuing lung injury in response to silica is the ongoing presence of PMN in the alveoli.[7]

Morphology and Cell Kinetics

Using good injection techniques, silica is well spread throughout the lung and a general inflammatory response is seen by light microscopy. After a few days, the tendency towards central clearance of free particles and cells, both PMN and macrophages, results in a more peribronchiolar location of inflammatory cells (Figure 4A). Cellular changes are observed

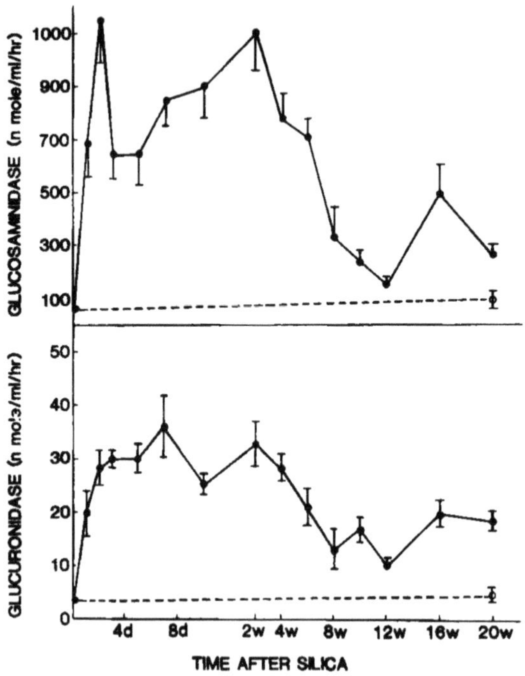

FIGURE 3. Lysosomal enzyme activity (±SE) in lavage fluid recovered at intervals after 2 mg silica. The control range is shown by the broken lines. (From Adamson, I. Y. R. and Bowden, D. H., *Am. J. Pathol.*, 117, 37, 1984. With permission.)

within the phagocytes in sections of lung tissue and in sections made from pellets of lavaged cells. Both PMN and macrophages phagocytize particles and both show increased cytoplasmic vacuolation (Figure 4B), although few necrotic macrophages are found. Cells containing silica are seen in the alveoli throughout the 20-week period.

Silica is occasionally seen by electron microscopy in vacuoles of type I epithelial cell cytoplasm (Figure 5). Some epithelial necrosis occurs and focal areas of intra-alveolar fibrin are observed.[8] Rapid repair of the epithelium is evident, particularly in autoradiographs where labeled type II cells, the progenitors of epithelial regeneration, are seen a few days after injecting silica.[7,8]

From 1 week onward, the PMN number is greatly reduced and aggregates of macrophages are found in air spaces adjacent to terminal bronchioles and in the perivascular interstitium. This latter group of cells appears to originate both from monocyte migration and by proliferation of interstitial macrophages.[3,4,8] They are particularly evident at areas where free particles cross the epithelium and the silica is subsequently phagocytized by interstitial macrophages (Figure 6).[8,18] These nodular aggregates of cells eventually form the peribronchiolar granulomas which are initially composed mostly of macrophages with a few PMN, but after 1 week contain increasing numbers of fibroblasts and macrophages (Figure 7). After a few weeks, the lesions appear less cellular and more fibrillar collagen is observed (Figure 8). By 6 weeks, much of the collagen forms a linear band around the bronchioles or is seen in fibrotic nodules, with some thickening of alveolar walls seen in places. The amount of fibrosis is difficult to estimate; some quantitation may be done in sections using morphometric methods, although measures of pulmonary function in larger animals or biochemical measurements are more reliable.

Autoradiographic studies provide more data on the reaction of specific cell types. When

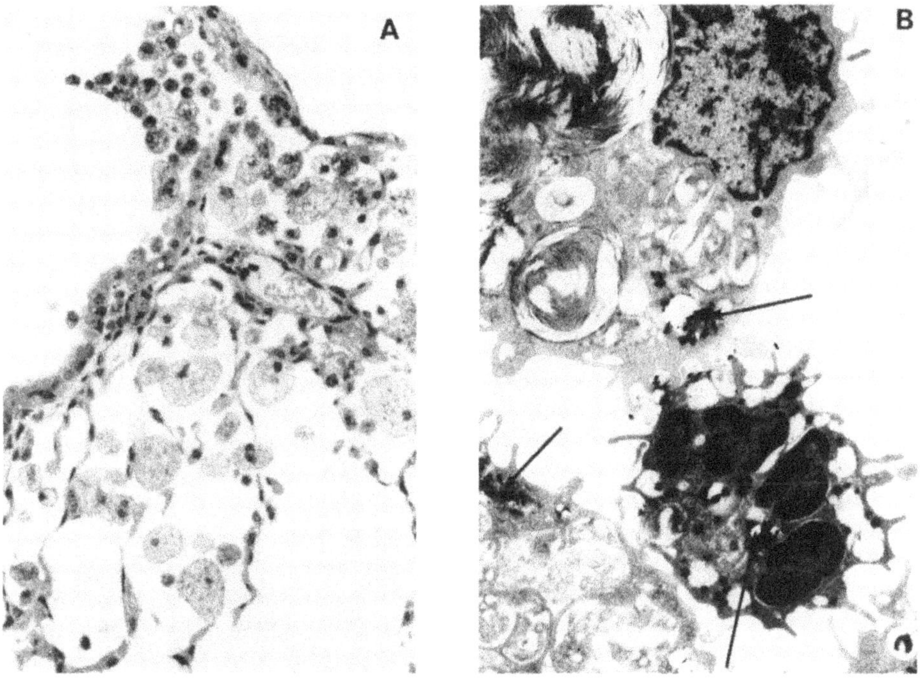

FIGURE 4. (A) At 1 week after silica, many PMN and large macrophages (AM) are concentrated in peribronchial areas. (Magnification × 150.) (B) By electron microscopy, both AM and PMN phagocytize silica (arrows). (Magnification × 6700.)

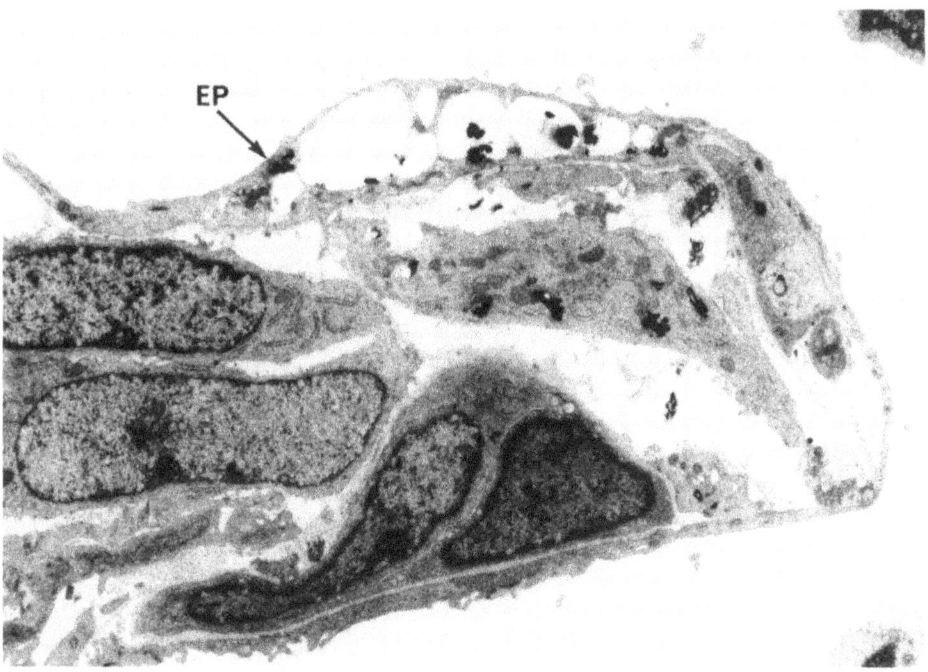

FIGURE 5. Alveolar wall of mouse lung 1 week after silica shows particles within vacuoles in the cytoplasm of type I alveolar epithelial (EP) cell. (Magnification × 9000.)

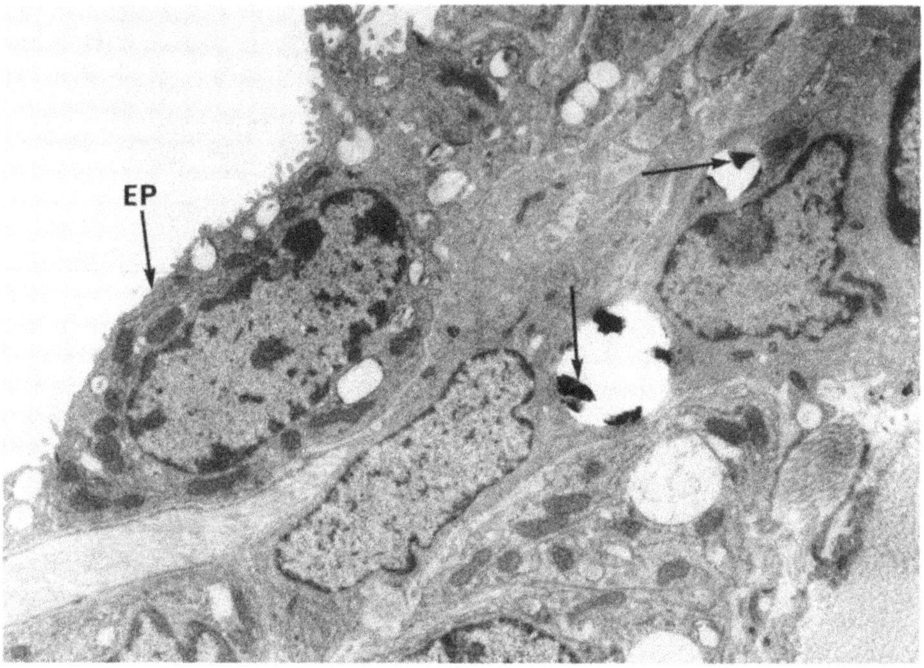

FIGURE 6. Some particles (arrows) are seen within interstitial macrophages 2 weeks after silica. The alveolar epithelium (EP) appears normal. (Magnification × 7000.)

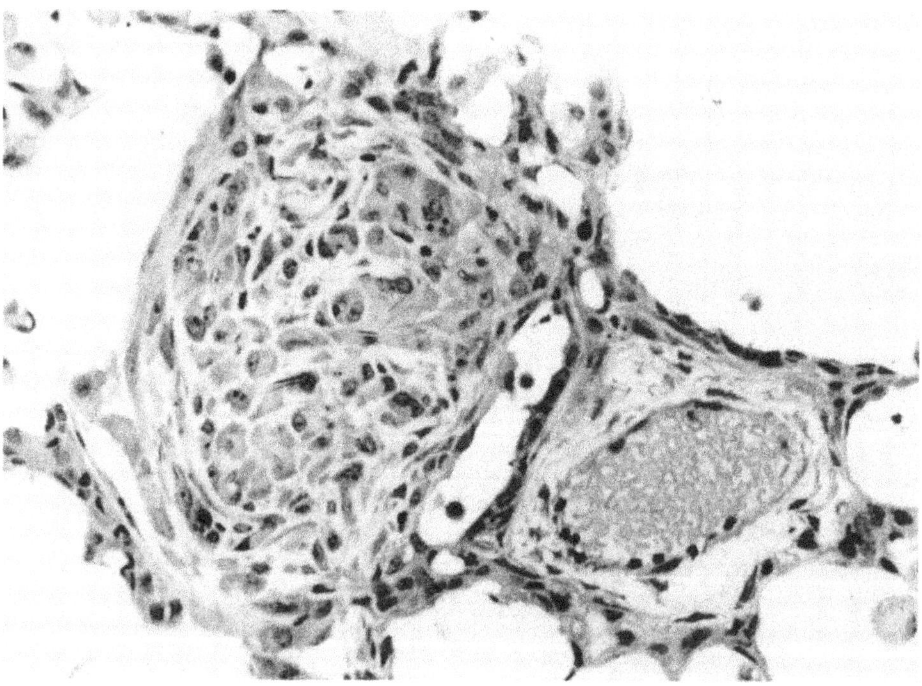

FIGURE 7. Pulmonary granuloma, composed mostly of macrophages and fibroblasts, 2 weeks after 2 mg silica. (Magnification × 350.)

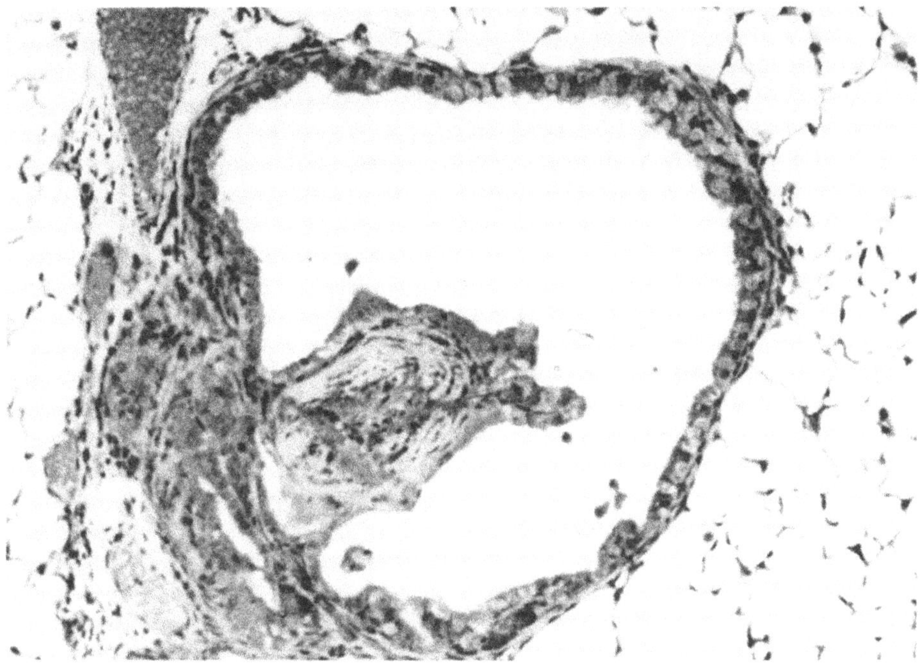

FIGURE 8. Granulomas are less cellular and areas of collagen deposition are seen 6 weeks after silica. (Magnification × 220.)

the overall percentage of the thymidine-labeled cells is calculated, it can be shown that there is a proliferative response in the lung between 2 d and 6 weeks (Figure 9). Because the lung is a multicellular organ, this peak does not indicate which cell types are responsible for enhanced DNA synthesis. However, the quality of the plastic section is sufficient to make a differential count of labeled cells by identifying each one as epithelial, endothelial, interstitial, intravascular, or unknown. By calculating the product of the overall labeling percentage and the differential percentage of each labeled cell type, a radiographic index can be calculated for each cell type.[7] The results show that there is an early peak of type II cell labeling corresponding to repair of the injured alveolar epithelium (Figures 9 and 10A). Later, the peak is due to labeling of cells in the interstitium; in the first 2 weeks this appears to be due to labeled macrophage precursors, whereas at 2 to 6 weeks the labeled cells are clearly fibroblasts (Figure 10B). Thus, the combined microscopy and cell labeling data show the patterns of cell injury and repair in the lung and indicate a prolonged phase of fibroblast proliferation.

Biochemical Studies

Some of the possible biochemical measurements have been described when considering lung lavage studies. However, crucial to establishing the model is the demonstration that fibrosis is produced. In addition to morphologic evidence, it is useful to make a biochemical determination of collagen to allow comparison with the various postinjection times, with other doses of silica, or with other injected or inhaled particulates. One of the simplest procedures is to measure the hydroxyproline (HYP) content of the tissue.

Either fresh or lyophilized lung samples are homogenized in cold water. Aliquots of the homogenate can be taken for measurements of DNA and protein content using standard methods. Another aliquot is hydrolysed by hydrochloric acid which is removed by boiling. HYP is then determined using a spectrophotometric method described by Woessner.[19] The

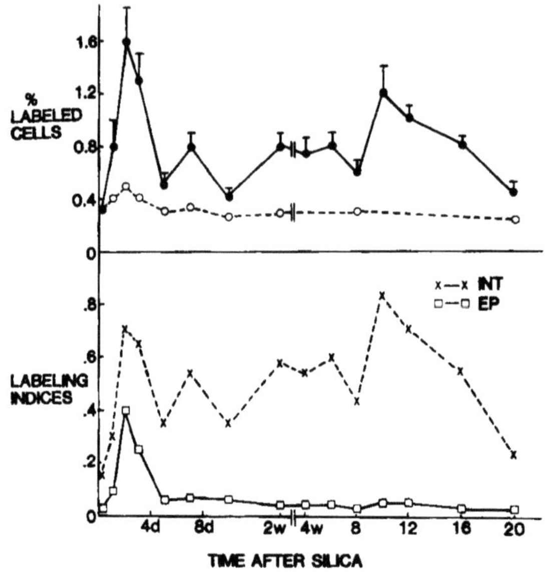

FIGURE 9. (Upper) Percentage of thymidine-labeled lung cells after silica (closed circles) and water (open circles). (Lower) Labeling index of interstitial cells (crosses) and type II epithelial cells (squares) after silica. Control values are the time zero numbers. (From Adamson, I. Y. R. and Bowden, D. H., *Am. J. Pathol.*, 117, 37, 1984. With permission.)

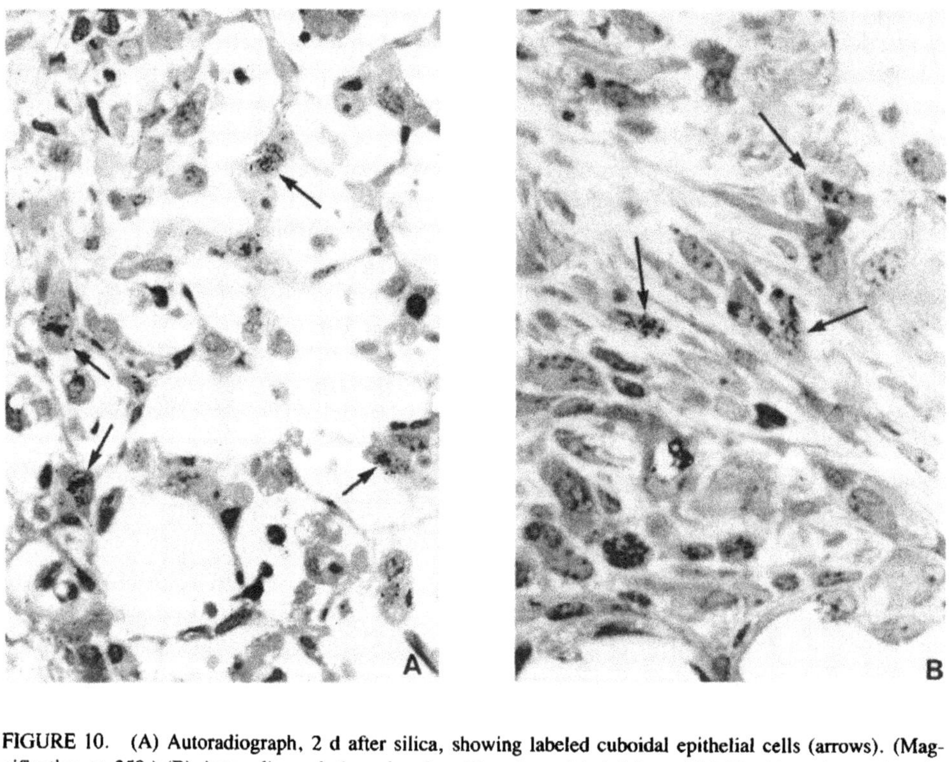

FIGURE 10. (A) Autoradiograph, 2 d after silica, showing labeled cuboidal epithelial cells (arrows). (Magnification × 350.) (B) Autoradiograph 4 weeks after silica shows labeled interstitial fibroblasts (arrows) at the edge of a granuloma. (Magnification × 500.)

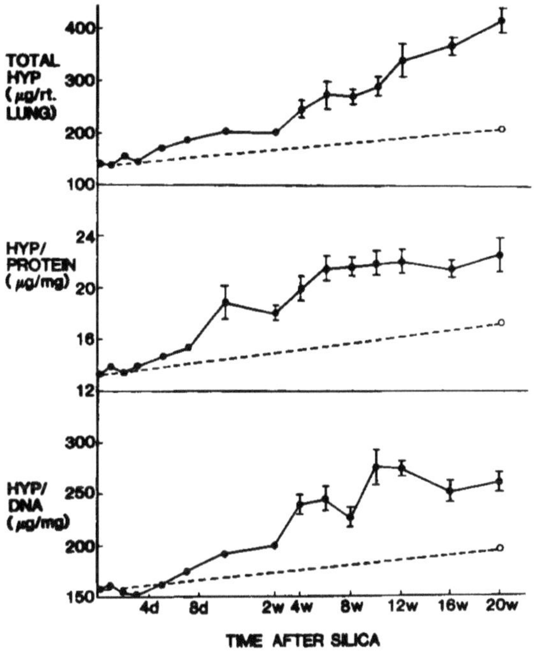

FIGURE 11. Levels of hydroxyproline (HYP) in the lung after silica (solid circles). Control values are marked by the broken lines. SE shown when greater than control ($p < 0.01$). (From Adamson, I. Y. R. and Bowden, D. H., *Am. J. Pathol.*, 117, 37, 1984. With permission.)

results can be expressed as total HYP per right lung in this case, or by HYP/DNA or HYP/dry weight. After silica, the increase in these parameters shows the progression of the lung to fibrosis (Figure 11). The ratio HYP/total protein appears more sensitive and shows the change in metabolism earlier than the measurements of total collagen. The amount of collagen in the lung normally rises with age, so it is important to use age-matched controls for comparison.

In several studies where the lipid components of the lung have been studied in relation to type II cell function and the observation of alveolar lipoproteinosis, samples of the lung homogenate can be used to examine phospholipid metabolism and surfactant content following lipid extraction. Phospholipid measurements can also be made on the lung lavage fluid.

MECHANISMS OF LUNG INJURY

The fibrogenic activity of silica has been attributed to toxic interactions with cell membranes and the release of fibroblast stimulating factors by macrophages.[5,20] In the initial phase of the pulmonary response, the rapid movement of particles from the periphery of the lung to the bronchioles and adjacent air sacs effectively clears most of the alveoli within 24 h. In this phase, clearance of silica is not different from inert particles such as carbon or polystyrene. Similarly, the initial efflux of macrophages and PMN is comparable for various particles, the critical factor being the number of particles reaching the alveoli. At this point the similarities cease. The reaction to carbon and polystyrene is limited to the exudation of inflammatory cells with transient focal edema and no evidence of epithelial necrosis. With silica, there is focal centrilocular necrosis of type I epithelial cells accompanied by a fibrinous exudate. The epithelium is rapidly repaired by proliferation of type II cells, which is generally regarded as the standard reparative reaction after type I cell injury. The fact that epithelial repair occurs so rapidly (Figure 9) makes it unlikely that this surface injury is the major reason for the stimulation of the underlying fibroblasts which occurs several weeks later.

The proximity of the epithelial injury to the exudate of PMN and macrophages suggests that release of proteolytic enzymes from these cells may be responsible for the injury. The relationship is unlikely to be direct because the instillation of particles such as carbon or polystyrene induces an equivalent inflammatory cell exudate with no evidence of accompanying cell injury.[3] In the case of silica, however, injury to lysosomal membranes has been demonstrated in macrophages. An examination of the electron micrographs in this experiment indicates a similar process occurring in the PMN, which is quantitatively a more potent source of proteases then the macrophage. The persistence of these cells for many weeks after the instillation of silica introduces the possibility of continuing protease-induced injury to pulmonary cells. On the other hand, collagenase secretion by PMN could inhibit the deposition of collagen.

The centripetal movement of silica crystals during the first 24 h brings a high concentration of particles into close contact with the epithelium of centrilobular ducts and air sacs. Direct membrane toxicity has been demonstrated, and it is not unlikely that the concentrated silica crystals directly alter the surface of the alveolar cells. Such particle-membrane interaction can change membrane permeability and increase calcium influx to the cell.[20] Whether or not the progression of cell injury to cell death involves excess calcium levels of proteolytic enzymes, damage to the epithelial barrier increases the likelihood of free particles passing from air sacs to interstitium, where they are ingested by phagocytes.[3,8,18] It is at these sites, predominantly in central regions of the lung, that the majority of fibrotic lesions are observed. Within the peribronchial granulomas, the predominant cells are macrophages and fibroblasts, with a smaller number of PMN probably arising from the chemotactic influences of tissue injury and macrophage-derived factors.

Most research studies indicate a central role for the alveolar macrophage in fibrogenesis. The progressive nature of silicosis is usually thought to be dependent upon successive waves of phagocytosis, cell destruction, and release of particles, a process that continues long after exposure to silica ceases.[5] However, several studies have shown that it is not the products of dead cells that activate fibrogenesis but rather a secretion of viable cells altered somehow by exposure to silica. When silica is incubated with normal macrophages *in vitro,* growth factors for fibroblasts have been demonstrated in supernatants[21] which also promote collagen synthesis.[22] Some authors have been unable to duplicate these results and have reported poor growth of fibroblasts with a tendency to inhibition on exposure to macrophage supernatants. Using different systems, mononuclear cell supernatants have been shown to increase fibroblast proliferation[23] or to inhibit growth.[16,24] This suggests that even under normal circumstances, pulmonary macrophages may play a role in controlling fibroblastic growth in the lung.

There is evidence that alveolar macrophages recovered several weeks after silica injection to the lung are altered and secrete greater amounts of a fibroblast growth factor when cultured for a short time.[15] This finding is not limited to silica exposure and seems to be a general finding for macrophages lavaged from lungs during a fibrotic reaction. Increased levels of fibroblast growth factor are secreted by macrophages in human idiopathic pulmonary fibrosis[25] and in animals with bleomycin-,[26] silica-,[15] and asbestos-induced lung disease.[27] Proteins of varying molecular weight have been isolated from supernatants and include macrophage-derived growth factor and interleukin-1.[25,28] Although stimulation of fibroblast growth has usually been shown in cells cultured in serum-free medium, this is not a universal finding. These results suggest that the activation of the macrophage and its link to fibrosis may be a general mechanism and not specific to the silica model of pulmonary fibrosis.

The studies on macrophages lavaged from the lung not only indicate a secretory function for these free cells, but may mirror the activity of interstitial macrophages. An active molecule secreted within the alveolar space is more likely to be inactivated by serum factors before it can cross the epithelial barrier to the connective tissue. In contrast, any growth-promoting substance released by macrophages containing silica within the pulmonary interstitium is already in close proximity to fibroblasts and seems more likely to be effective. For this reason, translocation of particles across the alveolar epithelium may be very important. Epithelial cell injury may be a key to the severity of the reaction because necrosis will allow more particles direct access to interstitial macrophages for phagocytosis at this site. In addition, there is evidence that a direct effect of translocated silica particles on the fibroblast may stimulate fibrogenesis, since in-culture incubation of fibroblasts with silica results in enhanced proliferation and collagen production.[29]

Macrophages within the interstitium tend to remain there for a long time, as evidenced by pigment-laden cells in the lungs of miners many years after exposure ceased. These interstitial macrophages may be cleared to alveoli or to lymphatics, but clearance is very much slower than for free alveolar cells. In addition, extra collagen deposition in the interstitium may further obstruct clearance of particle-laden macrophages and so perpetuate the localized secretion of any growth factor at peribronchiolar sites. The close interaction of the silica-laden macrophages and adjacent fibroblasts within the pulmonary interstitium may allow more-efficient transfer of a fibroblast growth factor. Collagen deposition at these peribronchiolar sites eventually results in an obstructive bronchiolar fibrosis. This pattern, with lesser involvement of alveoli, is similar to that described in man.

INHIBITION OF FIBROSIS

The most obvious method of inhibiting any reaction to an environmental hazard is to prevent exposure with the use of masks or filters. In experimental work, the belief that the pulmonary reaction to silica was due to its cytotoxic nature on macrophages led to the use of agents that would stabilize cell membranes, particularly of the lysosome. For example, the polymer polyvinylpyridine-*N*-oxide inhibited cytotoxicity, and when macrophage supernatants were tested, no stimulatory effect in collagen synthesis was found in fibroblast cultures.[2]

Agents that interfere with the late stages of collagen synthesis and secretion may have some value in reducing fibrosis and improving pulmonary function. Steroids have been used clinically and may act either to reduce the inflammatroy reaction or somehow modify the fibrotic process. Durgs that specifically inhibit collagen deposition have been used experimentally in various models of pulmonary fibrosis. Compounds such as D-penicillamine and *cis*-hydroxyproline block the production of cross-linked collagen molecules and inhibit fibrosis.[30]

These methods concentrate on the late stages of the disease process, whereas a more promising approach may be to modify macrophagic function. If specific growth factors are identified and shown to be responsible for the fibroblast stimulation, it should be possible to characterize these and develop strategies for their inhibition. The general nature of the macrophagic response to many agents including silica suggests that understanding the mechanisms of growth factor synthesis, secretion, and inactivation irrespective of the model of injury could have widespread implications for inhibiting pulmonary fibrosis.

REFERENCES

1. **Parkes, W. R.,** *Occupational Lung Disorders,* 2nd ed., Butterworths, London, 1982, chap. 7.
2. **Heppleston, A. G.,** Pulmonary toxicology of silica, coal and asbestos, *Environ. Health Perspect.,* 55, 111, 1984.
3. **Adamson, I. Y. R. and Bowden, D. H.,** Dose response of the pulmonary macrophagic system to various particulates and its relationship to transepithelial passage of free particles, *Exp. Lung Res.,* 2, 165, 1981.
4. **Adamson, I. Y. R. and Bowden, D. H.,** Chemotactic and mitogenic components of the alveolar macrophage response to particles and neutrophil chemoattractant, *Am. J. Pathol.,* 109, 71, 1982.
5. **Bateman, E. D., Emerson, R. J., and Cole, P.,** Mechanisms of fibrogenesis, in *Occupational Lung Diseases,* Weill, H. and Turner-Warwick, M., Eds., Marcel Dekker, New York, 1981, chap. 11.
6. **Phalen, R. F., Mannix, R. C., and Drew, R. T.,** Inhalation exposure methodology, *Environ. Health Perspect.,* 56, 23, 1984.
7. **Adamson, I. Y. R. and Bowden, D. H.,** Role of polymorphonuclear leukocytes in silica-induced pulmonary fibrosis, *Am. J. Pathol.,* 117, 37, 1984.
8. **Bowden, D. H. and Adamson, I. Y. R.,** The role of cell injury and the continuing inflammatroy response in the generation of silicotic pulmonary fibrosis, *J. Pathol.,* 144, 149, 1984.
9. **Morgan, A., Moores, S. R., Holmes, A., Evans, J. C., Evans, N. H., and Black, A.,** The effect of quartz, administered by intratracheal instillation, on rat lung. I. The cellular response, *Environ. Res.,* 22, 1, 1980.
10. **Reiser, K. M., Haschek, W. M., Hesterberg, T. W., and Last, J. A.,** Experimental silicosis. II. Long term effects of intratracheally instilled quartz on collagen metabolism and morphologic characteristics of rat lungs, *Am. J. Pathol.,* 110, 30, 1983.
11. **Dauber, J. H., Rossman, M. D., Pietra, G. G., Jimenez, S. A., and Daniele, R. P.,** Experimental silicosis: morphologic and biochemical abnormalities produced by intratracheal instillation of quartz into guinea pig lungs, *Am. J. Pathol.,* 101, 595, 1980.
12. **Dale, K.,** Late effects of large doses of quartz dust on pulmonary function and tissue. An experimental study on rabbits, *Scand. J. Respir. Dis.,* 54, 244, 1973.
13. **O'Neil, J. J. and Raub, J. A.,** Pulmonary function testing in small laboratory mammals, *Environ. Health Perspect.,* 56, 111, 1984.
14. **Adamson, I. Y. R.,** Cellular kinetics of the lung, in *Handbook of Experimental Pharmacology,* Vol. 75, Witschi, H. P. and Brain, J. D., Eds., Springer-Verlag, Berlin, 1985, chap. 11.
15. **Lugano, E. M., Dauber, J. H., Elias, J. A., Bashey, R. I., Jiminez, S. A., and Daniele, R. P.,** The regulation of lung fibroblast proliferation by alveolar macrophages in experimental silicosis, *Am. Rev. Respir. Dis.,* 129, 767, 1984.
16. **Gritter, H. L., Adamson, I. Y. R., and King, G. M.,** Modulation of fibroblast activity by normal and silica-exposed alveolar macrophages, *J. Pathol.,* 148, 263, 1986.
17. **Lugano, E. M., Dauber, J. G. H., and Daniele, R. P.,** Silica stimulation of chemotactic factor release by guinea pig alveolar macrophages, *J. Reticuloendothel. Soc.,* 30, 381, 1981.
18. **Brody, A. R., Roe, M. W., Evans, J. N., and Davis, G. S.,** Deposition and translocation of inhaled silica in rats. Quantification of particle distribution, macrophage participation and function, *Lab. Invest.,* 47, 533, 1982.
19. **Woessner, J. F.,** Determination of hydroxyproline in connective tissue, in *The Methodology of Connective Tissue Research,* Hall, D. A., Ed., Joynson and Bruvvers, Oxford, 1976, 227.
20. **Farber, J. L.,** How do mineral dusts cause lung injury?, *Lab. Invest.,* 49, 379, 1983.
21. **Bateman, E. D., Emerson, R. J., and Cole, P. J.,** A study of macrophage-mediated initiation of fibrosis by asbestos and silica using a diffusion chamber technique, *Br. J. Exp. Pathol.,* 63, 414, 1982.
22. **Heppleston, A. G. and Styles, J. A.,** Activity of a macrophage factor in collagen formation by silica, *Nature (London),* 214, 521, 1967.
23. **DeLustro, F., Sherer, G. K., and LeRoy, E. C.,** Human monocyte stimulation of fibroblast growth by a soluble mediator(s), *J. Reticuloendothel. Soc.,* 28, 519, 1980.
24. **Elias, J. A., Rossman, M. D., and Daniele, R. P.,** Inhibition of human lung fibroblast growth by mononuclear cells, *Am. Rev. Respir. Dis.,* 125, 701, 1982.
25. **Bitterman, P. B., Adelberg, S., and Crystal, R. G.,** Mechanisms of pulmonary fibrosis: spontaneous release of the alveolar macrophage-derived growth factor in interstitial lung disorders, *J. Clin. Invest.,* 712, 1801, 1983.
26. **Clarke, J. G., Kostal, K. M., and Marino, B. A.,** Bleomycin-induced pulmonary fibrosis in hamsters. An alveolar macrophage product increases fibroblast prostaglandin E_2 and cyclic adenosine monophosphate and suppresses fibroblast proliferation and collagen production, *J. Clin. Invest.,* 72, 2082, 1983.
27. **Lemaire, I., Rola-Pleszczynski, M., and Begin, R.,** Asbestos exposure enhances the release of fibroblast growth factor by sheep alveolar macrophages, *J. Reticuloendothel. Soc.,* 33, 275, 1983.
28. **Wahl, S. M.,** Inflammatory cell regulation of connective tissue metabolism, *Rheumatology,* 10, 404, 1986.

29. **Richards, R. J. and Curtis, C. G.,** Biochemical and cellular mechanisms of dust-induced lung fibrosis, *Environ. Health Perspect.*, 55, 393, 1984.
30. **Riley, D. J., Berg, R. A., Edelman, N. H., and Prockop, D. J.,** Prevention of collagen deposition following pulmonary oxygen toxicity in the rat by cis-4-hydroxy-L-proline, *J. Clin. Invest.*, 65, 643, 1980.

LUNG RESPONSE TO CIGARETTE SMOKE INHALATION: AN ANIMAL MODEL

Daniel H. Matulionis

INTRODUCTION

Recent concern of the scientific community, industry, and the general public about potential health risks related to utilization of tobacco products demands sustained research in the area of smoking and health. Although increased incidence of pulmonary disorders has been implicated with smoking,[1,2] due to regulatory restrictions, it is often difficult (if not impossible) to design experiments which evaluate the cause-effect relationships between smoking and disease in humans. In addition, direct evaluation of risk factors related to smoking in humans might not be possible due to unavoidable inherent variability in the human population. For these reasons, studies dealing with effects of smoking on human health are frequently not complete or are conflicting. For example, Hirayama[3] reported that passive smoking is a causal factor of lung cancer in nonsmoking wives of smoking husbands, while Garfinkel[4] found that little or no increased risk of cancer occurred in a similar smoking situation. For these and other obvious reasons, investigations dealing with the tobacco and health issue can best be carried out using animal models. Over the past several years, research in the author's laboratory has been directed toward developing an animal model that could be used to show a cause-effect relationship between smoke inhalation and lung pathology. Such a model could subsequently be used to study the onset, progression, and end stage of the disorder(s). With the availability of a model, it might be possible to determine the mechanism(s) involved in the manifestation of pulmonary abnormality and gain insight into how to approach the assessment of similar lung disorders in humans. This chapter describes the evolution of this animal model.

Briefly, it seemed a simple matter to show that cigarette smoke inhalation by an animal (mouse) would result in pulmonary disfunction since, especially at present, "everyone knows" smoking is harmful.[1,2] All that was necessary was to expose animals to cigarette smoke, characterize the disorder, and design experiments to determine the mechanism of the pathologic manifestation. The task of developing an animal model proved to be more difficult than anticipated, since cigarette smoke inhalation alone failed to induce expected lung abnormalities in normal young adult animals (mice).[5] However, it was noted that smoke inhalation caused a marked increase of the pulmonary (alveolar and septal) macrophage population size.[5] Later experiments revealed that pathology does not manifest in animals with an intact macrophage system. Only when the macrophage population increase that occurs in response to cigarette smoke inhalation is experimentally impeded, or when this increase is hindered normally by age, do lung abnormalities develop in mice.[6,7] The abnormalities noted conformed to descriptions of alveolar proteinosis and pulmonary fibrosis.[6,7]

This chapter presents procedures used to implement the animal model. Tissue preparation for morphologic assessment of lungs and pulmonary macrophages and morphometric and physiologic methods used to determine the nature of the disease are detailed. Data obtained are summarized. The possible mechanism for the development of the disorder, future direction of the research, and limitation of the model are discussed.

METHODS

Implementing the Model

In developing an animal model that potentially could provide useful information for assessing pulmonary integrity in humans subsequent to inhalation of cigarette smoke, the

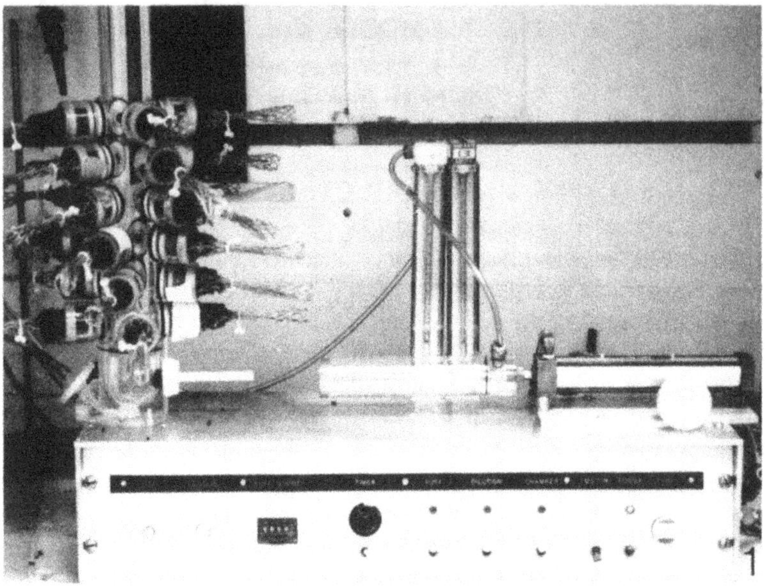

FIGURE 1. The Kentucky smoking machine. Mice are held in restrainers which allow the snouts of animals to protrude into the smoke exposure chamber.

response of a system in the animal should be as similar as possible to that of humans. The present model attempts to relate the significance of pulmonary macrophage system in maintaining lung integrity. Thus, the mouse rather than another species has been chosen because murine[5,8] and human[9] macrophage systems respond to smoke insult to an equal degree. Mice used should, in turn, be genetically resistant to spontaneous formation of lung abnormalities. C57BL/6 mice were used. To eliminate the potential differential effects of fluctuating reproductive hormones, only males were used.

All animals were housed in environmentally controlled (Bioclean) quarters maintained at 71°F and 48% relative humidity with a light/dark cycle of 12 h. Food and water were available to the animals at all times. The animals were purchased from a reliable vendor (Jackson Laboratories, Bar Harbor, ME) which assured disease-free mice. Mice should be quarantined for 10 d before the onset of the experiment and tested for *Mycoplasma pulmonis*, Sendai virus (D-Tec MM murine *M. pulmonis* and D-Tec MS murine Sendai virus antibody tests, Pitman Moore, Inc., Washington Crossing, NJ, respectively), and mouse hepatitis virus (MHV Organo Technicon antibody test, Charleston, SC). These serological tests should also be performed at the end of the experiment to insure that animals are disease free.

Nose-only exposure of animals to cigarette smoke should be carried out with smoking machines that generate and deliver smoke to mice simultaneously. Several smoking machines were evaluated during the development of the model.[10] It was determined that the Kentucky smoking machine[11] (University of Kentucky Tobacco and Health Research Institute, Lexington, KY) (Figure 1) was most effective in generating and delivering the smoke to elicit a marked pulmonary macrophage response. Briefly, smoke exposure of animals occurred in the following fashion.[11] A lit cigarette was placed in a holder connected to the smoke exposure chamber. Mice were restrained in specially constructed holders and their snouts were inserted into the exposure chamber. The puff was generated by an air cylinder that moved over the cigarette, creating a slightly above-ambient pressure around the cigarette. Smoke exited into the exposure column from the outlet of the cigarette holder and was diluted with air to a desired concentration (20% smoke). Smoke generation, dilution, and delivery occurred simultaneously. Following the smoke puff the air cylinder was retracted, allowing the cigarette to burn at atmospheric pressure. The smoke remained in the exposure chamber for 15 s; then the chamber was flushed with room air. The maximum age of smoke was 15 s. After 45 s, the next smoke puff was delivered into the exposure chamber. Animals were exposed once in the morning and once in the afternoon in the pulsatile fashion to ten puffs of smoke from one 2A1 University of Kentucky Reference Cigarette (University of Kentucky Tobacco and Health Research Institute, Lexington, KY). Sham treatment groups should parellel all smoke exposure groups to control the possible influence of stress resulting from restraint and manipulation. Sham treatment in the present model consisted of exposing animals to room air instead of smoke in machines used only for this purpose. Smoke exposure, sham treatment, and housing of animals occurred at the University of Kentucky Tobacco and Health Research Institute and were carried out by the Inhalation Core Facility personnel of the Institute. All mice tolerated the treatments well and exhibited no signs of undue discomfort. Appropriate number (six minimum) of absolute (room) control animals should accompany the smoke-exposed and sham groups. Smoke exposure time required to induce lung pathology depends on whether the smoke-induced increase in lung macrophage numbers is curtailed experimentally or allowed to occur naturally.

It is possible to impede the macrophage response in young (2 to 3 months old) adult C57BL/6 male mice by hydrocortisone (HCA) (hydrocortisone 21-acetate; Sigma Chemical Company, St. Louis) treatment.[6,12] The procedures to accomplish this are as follows. Seven experimental groups, each consisting of twelve animals, were used. The groups were (1) absolute room controls, (2) room controls injected with HCA, (3) room controls injected with a vehicle used to suspend the steroid, (4) mice subjected to sham treatment only, (5) sham-treated animals injected with HCA, (6) mice exposed to cigarette smoke only, and (7) smoke-exposed mice injected with HCA. Animals were exposed to smoke and sham treated for 35 d prior to HCA administration. HCA suspended in 0.1 ml of vehicle (0.5% carboxymethyl cellulose, 0.09% benzyl alcohol, and 0.4% polysorbate 80 in saline) was injected subcutaneously in the nuchal region.[13] Steroid dose administered was approximately 10 mg/kg body weight per day. HCA and vehicle treatments were maintained for 21 d, during which time animals continued to be exposed to smoke or sham-treated. Thus, 56 d of smoke exposure are required to induce lung pathology by experimentally impeding the smoke-related increase in numbers of pulmonary macrophages.

Chronic cigarette smoke inhalation by old C57BL/6 mice without additional intervention likewise results in lung abnormalities.[7] Implementation of the "aged mouse model" requires smoke exposure (twice daily) of 8- to 10-month-old C57BL/6 mice of 9 months. The old mice are not as resistant to manipulatin as the young animals while being loaded into and unloaded from the smoking or sham machines. During the last 3 or 4 months of the 9-month experiment, some animals developed skin lesions over the nasal bones anterior to the eyes and others sustained fractures of the upper or lower limbs. When skin lesions or fractures were noted, the animals were immediately sacrificed. In view of this, approximately 50% more old mice than needed should be alloted per treatment group. Groups of young animals (2 months of age) (smoke, sham, and controls) should parallel all old groups of animals to control age effect.

Evaluating the Model

At the end of the experiments, lung tissue should be collected for microscopic assessment. Tissue samples were obtained and processed as follows. Prior to autopsy of the lungs, mice were anesthetized by intraperitoneal injections of sodium pentobarbital (The Butler Company, Columbus, OH) (50 mg/kg body weight), weighed, and exsanguinated by severing the abdominal aorta. The left lung was dissected from the thoracic cavity, immersed in a 3.5% cacodylate-buffered (pH 7.2 to 7.4) glutaraldehyde fixative, and divided into cephalic two-thirds and caudal one-third parts. Care was taken to manipulate the tissues gently during these procedures. The cephalic portion of the lung was then fixed by immersion in a 10% neutral buffered formal in fixative for 24 to 48 h. After fixation, the tissue was prepared for light microscopy by routine paraffin methods and serially sectioned at 4 μm through approximately 360 μm of tissue (starting from the cut end of the lung). Three slides were stained with periodic acid-Schiff reagent-hematoxylin. The slides were chosen from different levels of the sectioned tissue. Two 1-mm slices from the caudal one-third of the left lung were dissected into 1-mm cubes while immersed in cacodylate buffered (pH 7.4) glutaraldehyde fixative, fixed by immersion, and processed for electron microscopy according to routine methods.

Tissues prepared for light microscopy were used to evaluate the general morphology of the lungs and to quantify morphometrically the volume fraction (percent volume) of alveolar space, parenchymal tissue (alveolar septa), other lung components (blood vessels, air passages, connective tissue, etc.), and mononuclear cell aggregations. For the purpose of gaining a more critical insight into the altered lung morphology, the parenchymal tissue and alveolar space were designated as normal or abnormal. Abnormal parenchyma was defined as markedly thickened (hypertrophied) alveolar septa that clearly deviated from the classically described normal state of these structures. Normal parenchyma was designated as that which conformed to the classical descriptions. Alveolar space was classified as abnormal if it was encompassed by abnormal parenchyma and normal when surrounded by normal parenchyma. If assessment of parenchyma or space, in regard to normalcy, was problematic, it was scored as normal. Such scoring procedures yield underestimates of the parameters identified as abnormal, which lends credibility to the morphometric observations since the quantification was intended to verify manifestation of anomalous conditions.

Quantitation of the identified parameters was done with light microscopy on transverse tissue sections of lungs at two levels within a 360-μm region starting at the inferior border of the cephalic two thirds of the left lung. The tissue sections and levels were randomly selected. Using a multipurpose grid,[14] more than 450 independent test points ("hits") that intercepted various lung components (i.e., alveolar space, parenchyma, etc.) were identified and recorded by superimposing the visual image of the tissue via a camera lucida on a data sheet containing the multipurpose grid. The volume fractions were estimated by the equation Vvi = Ppi,[14] where Vvi is the volume fraction (percent) occupied by a structure and Ppi is equal to the fraction of test points intercepted by the structure. The Vvi reported represent the percent volume of the assessed structures per unit volume of lung within the sampled region.

Tissue prepared for light microscopy was also used to quantify the number of pulmonary macrophages (alveolar and septal) per unit area (0.028 mm^2) of lung tissue. Macrophage profiles intersected by one half of the frame perimeter delineating the test area were included in the estimates, while those intersected by the other half were not. Initially, macrophages from one animal from each group were counted in ten randomly selected fields of tissue measuring 0.028 mm^2 from each of the three slides. Analysis of variance revealed no difference in macrophage numbers at three different levels over the 360 μm of lung tissues sampled. Subsequently, macrophages from one of the three randomly chosen slides of lung tissue from the remaining animals were counted. The number of pulmonary macrophages was determined using 10 fields of lung tissue (0.028-mm^2 area per field) since evaluation of data by analysis of variance obtained from 20, 30, or 40 fields revealed no reduction in the size of confidence interval or standard error. The number of macrophages per unit area was converted to number per square millimeter arithmetically. Each field was composed of the most normal appearing lung tissue only. The number of macrophages per square-millimeter area of lung tissue represents a certain fraction of the total (alveolar and septal) pulmonary macrophage population.

To assess whether differential shrinkage and/or collapse of lungs occurred following pneumothorax and tissue processing (as a result of different treatment of animals), the cross-sectional area of the above lung sections from all mice should be calculated using an image analysis system (Zeiss Video Plan Image Analysis System, Carl Zeiss Inc., Thornwood, NY). Lung sections for area analysis and subsequent macrophage counts should be the same cut at identical locations regarding the external pulmonary surface. If necessary, the number of macrophages per unit area in lungs of smoke-exposed and sham-treated mice must be normalized by the area of lung section of control animals (lung section area [square millimeter] of "smoked" or sham animal, times number of macrophages per unit area in "smoked" or sham mice, divided by lung section area of controls). Such a correction is necessary because the density of macrophages per unit area of lungs would artifactually be increased or decreased depending on whether the lungs shrank or expanded (respectively) during tissue processing as a result of their altered state due to different treatments.

In view of the fact that tobacco and health research is highly scrutinized, it is desirable to confirm morphological observations by other data. In the present model, structural lung abnormalities (pathology) were confirmed by evaluating pulmonary function of smoke-exposed, sham-treated, and control mice using the following procedures. The mice were anesthetized with an intraperitoneal injection of ethyl carbamate (Aldrich Chemical Company, Milwaukee) (0.5 to 0.75 g/kg). The trachea was exposed and cannulated with a modified hypodermic needle (blunt end, 2.5 cm × 18 gauge), which was secured firmly with silk sutures to prevent leakage. Body weights and rectal temperatures were recorded for use in subsequent calculations. Throughout the physiological testing period, cardiac electrical activity of each animal was monitored. After the tracheotomy was performed, the animals were placed supine on a solid hemicylindrical plexiglass shelf which slid into a cylindrical plexiglass chamber that was sealed at both ends and which served as a constant-volume, whole-body plethysmograph (I.D., 6 cm; lengh, 15.25 cm; volume with shelf in place 400 cm^3). The tracheal cannula was connected to a fitting which passed through one end of the plethysmograph, hence permitting spontaneous or controlled respiration to occur. The fitting was constructed to allow the attachment of a pressure transducer (Statham PM 131 TC ± 2.5 PSID; Statham Laboratories, Inc., Hato Rey, PR) to monitor airway opening pressure (P_{ao}) with a minimal contribution to respiratory dead space (0.18 cm^3). Total volume displacement within the plethysmograph was standardized to 40 cm^3 (assuming body tissue density = 1 g/cm^3) by the addition of small solid beads of plexiglass of known displacement into the plethysmograph which served to minimize the effect of varying animal size upon pressure calibration factors.

While the animals were breathing spontaneously, respiratory frequency and tidal voume were measured. Tidal volume was measured as the change in pressure within the sealed plethysmograph by a Statham PM 15 ± 0.4 PSID transducer connected to a side wall port in the plethysmograph. Subsequently, the animals were paralyzed with an intraperitoneal injection of gallamine triethiodide (Flaxedil; Davis and Geck, Princeton, NJ) (1.0 mg/kg) and artifically ventilated with a small animal respiratory (Ealing Bioscience, South Natwick, NJ). The volume and frequency of the artificial ventilation were adjusted to approximate the spontaneous respiratory rate of each animal and hence approximate alveolar P_{CO_2} and P_{O_2} levels.

Quasistatic pressure-volume (P-V) characteristics of the respiratory system were measured during controlled inflations and deflations of the lung at a constant flow of 6 ml/min. This maneuver was conducted using a reversible syringe pump (Model 351; Sage Instruments, Cambridge, MA). P-V loops (one inflation and deflation cycle) were measured using methods described by Sabo et al.[15] Briefly, lung volume was measured as the change in plethysmographic pressure by a Statham 283 ± 0.8 PSID transducer, and transthoracic pressure (P_{tt}) was measured as the differential pressure between P_{ao} and the pressure within the plethysmograph by a PM 131 TC ± 2.5 PSID transducer. The P-V loops were recorded from the relaxation volume (V_r, defined here as the lung volume in a paralyzed animal set by the static balance between opposing lung and chest wall recoil forces)[16] to total lung capacity (TLC, defined as the lung volume at P_{tt} = 25 cm H_2O).

Analysis of the P-V loops was performed by fitting inflation and deflation limbs separately to a modified logistic equation which was suggested by Paiva et al.[17]: $V = (TLC \times V_r)/(V_r + IC_e\text{-}kp)$, where: V = volume at a given P_{tt}, P = P_{tt}, TLC = total lung capacity, V_r = relaxation volume, IC = inspiratory capacity or TLC-V_r, and k is the volume independent shape constant of the P-V curve and as such is an index of lung elastic recoil.

Gas exchange properties of the lung were studied by a modified method[15] of Takazawa et al.[18] for the measurement of carbon monoxide diffusing capacity (DL_{co}) in small animals. This method allows the measurement of V_r and TLC by dilution of Ne gas which is incidental to the measurement of DL_{co} and the coefficient of diffusion (K_{co}).

All numerical data (morphometric and physiologic) were evaluated statistically using the F test (analysis of variance) and Fisher's multiple comparison procedures. A *P* value of ≤ 0.05 was required for significance.

EXPERIMENTAL FINDINGS

For clarity and convenience, data obtained using the present animal model is summarized in two sections. Section one deals with information obtained in the experiments in which mice were treated with hydrocortisone acetate,[6] and section two considers the findings of the study that assessed the pulmonary response in old mice following 9 months of cigarette smoke inhalation.[7]

Lung Response to HCA Treatment and Cigarette Smoke Inhalation

No ill effects from smoke exposure, sham manipulation, or HCA administration were perceived and no animal deaths attributable to the experimental treatments occurred during the course of the experiment. However, body weights of the smoke-exposed, HCA-treated mice (24.6 ± 0.64 g) were significantly lower ($p \leq 0.05$) that those of all control groups (range 26.5 ± 0.58 to 32.4 ± 1.1 g).

Lung morphology of Groups 1 to 6 (see "Implementing the Model" above) was similar and appeared normal (Figure 2). The alveolar spaces were distinct, patent, and free of debris. Interalveolar septa were then, clearly visible, and relatively uniform. Pulmonary tissue of HCA-treated, smoke-exposed mice (Group 7), on the other hand, deviated considerably from the norm (Figure 3). Alveolar spaces appeared reduced in size and were often filled with material. Thickened alveolar septa and highly cellular areas also characterized the lung of animals in this group. Electron microscopic examination confirmed the abnormal lung state noted with light microscopy and suggested that the conditions were similar to pulmonary fibrosis and alveolar proteinosis. It should be noted, however, that some areas of Group 7 lungs appeared normal or nearly normal. Morphometric assessment of volume fraction (percentage volume) of normal-appearing septa (normal tissue) (9.2 ± 2.4%) and air spaces surrounded by normal septa (normal space) (18.1 ± 4.2%) significantly ($p \leq 0.02$) decreased and that of abnormal tissue (37.1 ± 5.0%) and alveolar space surrounded by abnormal septa (abnormal space) (18.6 ± 1.3%) increased ($p \leq 0.03$) in smoke-exposed, HCA-treated mice when compared with that of all control group animals (range: normal tissue 17.5 ± 2.3 to 24.5% ± 2.3%, normal space 31.9 ± 9.0 to 51.7 ± 5.2%, abnormal tissue 24.0 ± 1.1 to 17.7 ± 1.3%, abnormal space 4.8 ± 0.6 to 16.3 ± 3.5%). Morphologic and morphometric assessment clearly indicated that the total amount of lung parenchymal tissue increased and that of alveolar space decreased in the HCA-treated, smoke-exposed animals.

The lung size in animals of different treatment groups appeared to be similar. Lung section

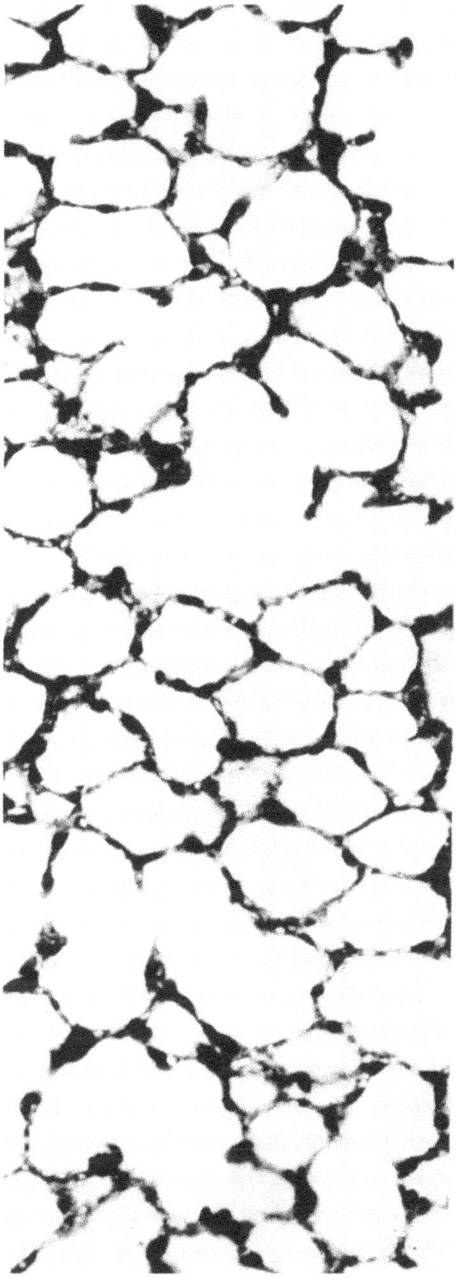

FIGURE 2. Light micrograph of lung tissue from a young C57BL/6 mouse exposed to cigarette smoke for 56 d. The tissue is normal in appearance. (Magnification × 420.)

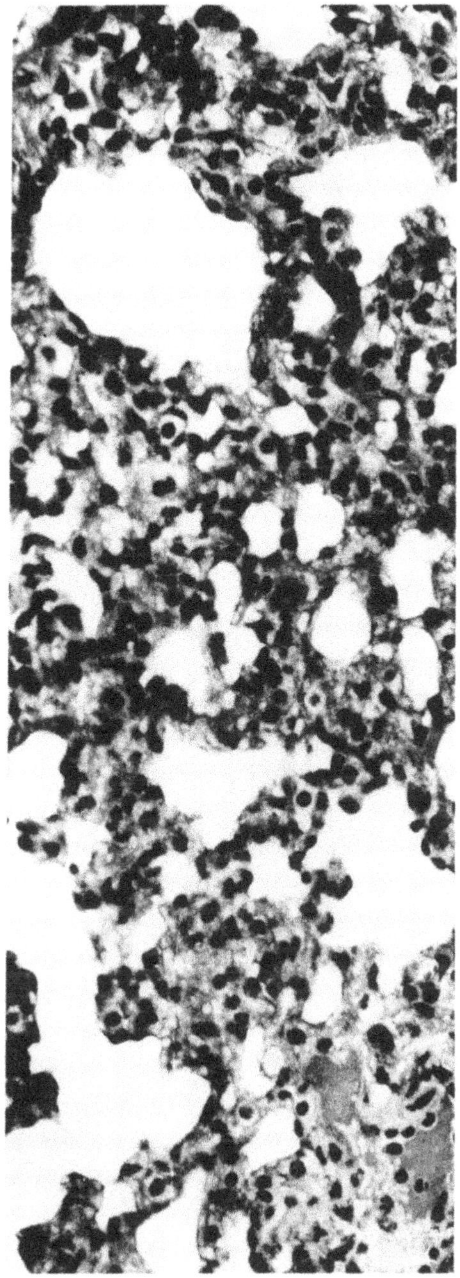

FIGURE 3. Light micrograph of lung tissue from a young C57BL/6 HCA-treated, smoke-exposed mouse. Morphology suggests a decrease in amount of air space and an increase in quantity of tissue. (Magnification × 420.)

area quantification substantiated this subjective impression. Therefore, lung shrinkage and/ or inflation was uniform ($p > 0.9$) and unaffected differentially by various experimental treatments. In view of the above, the altered numbers of pulmonary macrophages per unit area of lung tissue would reflect treatment effect rather than lung shrinkage. The number of macrophages (46.4 ± 1.9) in lungs of Group 7 mice (smoke + HCS) was more than

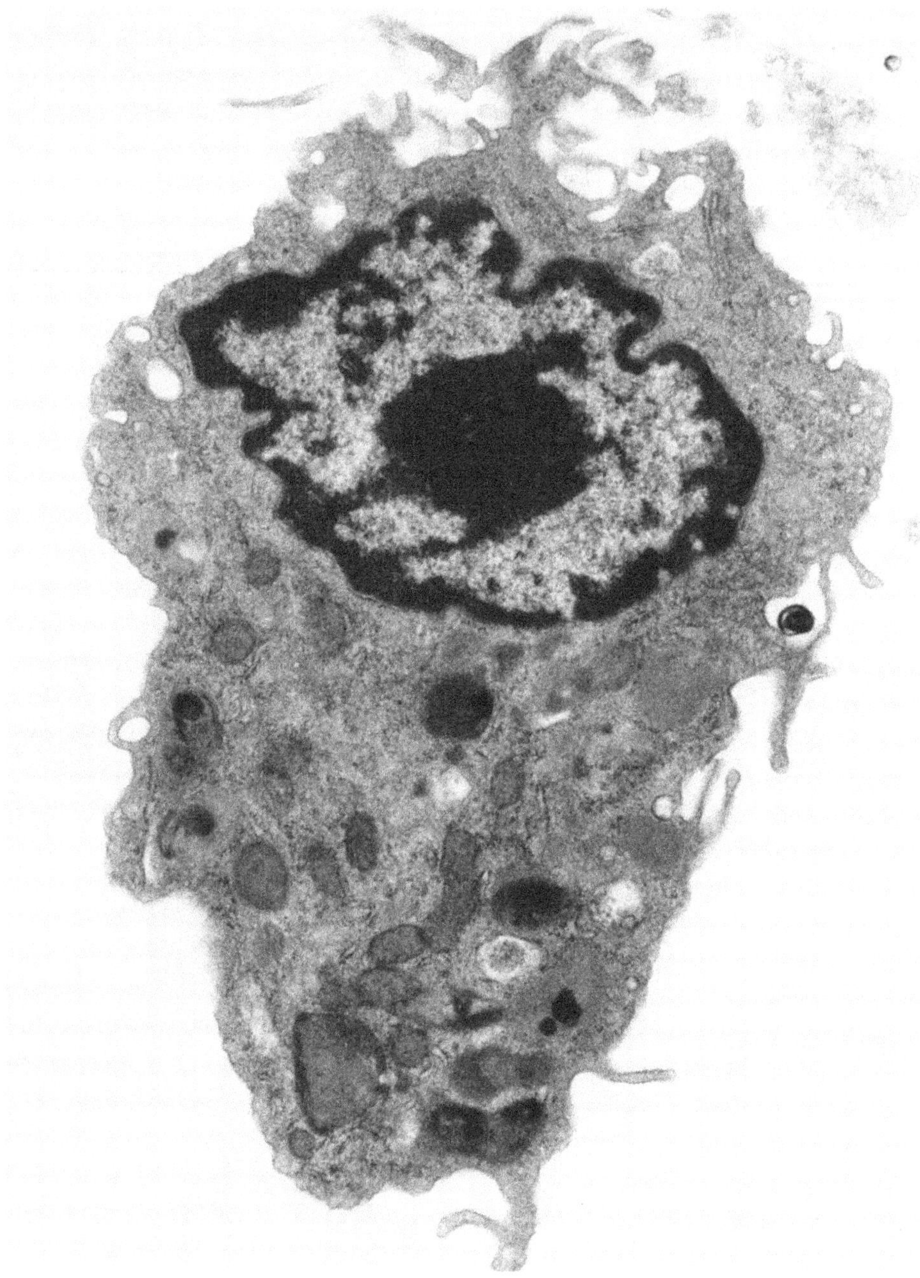

FIGURE 4. Electron micrograph of an alveolar macrophage from lungs of a Group 3 (vehicle injected) C57BL/6 mouse. Ultrastructure of the cell is normal. (Magnification × 14,100.)

six times less ($p = 0.0001$) than that found in lungs of smoke-exposed-only (292.7 ± 7.1) mice. Macrophage numbers (range 11.8 ± 1.4 to 21.4 ± 2.1) in lungs of Groups 1 to 5 were similar but significantly lower ($p = 0.0001$) than those of Groups 6 and 7. Clearly, HCA treatment depressed the macrophage response elicited by cigarette smoke inhalation.

Ultrastructure of pulmonary macrophages in lungs of animals in Groups 1 to 5 (Figure 4) was normal and appeared similar to that described by Matulionis and Traurig.[5] Phagocytes of Group 6 mice were generally similar to those of the above groups, although the cells differed slightly in two respects. There appeared to be an increased number of lysosomes and the lysosomes seemed to be more diverse in structure. Pulmonary macrophage ultrastructure of Group 7 mice deviated markedly from the norm. Many phagocytes appeared to be degenerating. The intact cells were considerably larger than those of control animals and their cytoplasm was replete with heterogenous lysosomes, some of which contained crystalline inclusions (Figure 5). The latter were also encountered free in the cytoplasm (Figure 5 insert) of the phagocytes. Macrophages in lungs of smoke-exposed-only mice were devoid of the crystalline material.

Pulmonary function of HCA-treated, smoke-exposed mice was reduced when compared with all other control groups.[6] Total lung capacity, relaxation volume, carbon monoxide diffusing capacity and CO diffusing capacity per unit volume of lung were significantly lower ($p \leq 0.02$) in this group of animals. Significantly compromised lung function in the smoke-exposed, HCA-treated mice indicated restrictive, space occupying lung disorders complementing morphologic and morphometric data which suggested the anomalous condition to be pulmonary fibrosis.

Lung Response of Old Mice to Chronic Cigarette Smoke Inhalation

The general health condition of the young (2 months old) C57BL/6 mice was good during the 9-month smoke-exposure period (Figure 6). Except for the skin lesions and fractions sustained (see above) by the old (8 to 10 months old) mice as a result of loading and unloading the animals into and out of the smoking and sham machine, their health condition was likewise good.

Microscopic analysis of lung tissue from young and old sham-treated and control animals revealed normal morphology. However, lungs of young and old smoke-exposed mice were characterized by perivascular and peribronchiolar accumulations of mononuclear cells. These lesions were composed of lymphocytes and macrophages. The latter were large, highly vacuolated, and contained a considerable amount of brown pigment. Such aggregations of these cells were never observed in lungs of any sham and control animals. Thus, these lesions indicate smoke but not age effect. In addition to the above abnormalities, which were common in both young and old smoke-exposed mice, lungs of the old smoke-treated animals contained highly cellular areas and thickened alveolar septa. In general, lung tissue morphology of old smoke-exposed mice was similar to that of HCA-smoke-treated animals (Figure 3). Lung structure of old smoke-exposed mice clearly suggested an increase in tissue and a decrease in alveolar space.

These visual impressions were confirmed by quantitative morphometric procedures. Smoke inhalation by old animals caused an increase in volume fraction of abnormal space and abnormal tissue and a decrease in volume of normal space and normal parenchymal tissue when compared with sham and control values.[7] On the other hand, neither smoke exposure, sham treatment, nor control conditions produced detectable effects regarding normality or abnormality of alveolar space and lung tissue in young animals ($p \geqslant 0.3$). Morphometric data indicated an interaction between smoke inhalation and age and suggested restrictive pathology in lungs of old smoke-exposed mice.

Macrophage response to smoke inhalation was different in young and old mice. However, the number of pulmonary phagocytes per square millimeter of lung tissue of old and young sham and control animals was statistically ($p = 0.08$) similar (17.1 ± 3.9 — old; 25.7 ± 3.2 — young). Lung macrophage population of young smoke-treated animals (279.9 ± 13.2 no./mm^2) was elevated more than ten times that of young sham and control values ($p = 0.0001$). Likewise, phagocyte numbers (89.3 ± 9.9 no./mm^2) of old smoke-exposed mice increased markedly (five times) over their control counterparts ($p = 0.0001$). However, the macrophage response in young smoke-exposed animals was more than three times that

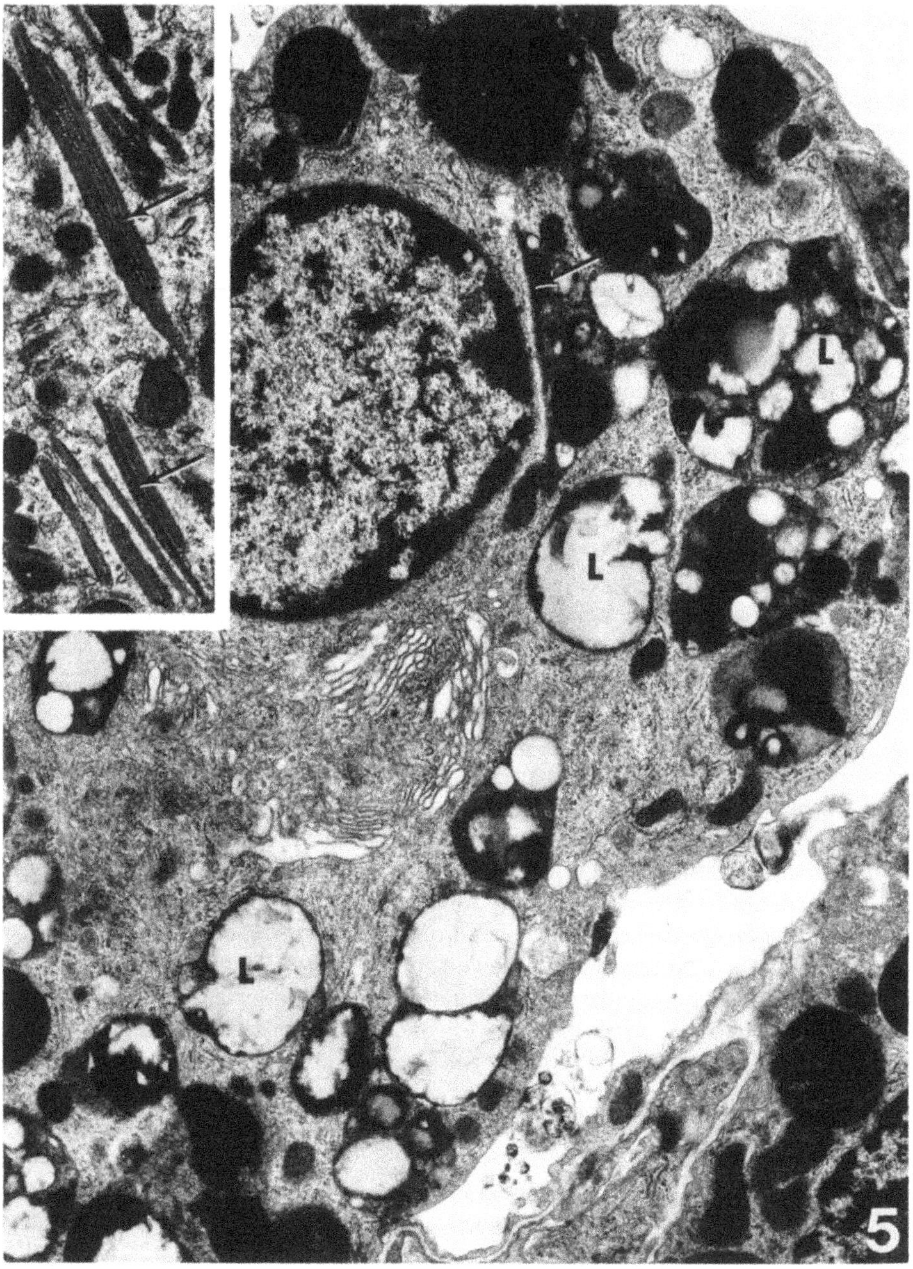

FIGURE 5. Electron micrograph of an alveolar macrophage from lungs of a Group 7 (HCA-treated, smoke-exposed) mouse. Note the numerous lysosome bodies (L) and crystal-like inclusions (arrow). (Magnification × 11,700.) Insert depicts crystalline material free in the cytoplasm (arrow). (Magnification × 22,200.)

of old mice that inhaled smoke ($p = 0.0001$). Could this decreased response be related to development of the restrictive lung lesion?

Pulmonary macrophage ultrastructure of young and old sham-treated and control animals were similar and normal in appearance. These cells contained a single elongate, often eccentric, nucleus with a prominent nucleolus. The most marked cytoplasmic organelles were lysosome-like bodies (primary and secondary lysosomes) and Golgi complex. Other

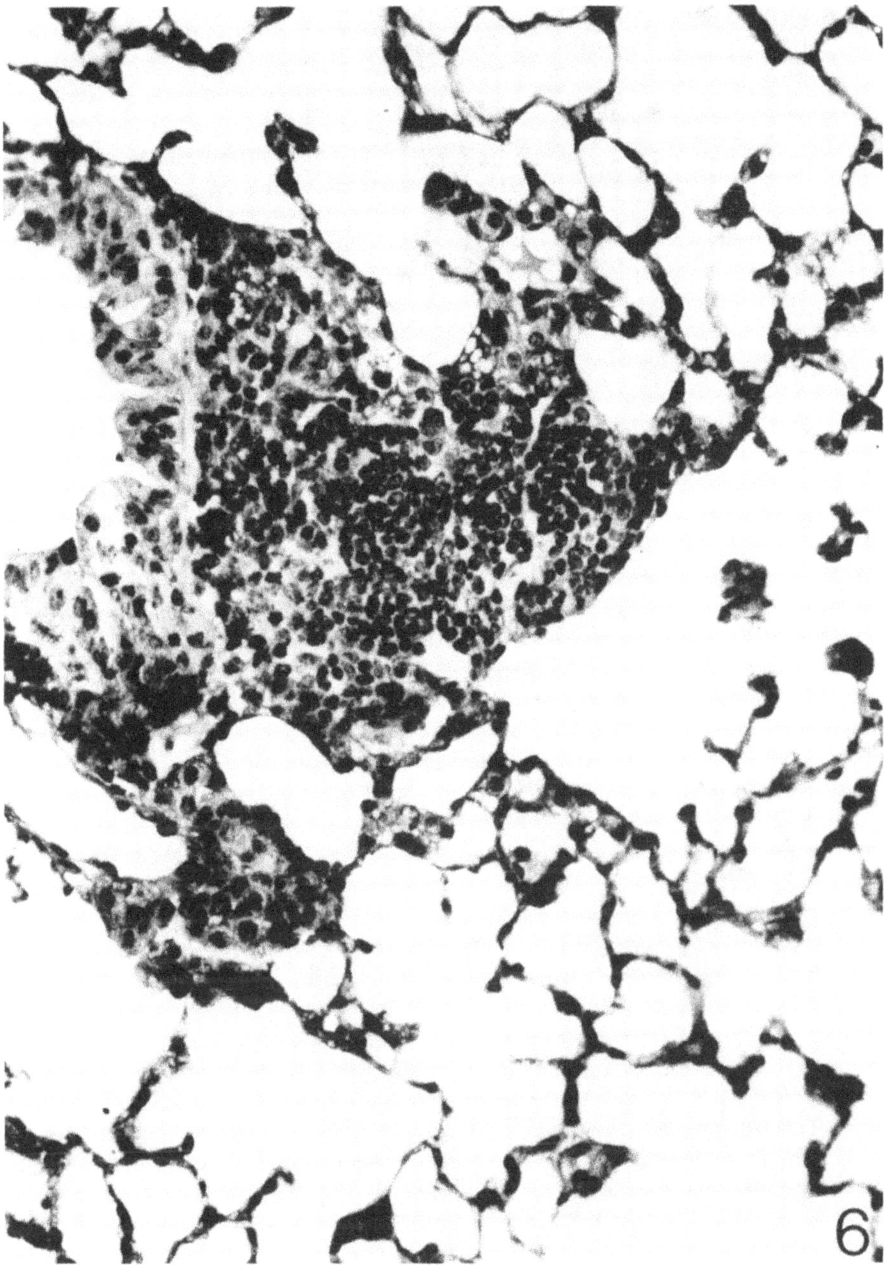

FIGURE 6. Light micrograph of lung tissue from a young C57BL/6 mouse exposed to smoke for 9 months. Note the mononuclear cell aggregation adjacent to a bronchiole. (Magnification × 420.)

organelles of the macrophages were not remarkable in distribution, number, or morphology. Ultrastructure of phagocytic cells in lungs of young and old smoke-exposed animals were more varied that that of sham and control animals. Some macrophages were similar in appearance to those of sham and control mice, other phagocytes appeared slightly larger and contained more heterogeneous lysosomes than those of control animals, and still others were large cells replete with heterogeneous lysosomes of various dimensions. The last type of macrophage (Figure 7) was encountered most frequently in old smoke-exposed animals.

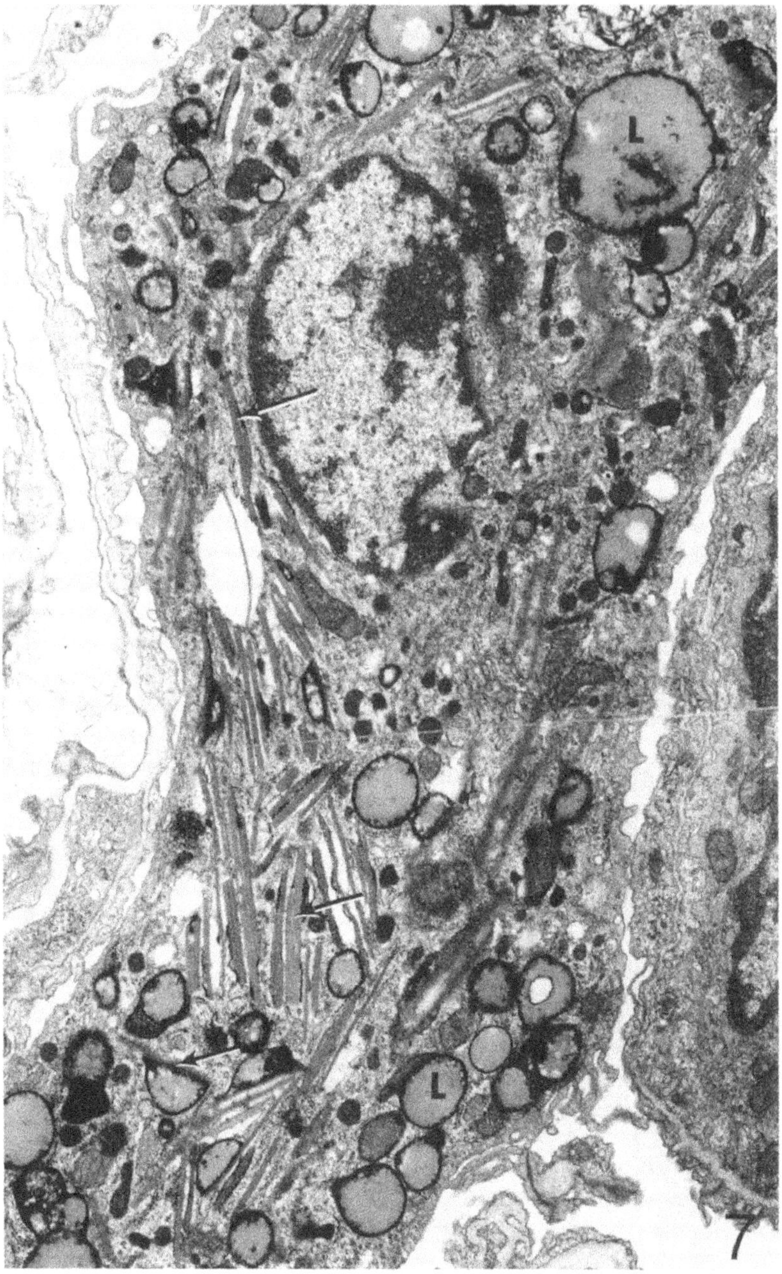

FIGURE 7. Electron micrograph of an alveolar macrophage from an old C57BL/6 mouse exposed to cigarette smoke for 9 months. The cell is replete with lysosome-like bodies (L) and crystalline inclusions (arrow). (Magnification × 10,800.)

Many inclusions resembling crystalline material were noted in these cells of smoke-exposed mice. The latter were associated with lysosomes or were free in the cytoplasm (Figure 7). Further, many phagocytes were noted in various states in degeneration.

The nature of the lesions was further confirmed by assessing pulmonary function. Total lung volume, rate of compliance change (a measure of lung elastic recoil), CO-diffusing capacity, and diffusing capacity per unit lung were significantly reduced in the old smoke-treated mice when compared to all other animal groups.[7] Smoke inhalation by young animals had no apparent effect on pulmonary functions. Slonim and Hamilton[19] reported that lung fibrosis is functionally accompanied by reduced total lung volume and diffusing capacity. Thus, physiologic data suggest, as do morphologic and morphometric data, that the anomalies observed in the old smoke-exposed animals were restrictive in nature conforming most closely to pulmonary fibrosis.

DISCUSSION

Possible Mechanism for Development of Fibrosis

A striking feature noted in macrophages of smoke-exposed, HCA-treated animals was a marked increase in number and diversity of lysosomes, some of which contained many conspicuous crystal-like membrane-bound inclusions. Similar crystalline inclusions have been observed in mice earlier,[5] although they were noted only after a 16-week smoke-exposure period. Likewise, similar crystals were present in lung macrophages of human smokers.[20] Chemically, the material has been identified as kaolinite (aluminum silicate).[5,20] This apparently originates from the soil during tobacco plant growth and harvest and is conveyed to the lungs as a particulate component of cigarette smoke.[20] The appearance of the kaolinite crystalline material in macrophages of smoke-exposed HCA-treated mice but not in the smoke-exposed-only group might be explained as follows. It has been shown that normal turnover rate of lung macrophages is markedly increased[21] by HCA treatment, resulting in a significant decrease in the population size of the phagocytes. Kaolinite material, delivered via tobacco smoke in apparently small quantities, is cleared undetectably under smoking conditions by the normally large pool of pulmonary macrophages. Macrophages of smoke-exposed animals treated with HCA, residing in the lungs for an extended time, would have an opportunity to engulf the keolinite material over longer periods, thus concentrating and making them discernible since phagocytic activity is not impeded by HCA treatment.[22] This idea gains further support from observations that macrophages containing silica reside for long periods of time within the lungs after a short exposure to silica material.[23]

The decreased macrophage response to cigarette smoke inhalation in old compared with young mice cannot be definitively explained at present and only a speculative idea can be proposed. This phenomenon might be related to an increase of steroid levels with advancing age. Such a notion is supported by observation of elevated corticosterone (a steroid similar in activity to HCA) serum levels in normal aging rats.[24] Increased steroid levels in old animals could have impeded the pulmonary macrophage response to cigarette smoke inhalation as did exogenous HCA. The macrophages of the old smoke-treated animals were also characterized by many heterogenous lysosomes and crystal-like kaolinite inclusions. Thus, like in the HCA-treated, smoke-exposed mice, macrophages in lungs of old animals phagocytose smoke-conveyed particulates for extended periods of time. The latter would therefore reside in the cells for a longer time. In general, macrophage morphology in lungs of smoke-exposed animals (HCA-treated and old) from both experiments suggested a progressive interiorization and accumulation of material including keolinite conveyed by smoke over time. Since kaolinite crystal-like inclusions were conspicuously present in lung macrophages of animals with fibrosis and decreased phagocyte response to smoke inhalation, the inclusions might be involved in the development of the pathology. Definitive mechanisms by which this compound induces the disorder is presently unknown; however, information exists to warrant speculation. Particulate challenge, in part by kaolinite crystal-like material, appears to overburden the macrophage clearance mechanism as suggested by cell size increase and/or degeneration of the phagocytes. Increased or prolonged phagocytic activity by macrophages of the old smoke-exposed animals is indicated by an increased lysosomal compartment size in old smoke-exposed mice.[25] It is conceivable that the macrophages might concentrate more on clearance of particles than on other activities such as secretion. In addition, becasue many phagocytes appeared degenerate and thus undoubtedly nonfunctional, secretory products required to maintain lung integrity could be reduced. Further, it has been reported that kaolinte interferes with amino acid incorporation into protein,[26] which is consistent with the possibility that the material might impede synthesis. Elias et al.[27] have shown that lung macrophages secrete a factor which inhibits fibroblast growth and collagen biosynthesis. Under the experimental conditions in the above studies, macrophages may not be capable of porducing sufficient amounts of this factor. Thus, fibroblasts could synthesize excess collagen, resulting in lung fibrosis.

Future Research

The present animal model indicates a relationship between cigarette smoke inhalation, the lung macrophage system, and pulmonary fibrosis and implicates kaolinite in the development of the disorder. However, reported studies have not directly addressed the mechanism by which kaolinite alters the phagocytic activity that leads to the pathology described. Perhaps the most significant contribution of the model is that it provides critical baseline information needed to assess the relationship of kaolinite to pulmonary fibrosis from a mechanistic aspect. Insight into the mechanistic action of kaolinite in the disease might be gained if future research addresses the following questions. Does kaolinite affect lung integrity independently or in a synergistic fashion with other smoke components? Is there a dose effect of the compound in regard to onset, progression, and degree of disorder? Is there a positive correlation between the severity of the disease and the amount of kaolinite in the lungs and/or macrophages? How much fibroblast inhibitory factor do kaolinite-ladened lung macrophages secrete? How much of the factor is necessary to maintain normal lung integrity? Answers to these and other questions obtained via animal experiments might provide insight into the assessment and management of pulmonary disorders in humans. Continued research in this area might also provide evidence to justify efforts by the tobacco industry to remove the residual kaolinite from tobacco prior to manufacturing commercial cigarettes. Significance of such action is obvious.

Limitations of the Model

Finally, a word of caution is warranted regarding the utilization of the animal model presented. Cigarette smoke inhalation experiments are quite complicated and costly; they necessitate many man-hours and require considerable technical expertise to "smoke" and "sham" the animals. Also, much time, effort, and knowledge is necessary to operate, monitor, and maintain the smoking machines. Malfunction of these and adjunct equipment may cause an untimely end to an experiment and result in lengthy data acquisition delay. Great care should be taken to assure an infectious-disease-free state of the animals. Data obtained from such animals might reflect the disease rather than experimental effects. Thus, all animals should be housed in environmentally controlled quarters and their health continuously monitored. In general, smoke inhalation research should be approached with the understanding that proper smoke and sham treatment of animals is a crucial prerequisite to data acquisition. The experiments must be planned carefully and designed in a fashion that prevents unforeseen problems from destroying the study. Although such research is quite demanding, results might provide significant information that has a considerable impact on understanding the relation between cigarette smoking and health.

REFERENCES

1. Report of the Surgeon General, Smoking and Health, U.S. Department of Health, Education and Welfare, Washington, D.C., 1979.
2. Report of the Surgeon General, The Consequences of Smoking — Cancer, U.S. Department of Health, Education and Welfare, Washington, D.C., 1982.
3. **Hirayama, T.,** Non-smoking wives of heavy smokers have a higer risk of lung cancer: a study from Japan, *Br. Med. J.*, 282, 183, 1981.
4. **Garfinkel, L.,** Time trends in lung cancer mortality among nonsmokers and a note on passive smoking, *J. Natl. Cancer Inst.*, 66, 1061, 1981.
5. **Matulionis, D. H. and Traurig, H. H.,** In situ response of lung macrophages and hydrolase activity to cigarette smoke, *Lab. Invest.*, 37, 314, 1977.
6. **Matulionis, D. H., Kimmel, E., and Diamond, L.,** Morphologic and physiologic response of lungs to steroid and cigarette smoke: an animal model, *Environ. Res.*, 36, 298, 1985.
7. **Matulionis, D. H.,** Chronic cigarette smoke inhalation and aging in mice. I. Morphologic and functional lung abnormalities, *Exp. Lung Res.*, 7, 237, 1984.
8. **Matulionis, D. H.,** Reaction of macrophages to cigarette smoke. I. Recruitment of pulmonary macrophages, *Arch. Environ. Health*, 34, 293, 1979.
9. **Pratt, S. A., Smith, M. H., Ladman, A. J., and Finley, T. N.,** The ultrastructure of alveolar macrophages from human cigarette smokers and nonsmokers, *Lab. Invest.*, 24, 331, 1971.
10. **Matulionis, D. H.,** Effects of cigarette smoke generated by different smoking machines on pulmonary macrophages of mice and rats, *J. Anal. Toxicol.*, 8, 187, 1984.
11. **Benner, J. F., Owens, S., Hancock, R., and Griffith, R. B.,** Smoking machine development and inhalation atmosphere monitoring, in Proc. University of Kentucky Tobacco and Health Research Institute Conference, Rep. No. 4, Lexington, KY, 1973, 494.
12. **Matulionis, D. H.,** Pulmonary tissue and cigarette smoke. I. Cellular response to hydrocortisone, *Environ. Res.*, 27, 361, 1982.
13. **Thompson, J. and Van Furth, R.,** The effects of glucocorticosteroids on kinetics of mononuclear phagocytes, *J. Exp. Med.*, 131, 429, 1970.
14. **Weibel, E. R.,** Point counting methods, in *Stereological Methods. Practical Methods for Biological Morphometry*, Academic Press, New York, 1979, chap. 4.
15. **Sabo, J. P., Kimmel, E. C., and Diamond, L.,** Effects of the Clara cell toxin, 4, ipomeanol, on lung function in rats, *J. Appl. Physiol. Respir. Environ. Exercise Physiol.*, 54, 337, 1983.
16. **Vinegar, A., Sinnett, E. E., and Leith, D. E.,** Dynamic mechanisms determine functional residual capacity in mice, *Mus musculus*, *J. Appl. Physiol. Respir. Environ. Exercise Physiol.*, 46, 867, 1979.
17. **Paiva, M., Yernault, J. C., Van Eerdeweghe, P., and Englert, M.,** A sigmoid model of statis volume-pressure curve of human lung, *Respir. Physiol.*, 23, 317, 1975.
18. **Takezawa, J., Miller, F. J., and O'Niel, J. J.,** Single-breath diffusing capacity and lung volumes in small laboratory animals, *J. Appl. Physiol. Respir. Environ. Exercise Physiol.*, 48, 1052, 1980.
19. **Slonim, N. B. and Hamilton, L. H.,** Lung volume and its subdivisions; and Diffusion of gases in the lung, in *Respiratory Physiology*, 4th ed., C. V. Mosby, St. Louis, 1981, chap. 4 and chap. 7.
20. **Brody, A. R. and Craighead, J. E.,** Cytoplasmic inclusions in pulmonary macrophages of cigarette smokers, *Lab. Invest.*, 32, 125, 1975.
21. **Van Furth, R. and Blusse Van Oud Alblas, A.,** The current view on the origin of pulmonary macrophages, *Pathol. Res. Pract.*, 175, 38, 1982.
22. **Slonecker, C. E. and Lim, W. C.,** Effects of hydrocortisone on the cells in an acute inflammatory exudate, *Lab. Invest.*, 27, 123, 1972.
23. **Brody, A. R., Roe, M. W., Evans, J. N., and Davis, G. S.,** Deposition and translocation of inhaled silica in rats: quantification of particle distribution, macrophage participation and function, *Lab. Invest.*, 47, 533, 1982.
24. **DeKosky, S. T., Scheff, S. W., and Cotman, C. W.,** Elevated corticosterone levels: a possible cause of reduced axon sprouting in aged animals, *Neuroendocrinology*, 35, 38, 1983.
25. **Matulionis, D. H. and Simmerman, L. S.,** Chronic cigarette smoke inhalation and aging in mice. 2. Quantitation of the pulmonary macrophage response, *Exp. Lung Res.*, 9, 309, 1985.
26. **Low, R. B., Leffingwell, C. M., and Bulman, C. A.,** Effects of kaolinite on amino acid transport and incorporation into protein by rabbit pulmonary macrophages, *Arch. Environ. Health*, 35, 217, 1980.
27. **Elias, J. A., Rossman, M. D., Zurier, R. B., and Daniele, R. P.,** Human alveolar macrophage inhibition of lung fibroblast growth, *Am. Rev. Respir. Dis.*, 131, 94, 1985.

Index

INDEX

A

B

C

D

E

F

H

O

P

V

W

X

www.ingramcontent.com/pod-product-compliance
Ingram Content Group UK Ltd.
Pitfield, Milton Keynes, MK11 3LW, UK
UKHW052220190926
475702UK00006B/1124
* 9 7 8 1 3 1 5 8 9 1 8 6 6 *